Computational Quantum Chemistry
An Interactive Guide to Basis Set Theory

Computational Quantum Chemistry
An Interactive Guide to Basis Set Theory

Charles M Quinn

*Department of Chemistry, The National University of Ireland, Maynooth,
Maynooth, Co. Kildare, IRELAND.*

ACADEMIC PRESS

An Elsevier Science Imprint

*San Diego   San Francisco   New York
Boston   London   Sydney   Tokyo*

Academic Press
An Elsevier Science Imprint
Harcourt Place, 32 Jamestown Road, London NW1 7BY, UK
http://www.academicpress.com

Academic Press
An Elsevier Science Imprint
525 B Street, Suite 1900, San Diego, California 92101-4495, USA
http://www.academicpress.com

ISBN 0-12-569682-5

CD ISBN 0-12-569683-5

Library of Congress Catalog Number: 2001091370

A catalogue record for this book is available from the British Library

Typeset by Laser Words, Chennai, India
Printed and bound in China by RDC

02 03 04 05 06 07 RD 9 8 7 6 5 4 3 2 1

Contents

Preface

'In most numerical work, a working sheet will be used for recording data and intermediate results of the calculation. A clear and orderly arrangement of this working sheet is a great help both in avoiding mistakes and in locating and correcting any that do happen to be made. Numerical work should not be done on odd scraps of rough paper, but laid out systematically and in such a way as to show how the intermediate and final results were obtained; and the number entered on the work sheet should be written neatly and legibly. Use of ruled paper is a help in keeping the layout of the work neat and clear. It is advisable to use loose sheets rather than a book since it is rather easy to make mistakes in copying from one page to another of a book; with loose sheets the number to be copied from one sheet, and the place to which it is to be copied on another, can more easily be brought close together, and the copy made and checked more easily.'

Professor Douglas R. Hartree, *Numerical Analysis*
[1952 The Clarendon Press, Oxford, England]

One can but wonder what Hartree and all his illustrious contemporaries would have thought of modern spreadsheet technology, in comparison to the tools available to them. Hartree listed these to be desk (*calculating*) machines, tables, the slide rule and graph paper. On a modern spreadsheet, not only are all of Hartree's requirements, for good practice, fulfilled in abundance, but also the design of a spreadsheet can be arranged so that a change, in any one feature of a calculation, is reflected very rapidly in the result.

This sets the context of this book. The power of the modern spreadsheet is applied to make familiar fundamental concepts of Basis Set theory, which is at the heart of modern Molecular Orbital theory. There are over 100 fully interactive EXCEL files on the CD supplied with this book. You can treat the CD as a learning aid and simply open and consult each file during your study of the various topics discussed in the text. There is a separate folder for each chapter and the corresponding figure numbers of the illustrations in the text identify the spreadsheets.

There is much more potential in the material on the CD. All worksheet cells, bordered in black, are unprotected. Thus, you can input different values for Slater exponents, Gaussian basis sets, bond lengths, scaling factors and other parameters involved in the calculations described. All the results of such changes in input data can be saved for later use or application. This versatility should make the material covered interesting to those preparing courses in the theory covered in the book.

If you copy the files from the CD to your hard-drive, they transfer in read-only form. This read-only attribute can be changed in the File/Properties menu of Windows Explorer for all the files copied folder by folder from the CD in one operation by using the Edit/Select all sequence. Note, that the Herman-Skillman program must be transferred onto the hard drive before it can be used, since READ/WRITE operations are carried out during its execution.

If you want to explore the design features of the various spreadsheets in detail, you will need to change the spreadsheets into unprotected form. Protection levels on open files are set in the TOOLS dialogue box and the universal password for the files on the CD is *hfsscf*. This acronym is formed from *Hartree-Fock-Slater-Self-Consistent-Field*.

The CD-ROM supplied contains spreadsheet files based on the Microsoft EXCEL97/2001 software. The CD-ROM is compatible with a wide range of Windows and Macintosh operating systems. The EXCEL files will open and perform properly on all Windows platforms. The EXCEL files can be accessed, too, using the Macintosh OS9.2 and OSX operating systems. In OS9.2 the files will open in EXCEL2001, while with the OSX system, the files open in classic mode. In order to run the Herman-Skilmann program it is necessary to copy the input files and executable program into a suitable folder, e.g. My Documents, on your hard drive. For Windows NT and 2000 platforms it is necessary, also, to login with Administrator privileges to avoid 'access violation' errors during execution of HS.EXE.

Finally, if you follow the design instructions in each exercise and make your own versions of the spreadsheets, I hope that you will learn not only about Electronic Structure theory and the methodology of calculations in Molecular Orbital theory, but, as well, many advanced features of EXCEL, which you can apply to illustrate other aspects of Chemistry and Physics or indeed your own work.

Charles M. Quinn,
National University of Ireland, Maynooth,
Maynooth,
Co. Kildare, IRELAND.
April 2001.

1

Essential atomic orbital theory

This chapter revises basic atomic orbital theory. The chapter begins with the exact results for the case of the hydrogen atom and the orbital concept for many-electron atoms. It is very important to understand these details about atomic orbitals, since the orbital concept is essential in the approximation to chemical bonding known as the Linear Combination of Atomic Orbitals — Molecular Orbital [LCAO-MO] theory. Only in the case of the hydrogen atom are these atomic orbitals, as exact solutions to Schrödinger's equation, available as functions.

However, for practical purposes, even the hydrogen radial functions are represented as linear combinations of other functions, which can be applied more easily in molecular orbital calculations. Nowadays, these functions and the approximations to the numerical radial functions for the other atoms of the elements of the Periodic Table derived from the hydrogenic ones are the basis sets of modern molecular orbital theory.

In this chapter you will learn:

1. how to use a spreadsheet to make plots of radial wavefunctions and radial distribution functions;
2. how to calculate numerical wave functions for many-electron atoms using the Herman–Skillman atomic structure programme;
3. how to apply Slater's rules to construct functions to model the numerical data for many-electron atoms;
4. how to use Gaussian functions to model these Slater functions.

1.1 ATOMIC ORBITALS FOR THE HYDROGEN ATOM

If you are good at mathematics, it is relatively straightforward to solve Schrödinger's equation for the hydrogen atom. The attractive potential, between the single extra-nuclear electron and the nucleus of one proton, depends only on the reciprocal distance apart of these charged particles. In the simplest solution, the nucleus is assumed stationary and this separation of the proton and the electron is a radial distance measured at the electron. So, it is most convenient to solve Schrödinger's equation by rewriting the equation in *spherical polar* coordinates, because, then, it divides into two distinct differential equations, the first depending only on the radial distance (r) and the second depending only on the polar (θ) and azimuthal angles (ϕ) of the new coordinate system, both with exact solutions.

Figure 1.1 defines the spherical polar coordinate system. The *radius vector, r*, is the distance of any point from the origin; θ is the polar angle of the radius vector with respect to the z-axis and ϕ is the azimuthal angle, the angle of rotation of the projection of the radius vector onto the xy plane, with respect to the x-axis. The Cartesian and polar coordinate systems relate to each other in the identities,

$$z = r \cos \theta \qquad\qquad 1.1$$

$$x = r \sin \theta \cos \phi \qquad\qquad 1.2$$

$$y = r \sin \theta \sin \phi \qquad\qquad 1.3$$

$$r^2 = x^2 + y^2 + z^2 \qquad\qquad 1.4$$

The exact solutions to the separate equations, which result from this coordinate transformation of the Schrödinger equation for the hydrogen atom, are the sets of functions known as the associated Laguerre polynomials, for the radial equation, and the spherical harmonics, for the angular equation. The quantum numbers, n,l and m arise naturally in the solution of Schrödinger's equation, and so the symbolic form, for the eigenfunction solutions to the H-atom problem, known as atomic orbitals, is

$$\phi_{nlm}(r, \theta, \phi) = R_{nl}(r{:}E)Y_{lm}(\theta, \phi) \qquad\qquad 1.5$$

The quantum numbers, l and m, define the angular properties of the electron distribution as the square of $Y_{lm}(\theta, \phi)$. The familiar icons for s, p and d atomic orbitals identify the phases (signs of the values of the functions) of the $Y_{lm}(\theta, \phi)$ by including the signs or using shading to identify the contours of positive phase. Table 1.1 presents the detailed forms of the first few of these functions defined in Equation 1.5, while standard chemist's pictures of these orbitals are shown in Figure 1.2.

If required, the quantum number s is used to complete the description of the electron state as the spin orbital ϕ_{nlms} (r, θ, ϕ) with δ the spin coordinate [see Chapter 5].

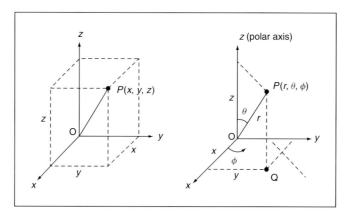

Figure 1.1   The standard definitions of the Cartesian and spherical polar coordinate systems. [From *The Chemist's Maths Book, Erich Steiner, OUP 1996,* © E. Steiner 1996].

Note, carefully, the entries for the normalization constants, the squares of which, render the integrals to be of unit value. Distinct normalization constants have been included for radial and angular parts. To get the overall constant, it is necessary to multiply the two partial constants together. The table has been constructed, in this manner, to draw attention to a possible confusion in basis set theory. Often, the normalization condition is not clear for particular basis sets. Moreover, only rarely are basis sets mutually orthogonal, one with respect to another. Thus, it will be important to check the normalization data in Table 1.1 as an exercise in using the numerical integration techniques developed in Chapter 2. Orthonormalization is the subject of Chapter 3 because, in the end, all calculations in quantum chemistry require the rendering of approximate wave functions mutually orthonormal.

The same procedure, of separation of variables, is possible for other atomic species, e.g. He^+, Li^{2+} etc., considered ionized to the extent that there is one extra-nuclear electron.

Table 1.1   The radial and angular components of the hydrogenic atomic orbitals with distinct normalization constants for the radial and angular functions. The parameter, $\rho = \left(\frac{2Z}{na_0}\right)$ extends the application of the functions in the table entries for non-hydrogen one-electron atomic species. Remember that the solutions to the angular equation in ϕ are $\exp(+/-im\phi)$ and the real forms given are obtained by taking the sums and differences of the expansions of the complex exponentials and then applying equations 1.1 to 1.3 to these results. The column headed '+/−' indicates the particular choices of sum when relevant.

Orbital	N_R	$R(r)$	$N_{Y(\theta,\phi)}$	$+/-$	$Y_{lm}(\theta,\phi)$
1s	$2\left(\frac{Z}{a_0}\right)^{3/2}$	$e^{-\rho r/2}$	$\left(\frac{1}{4\pi}\right)^{1/2}$		1
2s	$\left(\frac{1}{2(2)^{1/2}}\right)\left(\frac{Z}{a_0}\right)^{3/2}$	$(2-\rho r)e^{-\rho r/2}$	$\left(\frac{1}{4\pi}\right)^{1/2}$		1
$2p_x$	$\left(\frac{1}{2(6)^{1/2}}\right)\left(\frac{Z}{a_0}\right)^{3/2}$	$\rho r e^{-\rho r/2}$	$\left(\frac{3}{4\pi}\right)^{1/2}$	$+$	x/r
$2p_y$	$\left(\frac{1}{2(6)^{1/2}}\right)\left(\frac{Z}{a_0}\right)^{3/2}$	$\rho r e^{-\rho r/2}$	$\left(\frac{3}{4\pi}\right)^{1/2}$	$-$	y/r
$2p_z$	$\left(\frac{1}{2(6)^{1/2}}\right)\left(\frac{Z}{a_0}\right)^{3/2}$	$\rho r e^{-\rho r/2}$	$\left(\frac{3}{4\pi}\right)^{1/2}$		z/r
3s	$\left(\frac{1}{9(3)^{1/2}}\right)\left(\frac{Z}{a_0}\right)^{3/2}$	$(6-6\rho r+(\rho r)^2)e^{-\rho r/2}$	$\left(\frac{1}{4\pi}\right)^{1/2}$		1
$3p_x$	$\left(\frac{1}{9(6)^{1/2}}\right)\left(\frac{Z}{a_0}\right)^{3/2}$	$(4\rho r-(\rho r)^2)e^{-\rho r/2}$	$\left(\frac{3}{4\pi}\right)^{1/2}$	$+$	x/r
$3d_{z2}$	$\left(\frac{1}{9(30)^{1/2}}\right)\left(\frac{Z}{a_0}\right)^{3/2}$	$(\rho r)^2 e^{-\rho r/2}$	$\left(\frac{5}{16\pi}\right)^{1/2}$		$(3z^2-r^2)/r^2$
$3d_{xz}$	$\left(\frac{1}{9(30)^{1/2}}\right)\left(\frac{Z}{a_0}\right)^{3/2}$	$(\rho r)^2 e^{-\rho r/2}$	$\left(\frac{15}{16\pi}\right)^{1/2}$	$+$	xz/r^2
$3d_{yz}$	$\left(\frac{1}{9(30)^{1/2}}\right)\left(\frac{Z}{a_0}\right)^{3/2}$	$(\rho r)^2 e^{-\rho r/2}$	$\left(\frac{15}{16\pi}\right)^{1/2}$	$-$	yz/r^2
$3d_{xy}$	$\left(\frac{1}{9(30)^{1/2}}\right)\left(\frac{Z}{a_0}\right)^{3/2}$	$(\rho r)^2 e^{-\rho r/2}$	$\left(\frac{15}{16\pi}\right)^{1/2}$	$+$	xy/r^2
$3d_{x2-y2}$	$\left(\frac{1}{9(30)^{1/2}}\right)\left(\frac{Z}{a_0}\right)^{3/2}$	$(\rho r)^2 e^{-\rho r/2}$	$\left(\frac{15}{16\pi}\right)^{1/2}$	$-$	$(x^2-y^2)/r^2$

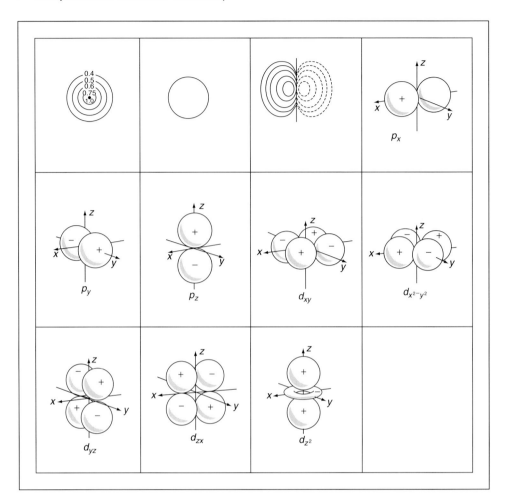

Figure 1.2   The chemist's traditional representation of the *shapes* of the atomic orbitals. The contour diagrams, in the first row, emphasize that the contours of mathematical functions are being represented. The orbital 'shapes' confirm the symmetries of these contours and include their phases as the signs on the lobes of the individual pictures. But, phase is not determined, since only the square of the wave function is of physical significance, as the probability density. And so, for example, the sign is not given for the s-orbital. [After *Coulson's Valence, R. McWeeny, OUP 1979* © Roy McWeeny 1979].

Thus, in the table, the results are generalized to take account of the numbers of protonic charges appropriate for the atomic number of different atoms ionized in this way.

Exercise 1.1.   Generation of an EXCEL chart to display the variation of the H_{1s} radial function with radial distance from the nucleus.

Figure 1.3 presents the graph for the 1s hydrogen atom radial function. All the spreadsheets on the CD are named fig*-*.xls to correspond to the diagrams in the text. If you wish and I recommend that you do, it is good practice to generate this

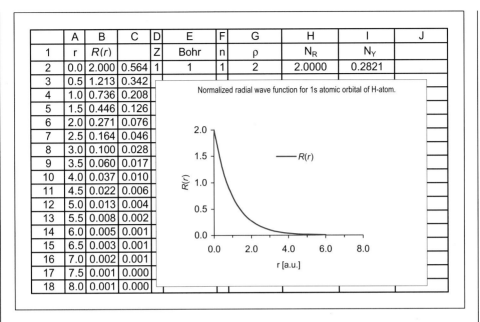

	A	B	C	D	E	F	G	H	I	J
1	r	R(r)		Z	Bohr	n	ρ	N_R	N_Y	
2	0.0	2.000	0.564	1	1	1	2	2.0000	0.2821	
3	0.5	1.213	0.342							
4	1.0	0.736	0.208							
5	1.5	0.446	0.126							
6	2.0	0.271	0.076							
7	2.5	0.164	0.046							
8	3.0	0.100	0.028							
9	3.5	0.060	0.017							
10	4.0	0.037	0.010							
11	4.5	0.022	0.006							
12	5.0	0.013	0.004							
13	5.5	0.008	0.002							
14	6.0	0.005	0.001							
15	6.5	0.003	0.001							
16	7.0	0.002	0.001							
17	7.5	0.001	0.000							
18	8.0	0.001	0.000							

Figure 1.3   The normalized [see Chapter 3] radial wave function, $R_1(r)$, for the 1s atomic orbital in the hydrogen atom [Z = 1] presented as an EXCEL graph, constructed as described in the text. Note, that since $Y_{00} = 1$ for s-orbitals, the only difference between this function and the total 1s atomic orbital is the factor $(1/4\pi)^{0.5}$ due to the normalization constant over the angular coordinates.

graph and all the others in the book using your own version of EXCEL. For the present example proceed as follows:

1.   On a new worksheet enter the labels as given in Figure 1.3 in cells $A$1 to $I$1.
2.   In cells $D$2, $E$2, $F$2 and $G$2, enter the appropriate parameter values for the hydrogen 1s atomic orbital. The choice $E$2 = 1 sets any calculations in Bohr units with $a_0 = 1$, because of the entry in $G$2.
3.   In cell $G$2 enter the formula for the parameter $\rho = 2*Z/(n*a_0)$ i.e.

$$\$G\$2 = 2*\$D\$2/(\$F\$2*\$E\$2)$$

Repeat this procedure to generate the entries for the normalization constants in cells $H$2 and $I$2, using the expressions given in Table 1.1.

4.   Select cell $A$2. Use the EDIT/FILL/SERIES command sequence to establish a radial mesh for the display of the hydrogen 1s radial wave function. To construct fig1-3.xls mesh intervals of 0.5 Bohr units were chosen and the maximum radius for the graph was set at 8.0 Bohr units.
5.   Enter the formula for the hydrogenic 1s radial function in cell $B$2 in the form

$$\$B\$2 = \$H\$2*\textbf{EXP}(-\$G\$2*\$A2/2)^{1}$$

[1]EXCEL intrinsic functions are in bold face type throughout this book.

Note the manner of identification of the cells containing the fixed parameters ρ and N_R. EXCEL-based absolute referencing has been used in this instruction. The use of the character '$' as a left-multiplier of a column or row address fixes that address in any further reference. Since much of the worksheet is constructed by using commands, which repeat a set of calculations for different parameter values, the '$' device enables fixed values to remain fixed. Thus, to repeat this formula with fixed parameter values in other cells of column B we need absolute referencing. It is good practice, too, to identify the column containing the radial data absolutely, thus $A*.

6. Select cells $B$2 to $B$18[2]. There is a variety of ways to do this in EXCEL. The simplest general procedure is to select the first cell, then to press F8 on the keyboard and select using the down arrow key for short runs down a column or along a row. To Select over larger ranges, select the first cell, press F8 and use the sequence EDIT/GOTO by entering the destination address, the final cell to be selected[3].

7. With the appropriate cells highlighted on the worksheet, issue the EDIT/FILL/ DOWN command [this can be done as a keyboard sequence CTL/D as indicated in the FILL menu].

8. Select cells $A$1 to $B$18 and in the INSERT menu opt to generate a graph of the two sets of radial data against the radial mesh. Follow the 'wizard' instructions to complete the reproduction of Figure 1.3[4].

Other examples of this procedure are given in Figures 1.4 and 1.5 for the 2s and 2p radial wave function for hydrogen.

Exercise 1.2.   Using the example of Figure 1.1 and the spreadsheet instructions above, reproduce Figures 1.4 and 1.5 as EXCEL charts.

Several familiar features of atomic wave functions are evident in these diagrams.

The maximum values of the radial wave functions for the 1s and 2s orbitals in hydrogen occur at the nucleus, the position $r = 0.0$. The 1s function of Table 1.1 tends to zero amplitude at about 6.0 Bohr units. The radial wave functions for atomic orbitals, with non-zero angular momenta, are of zero amplitude at the nucleus. This is evident in Figure 1.5 for the hydrogenic 2p radial function and can be verified for other virtual hydrogenic orbitals of interest by making appropriate charts for these functions, as listed in Table 1.1. Nodes appear in the graphs of the radial functions with increasing n. Since the principal (n) and azimuthal (ℓ) quantum numbers relate as $n \geq \ell + 1$, the number of nodes in a radial function go as $(n - \ell - 1)$. Thus, the 2s radial function passes through zero at the

---

[2]Command sequences are undone in EXCEL by pressing the ESC button twice.

[3]To make changes to already occupied cells in a column or row, the sequences CTl/SHIFT/↓, CTL/SHIFT/↑, CTL/SHIFT/→ and CTL/SHIFT/← select all the consecutively occupied cells in the directions indicated by the arrows.

[4]Non-adjacent sets of data can be entered at the wizard window offering confirmation of the ranges of cells with data for the chart. Click on the right 'black' arrow, select the first set of data, hold the control button down and select the other sets. If you know the end cell address, simply select a few cells, then amend the last address to the proper one, which procedure avoids the need to select over large areas.

To fill down a value or formula over the same range as an adjacent set of occupied cells, select the source cell and 'double-click' on the 'Fill Handle' in the corner of this cell.

	A	B	C	D	E	F	G	H
1	r	R(r)	Z	Bohr	n	ρ	N_R	N_Y
2	0.00	0.7071	1	1	2	1	0.353553	0.2821
3	0.50	0.4130						
4	1.00	0.2144						
5	1.50	0.0835						
6	2.00	0.0000						
7	2.50	-0.0506						
8	3.00	-0.0789						
9	3.50	-0.0922						
10	4.00	-0.0957						
11	4.50	-0.0932						
12	5.00	-0.0871						
13	5.50	-0.0791						
14	6.00	-0.0704						
15	6.50	-0.0617						
16	7.00	-0.0534						
17	7.50	-0.0457						
18	8.00	-0.0389						
19	8.50	-0.0328						
20	9.00	-0.0275						
21	9.50	-0.0229						
22	10.00	-0.0191						

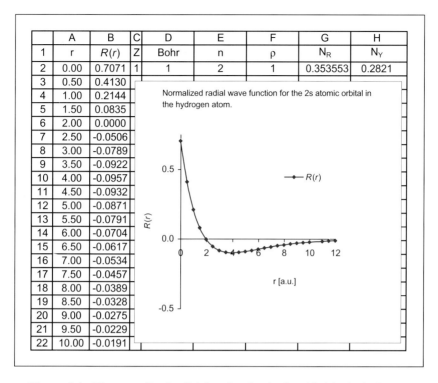

Normalized radial wave function for the 2s atomic orbital in the hydrogen atom.

Figure 1.4 The normalized radial function for the 2s orbital in the hydrogen atom.

radius satisfying the identity

$$2 = \left(\frac{Zr}{a_0} \right)$$

1.6

for the polynomial term in the Table 1.1, while the 3s function exhibits two nodes at the radii satisfying the quadratic identity,

$$6 + \left(\frac{Zr}{a_0} \right)^2 = \left(\frac{6Zr}{a_0} \right)$$

1.7

1.2 RADIAL DISTRIBUTION FUNCTIONS FOR THE HYDROGEN ATOM

Often, it is more meaningful physically to make plots of the radial distribution function, $P(r)$, of an atomic orbital, since this display emphasizes the spatial reality of the probability distribution of the electron density, as shell structure about the nucleus. To establish the radial distribution function we need to calculate the probability of an electron, in a particular orbital, exhibiting coordinates on a thin shell of width, Δr, between r and $r + \Delta r$ about the nucleus, i.e. within the volume element defined in Figure 1.6.

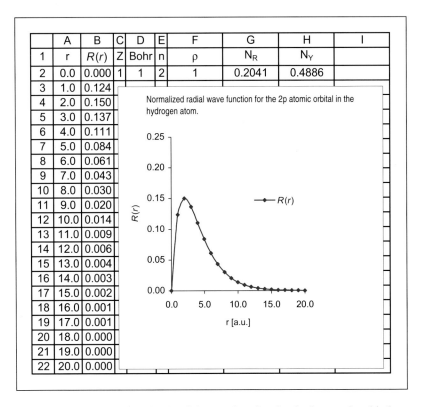

	A	B	C	D	E	F	G	H	I
1	r	$R(r)$	Z	Bohr	n	ρ	N_R	N_Y	
2	0.0	0.000	1	1	2	1	0.2041	0.4886	
3	1.0	0.124							
4	2.0	0.150							
5	3.0	0.137							
6	4.0	0.111							
7	5.0	0.084							
8	6.0	0.061							
9	7.0	0.043							
10	8.0	0.030							
11	9.0	0.020							
12	10.0	0.014							
13	11.0	0.009							
14	12.0	0.006							
15	13.0	0.004							
16	14.0	0.003							
17	15.0	0.002							
18	16.0	0.001							
19	17.0	0.001							
20	18.0	0.000							
21	19.0	0.000							
22	20.0	0.000							

Normalized radial wave function for the 2p atomic orbital in the hydrogen atom.

Figure 1.5   The normalized radial wave function for the 2p atomic orbital in the hydrogen atom.

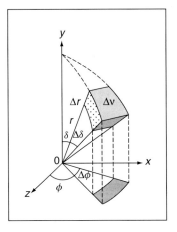

Figure 1.6   The integration volume element, Δv of the spherical coordinate system $r^2 \sin\theta \, \Delta\theta \, \Delta\phi \, \Delta r^1$. [From *The Chemist's Maths Book, Erich Steiner, OUP 1996* © E. Steiner 1996].

However, since the angular wave functions, the $Y_{lm}(\theta,\phi)$, are normalized, as defined in Equation 1.8, with

$$N_{\theta\phi}^2 \int_{\theta=0}^{\pi} \int_{\phi=0}^{2\pi} Y_{lm}(\theta,\phi) Y_{lm}(\theta,\phi) \sin(\theta)\, d\theta\, d\phi = 1 \qquad\qquad 1.8$$

and $N_{\theta\phi}$, as in Table 1.1, the radial distribution function reduces to[5]

$$P(r) = [R_{nl}(r)]^2 r^2 \int_{\theta=0}^{\pi} \int_{\phi=0}^{2\pi} Y_{lm}(\theta,\phi) Y_{lm}(\theta,\phi) \sin(\theta)\, d\theta\, d\phi = [R_{nl}(r)]^2 r^2 \qquad 1.9$$

Exercise 1.3.   EXCEL charts for the hydrogenic radial distribution functions.

The radial distribution function, for the 1s radial wave function of the hydrogen atom, is displayed in Figure 1.7. This EXCEL chart is constructed as follows. The broad design is the same as the previous example but now we need to use the appropriate cell entries for the components of equation 1.9 on the radial mesh.

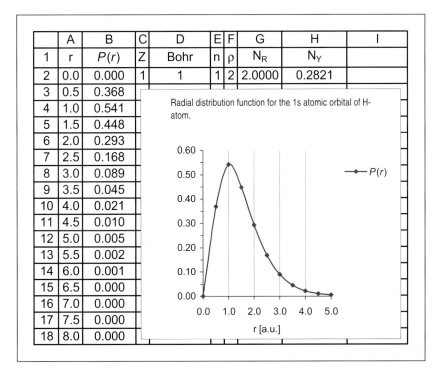

	A	B	C	D	E	F	G	H	I
1	r	P(r)	Z	Bohr	n	ρ	N_R	N_Y	
2	0.0	0.000	1	1	1	2	2.0000	0.2821	
3	0.5	0.368							
4	1.0	0.541							
5	1.5	0.448							
6	2.0	0.293							
7	2.5	0.168							
8	3.0	0.089							
9	3.5	0.045							
10	4.0	0.021							
11	4.5	0.010							
12	5.0	0.005							
13	5.5	0.002							
14	6.0	0.001							
15	6.5	0.000							
16	7.0	0.000							
17	7.5	0.000							
18	8.0	0.000							

Figure 1.7   The radial distribution function, $P(r)$, for the 1s atomic orbital in hydrogen. Note the maximum occurs at r = 1 Bohr unit.

[5]Note, the surface area of a spherical shell at radius r is $4\pi r^2$, but the normalization constant for the angular wave function reduces this to the r^2 factor in equation 1.14.

1. Make a copy of fig1-3.xls and rename this file fig1-7.xls. Such a copy of fig1-3.xls is made directly using the COPY/PASTE commands in Windows Explorer. The copy is renamed by 'clicking' with the right mouse button on the filename and choosing RENAME, in the dialog box, which appears or by applying a *gentle* click on the filename!

2. Substitute the formula[6]

$$\$B2 = \mathbf{POWER}(\$G\$2^*\mathbf{EXP}(-\$F\$2^*\$A2/2)^*\$A2,2)$$

in cell $B2 of the copied file and fill this new relation down the appropriate cells of column B, using the select and copy down sequences as before. The existing chart in the copied file undergoes automatic revision and the radial distribution function graph appears as in Figure 1.7.

3. The similar function for the 2s radial wave function or atomic orbital of the hydrogen atom follows if a copy of fig1-4.xls is made and the formula entries of column B are changed appropriately. Thus, enter

$$\$B2 = \mathbf{POWER}(\$G\$2^*(2-\$F\$3^*\$A2)^*\mathbf{EXP}(-\$F\$2^*\$A2/2)^*\$A2,2)$$

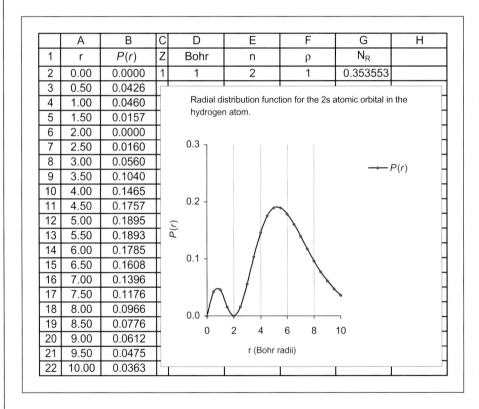

	A	B	C	D	E	F	G	H
1	r	P(r)	Z	Bohr	n	ρ	N_R	
2	0.00	0.0000	1	1	2	1	0.353553	
3	0.50	0.0426						
4	1.00	0.0460						
5	1.50	0.0157						
6	2.00	0.0000						
7	2.50	0.0160						
8	3.00	0.0560						
9	3.50	0.1040						
10	4.00	0.1465						
11	4.50	0.1757						
12	5.00	0.1895						
13	5.50	0.1893						
14	6.00	0.1785						
15	6.50	0.1608						
16	7.00	0.1396						
17	7.50	0.1176						
18	8.00	0.0966						
19	8.50	0.0776						
20	9.00	0.0612						
21	9.50	0.0475						
22	10.00	0.0363						

Figure 1.8   The radial distribution function, $P(r)$, for the 2s atomic orbital in hydrogen.

[6]The form $a\hat{\ }n = a^n$ is interchangeable with the **POWER** (a, n) function. Both expressions return the 'a' raised to the power of n.

in cell \$B2 and use the select and CTL/D sequence to fill this relationship down the column in the selected cells. The existing chart in the copy file again undergoes automatic revision and the radial distribution function $P_{2s}(r)$, Figure 1.8 is displayed.

4. Select the chart and then open the CHART menu. Choose CHART OPTIONS/ GRIDLINES and opt to display the major gridlines for the abscissa, shown on the figure.

The radial distribution function data in Figures 1.7 and 1.8 establish the concept of shell structure in atomic theory. We can distinguish the positions of highest probability for the 1s and 2s electronic states to be the nucleus, but the most probable positions to be within the K shell in the 1s state and the L shell for the electron excited into the 2s state, since these include the extra factor of the surface area, $4\pi r^2$, at each radial distance. Note, the coincidental agreement with the Bohr theory, in the result for the electron in the 1s state, the ground state electronic configuration of the hydrogen atom. The most probable distance of the electron from the proton is the radius of the first Bohr orbit of hydrogen, which is the reason that the atomic unit of length is identified often as the Bohr radius!

1.3   RADIAL WAVE FUNCTIONS FOR MANY-ELECTRON ATOMS

The term '*many-electron atoms*' includes all atoms and atomic ions with more than one extra-nuclear electron. Helium is the simplest many-electron atom. Even in this case the Schrödinger equation cannot be solved analytically. The helium atom, with two electrons and one nucleus, is an example of the '*Three-body Problem*', the equation of motion of which remains unsolved also in Classical Mechanics. The difficulty is that the motion of every particle, in a many-body problem, is coupled to the motion of all the other particles.

For many-electron atoms, the Schrödinger equation can be solved approximately, but very accurately nowadays, by transforming the difficult electron–electron repulsion terms into a spherically averaged form. Then, each electron can be considered to exhibit independent motion in the potential field due to the nucleus and the averaged repulsive interaction with the other electrons in the atom. This simplification leads to the '*Independent Particle Approximation*' preserving the orbital concept and first proposed by Hartree (1) for the simplest many-electron atom, helium, as the product

$$\Phi(1,2) = \phi(1)\phi(2) \qquad\qquad 1.10$$

with ϕ the 1s atomic orbital for the helium atom.

However, this is not a satisfactory description, especially, for electronic configurations in which there are unpaired electron spins, since this, so-called *Hartree* product form, does not comply with the anti-symmetry requirement of the Pauli Principle. In general, many-electron wave functions are written as Slater Determinants, which do exhibit the necessary anti-symmetry properties for electron exchange[7].

---

[7] See Chapter 5.

Thus, the separation of variables possible, in the case of the hydrogen atom, can be performed again. The solutions to these one-electron equations, also, are called 'atomic orbitals' and are similar in form to those for the hydrogen atom,

$$\phi_{nlm}(r, \theta, \phi) = R_{nl}(r:E)Y_{lm}(\theta, \phi)$$ 1.5

There are no analytical forms for the radial functions, $R_{nl}(r)$, as solutions of the radial wave equation. Hartree[2], in 1928, developed the standard solution procedure, the *self-consistent field method* for the helium atom by using the simple product forms of equation 1.10 to represent the two-electron wave function[8]. Herman and Skillman (4) programmed[9] a very useful approximate form of the Hartree method in the early 1960s for atomic structure calculations on all the atoms in the Periodic Table. An executable version of this program, based on their FORTRAN code, modified to output data for use on a spreadsheet is included with the material on the CDROM as hs.exe.

The primary wave function output data from the Herman–Skillman program are the products $rR(r)$, which are known as the numerical radial functions. The radial wave function itself can be recovered on division by the radial distance, r, and approximately near the origin by extrapolating to avoid the infinity. There is one other detail. For the purposes of the numerical integration procedure in the Herman–Skillman procedure the radial data are defined on a non-uniform grid, $\{x\}$, known as the Thomas–Fermi mesh (4). These are converted, in the output from hs.exe, to radial arrays specific to each atom, with

$$r(Z) = \left(\frac{1}{2}\right)\left(\frac{3\pi}{4}\right)^{2/3} Z^{-1/3}x = 0.88534138Z^{-1/3}x$$ 1.11

with Z the atomic number and both x and r expressed in atomic units.

To make an atomic structure calculation, using the program hs.exe, it is necessary to prepare a data input file which is read at the start of the program. For helium, for example, the input file has the following form

```
Helium
    1.000000
    2.  0   1 0.000000
    100 2.000000 −01.721
    −1
    01234567890123456789
```

[8]The Russian, Fock (2,3) showed how to extend this approach to the more general case, in which the self-interaction of the electron [the exchange interaction] is considered and for which the Slater determinantal form of wave function,

$$\Phi(1, 2) = \begin{vmatrix} \phi_{1s\alpha}(1) & \phi_{1s\alpha}(2) \\ \phi_{1s\beta}(1) & \phi_{1s\beta}(2) \end{vmatrix} = \phi_{1s\alpha}(1)\phi_{1s\beta}(2) - \phi_{1s\alpha}(2)\phi_{1s\beta}(1)$$

is appropriate. The Hartree–Fock–Slater method can be applied to atomic and molecular systems as a first approximation to accurate electronic structure information.

[9]The program uses the simplified $X\alpha$ exchange term suggested by Slater (5) so that the Fock term is replaced by an average exchange potential. Moreover, it is assumed that the many-electron wave functions can be represented well by single Slater determinants. The Slater approximation involves the replacement of the exchange potentials for different orbitals by a universal exchange potential formed by averaging the individual terms. The full code is available at the XANES project site, *http://giotto.phys.uwm.edu/~guest/projects/xanes/XANES.html*, wherein there is, too, a detailed discussion of the Herman–Skillman method.

The Herman–Skillman program is bundled on the CDROM with the permission of Professor Frank Herman, Distinguished Professor of Physics, San Jose State University, San Jose, California, USA.

Unlike most modern computer programs, the Herman–Skillman code, written in the FORTRAN language[10], uses *formatted* input and output statements. Thus the 'white' spaces in the text lines of input count to separate each input parameter.

For the helium calculation, apart from the title card[11], the formatted input, above, is constructed using the spacebar to introduce these white spaces. The first of the atomic data cards, card #2, is written in the format f10.6, which means that this input, the value of the alpha parameter in the exchange term, is a floating-point number with six decimal places, occupying in all ten spaces. The default value is 1.000000. The third card has the format f4.0, 2i3 and f10.6. This code means that we expect a floating-point number [f-type] with zero numbers after the decimal point[12] in the first four columns of the input line and it is good practice to include the decimal point, although some compilers tolerate its absence. Then, there are two integers [i-type], right justified, in the next two sets of three columns spaces and finally another floating-point number occupying the next 10 column spaces with six numbers after the decimal point. For helium, this input is the atomic number, 2, the number of core orbitals, 0, the number of valence orbitals, 1, and finally the charge, if any, on the atom, here none and so 0.000000 for the helium atom. The final card of atom-specific information for the helium atom identifies the electron configuration and the format is i3, 2f10.6. We have the compound symbol for the n, l and m quantum numbers, the occupation number and the orbital energy for the 1s [100] orbital. The actual last card is the 'termination' card containing the input -1, with the integer 1 in column 4. In FORTRAN this format is known as i4. The extra line of text, in red, has been added to help the identification of these formats.

Exercise 1.4.   Generation of the numerical radial wave function for the helium atom '1s' atomic orbital.

This exercise involves the calculation of the numerical radial wave function for the helium atom in the form $rR(r)$ output by the Herman–Skillman program for the helium atom and its processing into the equivalent of a radial wave function.

1. Run the Herman–Skillman program for the helium atom calculation. Select hs.exe and pressing the return button, or simply *double click* on the file icon. The program runs in interactive mode. When asked for the input file, type he.in, press return and type c:\he.out when asked to name the output file. The program runs in DOS mode and returns to WINDOWS after execution is complete[13].

---

[10]The MICROSOFT FORTRAN POWERSTATION is an excellent package for writing and running FORTRAN programs on a PC. The modifications of the Herman–Skillman program were written and compiled using this software, but the executable version, of course, will run on any PC with a background DOS environment. A good source for information on FORTRAN coding is *http://www.npac.syr.edu/hpfa/fortgloss/fortgloss.html*.

[11]In the 'early days' each line of computer code was typed on Hollerith cards and fed into the hopper on a computer!

[12]Floating-point numbers [f*.* formats] require only the decimal point to identify the field. But be careful to leave 'white' spaces if other data are to be entered on the same card.

[13]It is good practice to make new directories for each kind of application, so it is best perhaps to locate a copy of hs.exe in a directory named 'Herman–Skillman' or whatever.

2. Open the output file, he.out, generated by the program HS.EXE. If this is the first time you have tried to open this kind of file and the option box for the software needed appears opt to open such files always as text files in NOTEPAD or WORDPAD.
3. Select all the text in the file and COPY/PASTE into cell $A$3 of a new worksheet, fig1-9.xls. With luck, the EXCEL software will parse the pasted information into columns A and B of the worksheet. If not, open the TOOLS dropdown on the EXCEL toolbar and select the TEXT to COLUMNS option. Follow the wizard's instructions and the radial mesh data will separate in column A from the numerical wave function data, which will appear in column B.
4. Enter the appropriate labels in the header cells in rows 1 to 3.
5. To avoid the infinity problem and the appearance of the error message #ERROR the division by r is from the second radial mesh point only, enter

$$\$C5 = \$B5/\$A5$$

in cell $C5 and fill this formula down the column to match the radial data in A.
6. Select cell $C$4 and enter the formula bar by typing '='. Select 'f_x' on the toolbar and following the wizard through the *statistical* library choose **INTER-CEPT**. Opt for the select arrow on the right-hand side of the **INTERCEPT** window and using the mouse select cells $B$5 to $B$10 for the *y-values*

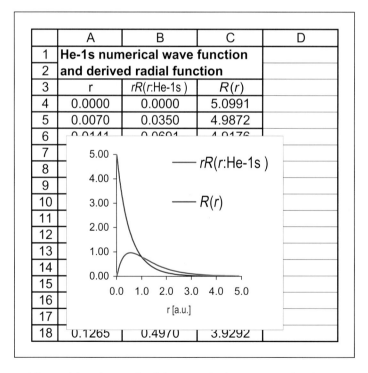

	A	B	C	D
1	He-1s numerical wave function			
2	and derived radial function			
3	r	rR(r:He-1s)	R(r)	
4	0.0000	0.0000	5.0991	
5	0.0070	0.0350	4.9872	
6	0.0144	0.0601	4.0176	
7				
8				
9				
10				
11				
12				
13				
14				
15				
16				
17				
18	0.1265	0.4970	3.9292	

Figure 1.9   The graph of the numerical $rR(r)$ data for the 1s atomic orbital of the helium atom and the derived results for $R(r)$ by using the **INTERCEPT** function.

and $A$5 to $A$10 for the *x-values*. Press enter and the software performs linear regression on the selected data to return the intercept value $R(0)$ in cell $C$4. The formula entry in cell $C$4 takes the form

$$= \textbf{INTERCEPT}(\$C\$5:\$C\$10, \$A\$5:\$A\$10)$$

7. Select cells $A$3 to $C$65. Open the INSERT menu on the toolbar and choose CHART.
8. Follow the wizard instructions to construct the *xy*-scatter chart shown in Figure 1.9 of $rR(r)$ and $R(r)$ against r.

Note, that the Herman–Skillman output, the numerical radial wave functions, $rR(r)$, are normalized in the form of equation 1.9.

Exercise 1.5. Generation of the numerical radial wave functions Li$_{1s}$ and Li$_{2s}$ orbitals.

Figure 1.10 presents the similar transformation, to Figure 1.9 of the helium data, for the numerical $rR(r)$ data for the lithium 1s and 2s orbitals. Construct the spreadsheet as follows.

1. Make a copy of the file he.in [use COPY and PASTE commands in EXPLORER]. Click on the file name with the right mouse button and change the filename to li.in. Then open this new file, LI.IN, and change the atom-specific data cards using the information for lithium in Table 1.2.
2. Run HS.EXE. Open the output file li.out. This time NOTEPAD should open the file automatically. Note, that in this case there are two sets of numerical data, the one for the 1s orbital and the second for the 2s orbital in lithium.

Table 1.2 The orbital energies, in Rydbergs, for the atoms of the first short period of the Periodic Table (1 Rydberg = 13.605 eV = 0.8000 Hartrees) returned by the Herman–Skillman program.

	1s	2s	2p
He	−1.721		
Li	−4.398	−0.4039	
Be	−8.698	−0.6012	
B	−14.373	−0.9259	−0.4598
C	−21.378	−1.2895	−0.6603
N	−29.737	−1.6959	−0.8445
O	−39.456	−2.1440	−1.0409
F	−50.536	−2.630	−1.2502
Ne	−62.990	−3.168	−1.471

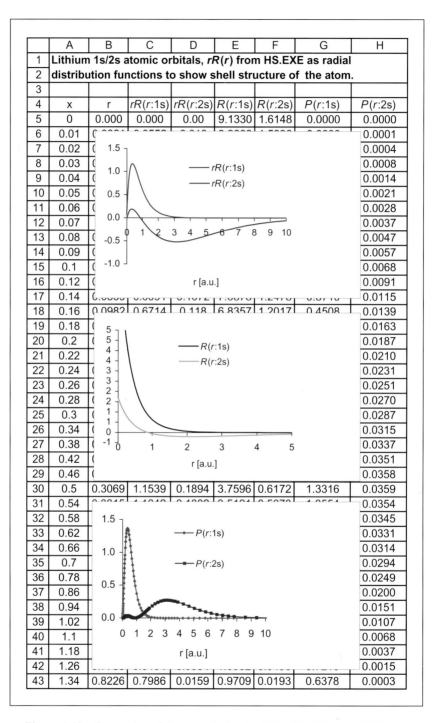

	A	B	C	D	E	F	G	H
1	Lithium 1s/2s atomic orbitals, $rR(r)$ from HS.EXE as radial							
2	distribution functions to show shell structure of the atom.							
3								
4	x	r	$rR(r{:}1s)$	$rR(r{:}2s)$	$R(r{:}1s)$	$R(r{:}2s)$	$P(r{:}1s)$	$P(r{:}2s)$
5	0	0.000	0.000	0.00	9.1330	1.6148	0.0000	0.0000
6	0.01							0.0001
7	0.02							0.0004
8	0.03							0.0008
9	0.04							0.0014
10	0.05							0.0021
11	0.06							0.0028
12	0.07							0.0037
13	0.08							0.0047
14	0.09							0.0057
15	0.1							0.0068
16	0.12							0.0091
17	0.14							0.0115
18	0.16	0.0982	0.6714	0.118	6.8357	1.2017	0.4508	0.0139
19	0.18							0.0163
20	0.2							0.0187
21	0.22							0.0210
22	0.24							0.0231
23	0.26							0.0251
24	0.28							0.0270
25	0.3							0.0287
26	0.34							0.0315
27	0.38							0.0337
28	0.42							0.0351
29	0.46							0.0358
30	0.5	0.3069	1.1539	0.1894	3.7596	0.6172	1.3316	0.0359
31	0.54							0.0354
32	0.58							0.0345
33	0.62							0.0331
34	0.66							0.0314
35	0.7							0.0294
36	0.78							0.0249
37	0.86							0.0200
38	0.94							0.0151
39	1.02							0.0107
40	1.1							0.0068
41	1.18							0.0037
42	1.26							0.0015
43	1.34	0.8226	0.7986	0.0159	0.9709	0.0193	0.6378	0.0003

Figure 1.10   Conversion of the numerical output (a) of the Herman–Skillman program to (b) 1s and 2s radial wave functions and (c) 1s and 2s radial distribution functions for lithium following the instructions in Exercise 1.5.

3. As before, select one set of data and COPY/PASTE into cell $A5 of a new worksheet and convert the text data into two columns of data as before.
4. Repeat this procedure for the second set of data for the 2s orbital using cell $C5 as the starting point.
5. Delete the redundant column B of radial data.
6. As before, convert the $r R(r)$ data into $R(r)$ over the radial mesh in both cases.
7. Then, following the instructions in Exercise 1.3, generate the radial distribution functions for the lithium 1s and 2s orbitals in columns G and H.
8. Finally, convert the results into xy-scatter charts to complete Figure 1.10.

The clear suggestions from this analysis are that the atomic orbitals for many-electron atoms are similar to the equivalent hydrogenic ones and so it might be possible to discover mathematical functions to represent analytically and continuously the essential details of the radial distribution functions. Our appreciation of how to achieve this becomes clearer if we construct the radial distribution functions as in the last diagram of Figure 1.10.

It is clear from the radial distribution function for the lithium 2s atomic orbital that the lithium 2s electron has the greatest probability of being outside the $1s^2$ core, the K shell region of the atom. Indeed, it is as if, for the most part, the $1s^2$ core electron density screens the single 2s electron from the attraction due to two protons. On the other hand, there is still a finite significant probability that the 2s electron can be found closer to the nucleus, so the screening is not that perfect and the 2s electron is said to penetrate the electron core of the atom. This differential penetration of different radial distribution functions, of course, is the reason, too, for the change in the orbital energy sequence for many-electron atoms compared to the one observed in the case of the hydrogen atom. The effect provides an explanation for much of the chemistry of different atoms in terms of their electronic configurations, as is explained in modern textbooks on inorganic chemistry.

The availability, only, of numerical data for the electron distributions in atoms other than hydrogen and the increasing complexity of that data with the atomic number of the atom, would be a serious limitation on our comprehension of atomic and molecular theory. In Chemistry the 'orbital' is fundamental to the understanding of all the body of data that can be catalogued using the modern Periodic Table. It is an essential concept, too, in modern bonding theory, because general rules can be established, based on orbital interactions.

So, let us concentrate on the matter of the fitting of mathematical expressions to the numerical data for the radial distributions for the electrons in many-electron atoms. Maybe it is appropriate to use an effective atomic number, Z_{eff}, rather than the full Z in the equations in Table 1.1. Such a change is consistent with the view that the other electrons screen some, but not all, of their nuclear protonic charges in many-electron atoms.

1.4   SLATER-TYPE ORBITALS

In 1930, J.C. Slater (6,7) proposed that this kind of modification of the mathematical functions for the hydrogen orbitals could be used to fit many-electron radial data[14]. Slater

---

[14] Slater worked out his rules empirically, on the basis of matching results of calculations to X-ray energy levels and other considerations, although he suggested also, in an earlier paper (6) that a variation-principle based approach might be appropriate.

suggested that the radial functions might be represented in the form[15]

$$R(r:n\zeta) = (2\zeta)^{n^*+(1/2)}[(2n^*)!]^{-1/2}r^{n^*-1}\exp(-\zeta r) \qquad 1.12$$

and that, if necessary, several Slater functions or orbitals could be combined to return a good fit to radial numerical data.

Notice the details of the form of Slater-type orbitals [STOs]. The polynomial terms of the radial functions in hydrogen, Table 1.1, are simplified to include the radius raised only to the power of the principal quantum number less one. In addition, an effective principal quantum number, n^*, is used.

The 'best' values of the effective principal quantum number, n^*, and this effective atomic number, ξ, follow from Slater's rules, namely,

1. For the principal quantum number, n, equal to 3 or less, n^* is equal to n.
2. $\zeta = (Z - S)/n^*$, with Z the atomic number and S the screening parameter for one-electron due to the others in the atom.
3. These parameters are the same in the groups of orbitals, 1s; 2s, 2p; 3s, 3p; 3d; and so on.
4. For these groups of orbitals then S is determined on the basis;

    i.   there is no contribution due to any electron in a group higher than the one under consideration;
    ii. 0.35 is added to S for every other electron in the group, but in the case of 1s electrons, only 0.30 is added;
    iii. 0.85 is added for each electron of an sp group with principal quantum number one less than that of the electron being considered, 1.0 is added for every other electron in groups nearer the nucleus and 1.0 for every inner electron to a d-group electron.

For Li, for example, the Slater's rules exponents, ζ, for the 1s and 2s orbitals are 2.70 and 0.65. Thus, we can imagine that the second 1s electron screens the first electron and vice versa from 0.3 of the three nuclear charges, while the $1s^2$ core screens the single electron in the STO equivalent to the 2s hydrogenic orbital from almost two nuclear charges.

Other prescriptions for the exponents, ζ, have been advanced over the years. Clementi and Raimondi (8) proposed in 1963 that the 'best' exponents should be based on the criterion that the atomic energy should be minimized. Clementi, too, (9,10) and others (11) have investigated the use of more than one Slater function to obtain a better representation of the radial wave functions for many-electron atoms.

These proposals, to use more than one function and thereby achieve extra flexibility in the matching of the approximate functions to the numerical data and to apply a variation principle condition to determine the 'best' approximations, remain essential considerations in the development of modern basis set theory. Table 1.3 lists Clementi's *double-zeta* sets for the atoms of the elements from He to Ne.

This approach, the making of a linear combination with coefficients subject to optimization for particular purposes is simpler, in principle, than depending on changes in the

---

[15]Note, with reference to Figure 1.6, the pre-exponential factor in Slater orbitals includes normalization only over the radial coordinate.

Table 1.3 The optimized *double-zeta* STO-exponents obtained by Clementi from accurate Hartree–Fock calculations on the atoms from He to Ne.

Atom		ζ_{1s}	ζ_{2s}	ζ_{2p}
He	Z = 2	1.4461		
		2.8622		
Li	Z = 3	2.4331	0.6714	
		4.5177	1.9781	
Be	Z = 4	3.3370	0.6040	
		5.5063	1.0118	
B	Z = 5	4.3048	0.8814	1.0037
		6.8469	1.4070	2.2086
C	Z = 6	5.2309	1.1678	1.2557
		7.9690	1.8203	2.7263
N	Z = 7	6.1186	1.3933	1.5059
		8.9384	2.2216	3.2674
O	Z = 8	7.0623	1.6271	1.6537
		10.1085	2.8216	3.6813
F	Z = 9	7.9179	1.9467	1.8454
		11.0110	3.0960	4.1710
Ne	Z = 10	8.9141	2.1839	2.0514
		12.3454	3.4921	4.6748

exponents of exponential functions. This is particularly true when the linear combinations involve the quadratic exponential Gaussian approximation described in the next section. Note, too, that a distinction is made between the exponent values for the 2s and 2p radial functions. This feature of the *double-zeta* Slater approach increases the computational difficulty by a factor of two in the numbers of integrals to be calculated and it was common practice to use the same exponent for valence s and p radial function approximations in early applications of Gaussian basis set theory, since in Gaussian basis set theory the approximate functions are linear combinations over, at least, several Gaussian functions.

Slater based his original rules for the screening constant, s, for any electron in an atom, on the matching of data to X-ray energy levels and other considerations, although he did suggest as well the possibility of applying a variation principle. This, in fact, is the modern basis for the current Slater rules (12,13) in which the screening for a particular electron follows from the electronic configuration as

$$s(1s) = 0.3000(\#_{1s} - 1) + 0.0072(\#_{2s} + \#_{2p}) + 0.0158(\#_{3s} + \#_{3p} + \#_{3d} + \#_{4p})$$

$$s(2s) = 1.7208 + 0.3601(\#_{1s} - 1 + \#_{2p}) + 0.2062(\#_{3s} + \#_{3p} + \#_{3d} + \#_{4p})$$

$$s(2p) = 2.5787 + 0.3326(\#_{2p} - 1) - 0.0773(\#_{3s}) - 0.0161(\#_{3p} + \#_{4s}) - 0.0048(\#_{3d})$$

$$+ 0.0085(\#_{4p})$$

$$s(3s) = 8.4927 + 0.2501(\#_{3s} - 1 + \#_{3p}) - 0.0778(\#_{4s}) + 0.3382(\#_{3d}) + 0.1978(\#_{4p})$$

$s(3p) = 9.3345 + 0.3803(\#_{3p} - 1) + 0.0526(\#_{4s}) + 0.3289(\#_{3d}) + 0.1558(\#_{4p})$

$s(3d) = 13.5894 + 0.2693(\#_{3d} - 1) - 0.1065(\#_{4p})$

$s(4s) = 15.505 + 0.0971(\#_{4s} - 1) + 0.8433(\#_{3d}) + 0.0687(\#_{4p})$

$s(4p) = 24.7782 + 0.2905(\#_{4p} - 1)$

In these identities the symbols $\#_{nl}$ are the occupation numbers of the orbitals specified in the electronic configuration for the particular atom. In general, for many-electron atoms other than the simplest ones, these definitions of the screening constants lead to much better agreements with Hartree–Fock–Slater numerical results such as the output of hs.exe, for the atoms of the typical elements in the Periodic Table.

Exercise 1.6.   Slater-type orbitals and the corresponding radial distribution functions.

To construct radial functions and radial distribution functions, based on the Slater approximation, only straightforward modifications to the designs of previous spreadsheets are required. The spreadsheet file, fig1-11.xls, and Figure 1.11, based on lithium 2s data, are based on the following instructions.

1. Run hs.exe and copy the output data for the lithium 2s radial function onto columns A and B of a new spreadsheet. This means that the Thomas–Fermi mesh will be the radial mesh for the remainder of the exercise. For simplicity delete column A the dimensionless x-mesh. Label the remaining columns of translated data from the Herman–Skillman output file, r[TF], $rR(r\_Li$-2s) and enter the headers RDF-Li_2s and RDF-Slater in cells $C$5 and $D$5.
2. In cell $C$6 enter

$$\$C6 = \textbf{POWER}(\$B6, 2)$$

to convert the numerical radial function to the radial distribution function and fill this formula down the active cells of column C over the radial mesh in column A.
3. Calculate the Slater exponent for lithium 2s using Slater's rules and enter this value in cell $D$4.
4. For each of the radial mesh points in column A calculate the Slater function equivalent of the radial distribution function in the cells of column D, with, for example,

$$\$D\$6 = \textbf{POWER}(\textbf{POWER}(2^*\$D\$4,2.5)^*\textbf{SQRT}(1/\textbf{FACT}(4))^*$$
$$\$A6^*\textbf{EXP}(-\$D\$4^*\$A6),2)^*\textbf{POWER}(\$A6,2)$$

using equation 1.12.
5. Finally, construct the xy-scatter chart of the variation in the derived data with the radial distance as in Figure 1.11.

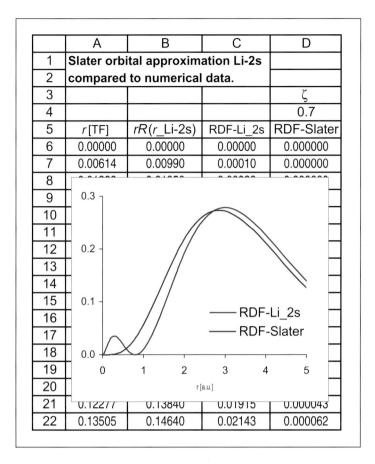

	A	B	C	D
1	Slater orbital approximation Li-2s			
2	compared to numerical data.			
3				ζ
4				0.7
5	r[TF]	$rR(r\_Li$-2s$)$	RDF-Li_2s	RDF-Slater
6	0.00000	0.00000	0.00000	0.000000
7	0.00614	0.00990	0.00010	0.000000
21	0.12277	0.13840	0.01915	0.000043
22	0.13505	0.14640	0.02143	0.000062

Figure 1.11 Comparison between the radial distribution function, calculated with HS.EXE, for the Li-2s orbital and from the
simple Slater function.

As you can see a fair degree of agreement has been achieved between the two distributions, in the outer 'valence' region of the lithium atom. But, because the Slater functions
are nodeless, the approximation fails within the K shell region of the atom. Since the
principal purpose of these approximations to the numerical HFS data was to provide only
a starting function for molecular calculations the matter of very close agreement is not
important nowadays.

In the 1960s, this last statement was not acceptable and, indeed, it was to minimize such
discrepancies that prompted Clementi to propose his optimized *double-zeta* sets. That better
results can be obtained using this approach is shown in Figure 1.12, wherein the comparisons are made between the optimized Clementi *double-zeta* function and the numerical
results. The detail of this calculation requires the involvement of the variation principle
to determine the relative weightings of the two components of the *double-zeta* basis by
minimizing the calculated 2s orbital energy. This requires that the 2s function be rendered
orthonormal [Chapter 3] to the 1s function in lithium. Thus, all four Slater functions in
the Clementi *double-zeta* basis, the two for the 1s and the two for the 2s, contribute to

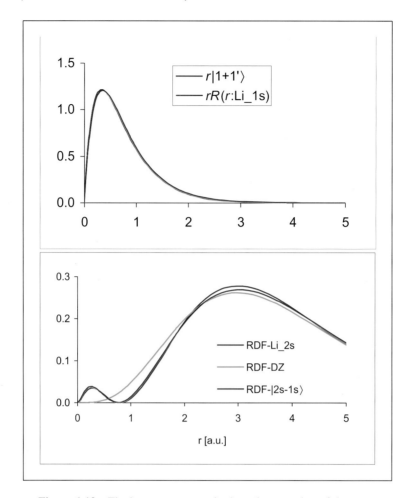

Figure 1.12   The better agreement in the valence region of the atom with the numerical data for the Li_{2s} radial wave function, which follows for the Clementi *double-zeta* basis sets for Li_{1s} and Li_{2s} mutually orthogonal.

the overall approximate function for the 2s radial wave function and lead to the much greater agreement with the numerical data evident in Figure 1.12.

In any event, the objective is to generate the 'best' energies of the atomic and molecular orbitals and increasingly, the total energy. Thus, there are many excellent calculations in the literature of molecular orbital theory using the Slater approach and there are many early successes of molecular orbital theory based on calculations involving Slater-type orbitals.

However, the major problem with Slater-type and hydrogenic orbitals is a computational one. For all but the simplest molecules, the evaluation of electron–electron repulsion integrals is a formidable problem, which is not simplified when the atomic orbitals are approximated using Slater functions. This is especially the more so when linear combinations of Slater orbitals are used, because the number of multi-centre integrals increase with the fourth power of the number of primitive orbital components.

Finally, it is useful to clarify some basis set notation. A minimal basis set of Slater orbitals for the atoms of the first short period of the Periodic Table comprises a single Slater function for the 1s, 2s and the three 2p orbitals except for hydrogen and helium for which only the 1s orbital counts. Thus, for example, for ethene, C_2H_4, we identify the minimal basis as (2s1p/1s), which translates to 1s, 2s and 2p on non-hydrogen atoms and 1s only on each hydrogen. In a *double-zeta* Slater basis, two Slater-type functions are used for each orbital, while in a *split* basis, this prescription applies for the valence orbitals but the core orbitals are represented by only one Slater function [this is distinguished in modern usage as a *split-valence* basis set]. It is common, too, to augment the basis to model polarization effects in molecular environments. For example, 'p-type' character is added to the hydrogen basis by including in the linear combination $2p_x$, $2p_y$ and $2p_z$ Slater orbitals, while 'd' type character is added to the atoms from Li to F by including the five 3d Slater orbitals. Such a *double-zeta polarized* basis set is identified by the notation (4s,2p,1d/2s1p). This notation is carried over to identify Gaussian basis sets.

1.5 GAUSSIAN-TYPE FUNCTIONS — THE |sto-3g⟩ MINIMAL BASIS SET

In 1950 Boys (14) proposed the use of Gaussian functions to approximate the radial components of atomic orbitals. Such functions had been used independently by McWeeny (15,16) who later (17) obtained Gaussian approximation (*contracted Gaussians*) to Slater orbitals of 1s, 2s and 2p types and used them in calculating X-ray Scattering factors. Boys' famous paper showed how the complicated integrals over Slater functions, needed for the solution of atomic or molecular Schrödinger equations, become simplified and easily soluble when these Slater functions are substituted for with Gaussian-type functions. In the 1950s there was relatively little further development of Boys' and McWeeny's ideas undoubtedly because the calculation of large numbers of integrals was a formidable problem in a world without computers as we know them today (18). However, the Gaussian approach was applied increasingly after 1960 (19–33) to calculations of atomic and molecular electronic structures and properties. During that period, too, work continued on the Slater functions integral problem, as well, for example (34), but by 1973, with the appearance of the GAUSSIAN 70 program[16] in the QCPE catalogue, the Gaussian methodology was established as the standard approach for the future.

The general form of the Gaussian-type function is

$$g(r{:}n, \alpha) = N_g r^{n-1} e^{-\alpha r^2} \qquad 1.13$$

The normalization constant, N_g, for the pure Gaussian function over the radial coordinate, only, is

$$N_g = \left[\frac{2^{2n+3/2}}{(2n-1)!!\pi^{1/2}} \right]^{1/2} [\alpha]^{(2n+1)/4} \qquad 1.14[17]$$

However, a normalized '1s' Gaussian function, $g(r{:}1s, \alpha)$ generally is written

$$g(r{:}1s, \alpha) = (2\alpha/\pi)^{0.75} \exp(-\alpha r^2) \qquad 1.15$$

---

[16]GAUSSIAN 70: Ab initio SCF-MO calculations· on organic molecules, W.J. Hehre, W.A. Latham. R. Ditchfield, M.D. Newton and J.A. Pople, *Quantum Chemistry Program Exchange [QCPE]*, **11** (1973) 236.

[17]$n!!$ identifies the *double factorial*. For n odd this is $n(n-2), (n-4), \ldots (3), 1$. For n *even* this is $n(n-2)(n-4)\ldots(4), 2$.

Note, carefully, the pre-exponential normalization factors in these equations. The N_g in equations 1.13 and 1.14 normalize only over the radial coordinate defined in Figure 1.6 and equation 1.9. In Gaussian basis set theory, it is normal to write the pre-exponential normalization factor including the angular terms in Table 1.1. Thus, to obtain the normalization constant in equation 1.15 we need to reduce equation 1.14 for the case $n = 1$ and then multiply by the angular normalization factor for 1s $(1/4\pi)^{1/2}$.

The Gaussian representation, of the numerical radial wave functions in atoms, is at the heart of modern molecular orbital theory. The different formulations are known as basis sets and the most common and proven basis sets are included in the various packages for molecular orbital theory calculations[18]. In this book, the symbol |sto-ng⟩ identifies a linear combination of n Gaussian orbitals[19] used, in the present context, to model the corresponding Slater function.

The great advantage of the Gaussian representation, based on linear combinations of Gaussians [*Gaussian primitives*], is that many of the difficult integrals in molecular orbital theory become relatively straightforward, when transformed following Boys' prescription (14). The fundamental result is that an integral over a product of 1s Gaussians centred about two positions, reduces to a single integral over a third Gaussian about an intermediate position. Boys pointed out, too, that more complicated integrals over higher order Gaussians, for example p and d-type Gaussians can be derived from this basic result, because differentiation of the basic product integrals with respect to the coordinates returns integrals over increasingly higher order functions.

Thus, all the multi-centre integrals required to solve the Schrödinger equation, using a Gaussian basis, can be reduced eventually to the standard form. Much effort has been devoted, over the intervening years (35–37) to the design of efficient computational procedures and much faster integration routines have been developed in recent years (38). In the end, however, all these methods depend on the original Boys' procedure for the transformation of the integral over Gaussian products.

Thus, for two 1s Gaussian functions in the form of equation 1.15, about two centres R_1 and R_2, the product Gaussian

$$K G(r - R_p : \alpha_p) = G(r - R_1 : \alpha_1) G(r - R_2 : \alpha_2) \qquad 1.16$$

with

$$K = \exp\left(-\frac{\alpha_1 \alpha_2}{(\alpha_1 + \alpha_2)} |R_1 - R_2|^2\right) \qquad 1.17$$

$$R_p = \frac{(\alpha_1 R_1 + \alpha_2 R_2)}{(\alpha_1 + \alpha_2)} \qquad 1.18$$

and

$$\alpha_p = \alpha_1 + \alpha_2 \qquad 1.19$$

simplifies all further calculation.

---

[18]For example, GAUSSIAN 98 *http://www.Gaussian.com/*, SPARTAN *http://www.wavfun.com/*, Q-CHEM, *http://www.q-chem.com/*, MOLPRO, *http://www.tc.bham.ac.uk/molpro/* and the GAMESS program, available free, for personal use from *http://www.msg.ameslab.gov/pcgames*.

[19]Called *primitives* and remember to check the normalization for the defining conditions!

It is straightforward to demonstrate the validity of equation 1.16 using a spreadsheet. All we need to do is plot the projections of the individual Gaussian functions and their products on a suitable radial mesh using the CHART wizard. The spreadsheet display is shown in Figure 1.13 and as you see provision is made for the variation of the locations of the Gaussian functions, the entries in cells $F$2 and $G$2, on the axis defined by the positions in column A. It is instructive to vary the positions and also to change the exponents, the entries in cells $F$4 and $G$4, to make this product relationship familiar. Finally, note the inclusion of equation 1.16 in the display as the entries in column D and the graph labelled 'G'. There is an exact agreement between the projections of this function and the product drawn as g1g2 on the chart from the data in column C. The design of fig1-13.xls is left as an exercise, but all the details are available on the spreadsheet.

There are disadvantages associated with the substitution of Slater functions by Gaussian functions. More than one Gaussian functions is required to model Slater functions accurately. In fact, about two to five times as many Gaussian functions are required to produce a given accuracy returned in calculations with Slater functions. For example (39) ten Gaussian functions must be taken in linear combination to return six-figure accuracy for helium and beryllium, while some sixteen functions are required (as ten 1s Gaussians and six 2p Gaussians) to return good energies for second-row atoms of the Periodic Table. For the elements aluminium to argon the use of twelve 1s and nine 2p-type Gaussian functions returns results as good as those obtained using a *double-zeta* Slater basis.

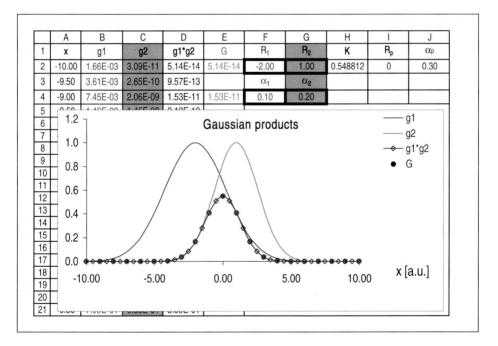

Figure 1.13 Demonstration of the validity of equation 1.16 for the Gaussian product of column D formed by multiplication of the projections of the Gaussians in columns B and C following equations 1.16 to 1.19. Note, the cells bordered in black in fig1-13.xls can be changed even with the spreadsheet 'locked'.

Table 1.4   The Pople, Hehre and Stewart minimum basis sets, |sto-3g⟩, for the 1s, 2s and 2p [Slater] functions of the hydrogen atom (33). Note the use of the same exponent in the basis sets to approximate the 2s and 2p Slater functions.

	α_{1s}	d_{1s}	α_{sp}	d_{2s}	d_{2p}
|sto-3g⟩	0.109818	0.444635	0.0751386	0.700115	0.391957
	0.405771	0.535328	0.2310310	0.399513	0.607684
	2.227660	0.154329	0.9942030	−0.0999672	0.155916

The taking of such linear combinations is known as 'contraction' in Gaussian basis set theory. Because the individual coefficients are fixed in calculations, the linear combination of primitive functions can be taken as one overall function in calculations. These simple linear combinations can be refined further to make, for example, *split-basis* or *split-valence* sets, in which the linear combination is divided into distinct parts. For calculations involving the atoms of the first short period of the Periodic Table, the minimum basis Gaussian sets model only the 1s function for hydrogen and helium and the 1s, 2s and 2p functions for the other atoms Li to F. Table 1.4 lists the coefficients and exponents of the |sto-3g⟩ minimum basis for the 1s, 2s and 2p Slater functions in hydrogen, determined by Hehre, Stewart and Pople (33). As you might expect, now, such minimal basis sets do not provide very good results in calculations.

Exercise 1.7.   The |sto-3g⟩ representation for H_{1s}.

1.   On a new spreadsheet, enter the basic data required to construct the graphs of the 1s radial function and the |sto-ng⟩ set. In this example, the |sto-3g⟩ data listed in Table 1.4. appears between cells $B$4 to $D$5.
2.   Provide for the multiplication by the normalization constant of equation 1.15, with, for example, in cell B$6

$$B\$6 = \textbf{POWER}((2^*\$B\$4/\textbf{PI}(\,)),0.75)$$

and copy this formula into $C$6 and $D$6.
3.   Following the procedure of Exercise 1.1, lay out a suitable radial mesh and corresponding column of radial function values for the hydrogen 1s atomic orbital, which is the same for this case only as the Slater function. Start with cell $A$5.
4.   For each of the three Gaussian primitives in Table 1.4, construct their projections on the radial mesh in cells $B$9 to $D$29. For example, the |g1⟩ entry in cell $B$9 is
$$\$B9 = \$B\$6^*\textbf{EXP}(-\$B\$5^*A9^2)$$

5.   Use the EDIT/FILL/DOWN procedure after selecting the appropriate cells over these columns for the radial mesh to propagate these formulae down the active cells.

6. In the next blank cell, $G9, in row G to the right of these entries, sum the contributions from the |sto-ng⟩ primitives. For example, in the present case, enter

$$\$G9 = \mathbf{SUM}(\$B9\!:\!\$D9)$$

As before, extend this relation over the radial mesh, using the EDIT/FILL/DOWN sequence.

7. Complete the construction of Figure 1.14 by generating two xy-scatter charts the first, of the individual primitives over the radial mesh and then, the second, the |sto-3g⟩ approximation to the hydrogen 1s radial function, which is also the Slater function, in this case.

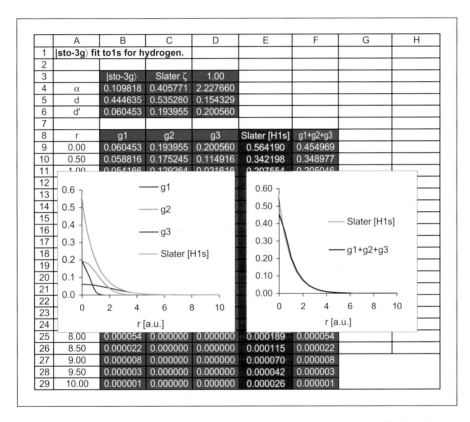

Figure 1.14   The primitives in the |sto-3g⟩ representation of the H_{1s} radial function in the hydrogen atom, and the 1s radial function itself, which, of course, is the Slater function in this 1s case. The second chart shows the degree of fit achieved for the minimal basis of Table 1.4.

Unlike the basic Slater function the basic Gaussian function provides a very deficient approximation to a 1s atomic radial function. Whereas the Slater function has a non-zero derivative at the nucleus and so can model the *cusp* of the radial function, the Gaussian function exhibits a zero derivative at the origin [compare the derivatives of equations 1.12 and 1.15 at $r = 0$]. The behaviour of the quadratic exponential Gaussian function is deficient, too, at large radial distances from a nucleus, since hydrogenic radial functions decay exponentially.

As you will see, in the remainder of this chapter, in the intermediate region of radial distance away from the nucleus, the Gaussian and the Slater representations are comparable and the majority of the Gaussian primitives are required to model the near nucleus behaviour. But, the critical behaviour problem with Gaussian bases has physical consequences. Thus, (40) the long-range overlap between atoms, important in weak interactions, is always underestimated in a Gaussian basis and equally, properties dependent on near nuclear behaviour of the wave function, such as charge and spin density at the nucleus [*the Mössbaüer effect, for example*] will be in error. While similar observations can be made concerning the performance of Slater functions, the deficiencies, which are found, are not all pervasive in this manner.

1.6 |sto-ng⟩ BASIS SETS

There is a need to clarify certain important details probably nowadays submerged in the history of the development of basis set theory. In the beginning, it was not universal, as now, to design Gaussian sets determined by atomic orbital energy criteria. The benefits, in calculations, too, of using the simplest 1s, 2p and 3d-based Gaussian representations had not been recognized.

In early important work, Reeves (22–24) worked out the minimum basis sets for the hydrogen orbitals listed in Table 1.5. Reeves' work is a good starting point to explore the various complexities of detail in Gaussian set theory. The data in the table consist of complete pre-exponential factors and exponents for the Gaussian linear combinations |sto-ng⟩. There are no other considerations in the construction of the approximate matches to the hydrogenic orbitals and this agreement and Figure 1.15a, b and c display the agreement for the 1s, 2s and 2p cases. Note that Reeves worked out different Gaussian exponents

Table 1.5   The Gaussian basis sets proposed by Reeves to represent the hydrogenic radial functions. The table entries, for each basis set, are the exponents, α, of the primitive Gaussians and then in the second columns the complete normalized coefficients, d, of the linear combinations.

	H_{1s}		H_{2s}		H_{2p}	
	α	d	α	d	α	d
|sto-1g⟩	0.28294	0.65147			0.04527	0.02976
|sto-2g⟩	0.2015	0.1760	0.01630	−0.03335	0.03240	0.01539
	1.3320	0.2425	0.32110	0.14093	0.1393	0.03952
|sto-3g⟩	0.1483	0.1084	0.01934	−0.03909	0.02465	0.00803
	0.6577	0.2164	0.2586	0.10736	0.07970	0.02870
	4.2392	0.1575	1.6844	0.08625	0.3363	0.03369
|sto-4g⟩	0.1233	0.0756	0.02006	−0.04064	0.02018	0.00452
	0.4552	0.1874	0.2276	0.09059	0.05572	0.02042
	2.0258	0.1620	1.0334	0.08108	0.1743	0.02944
	13.7098	0.0947	7.0420	0.04646	0.7342	0.02544

for the 2s and 2p hydrogenic radial functions, but note, most particularly, that Reeves' parameters are based on the making of linear combinations of 1s Gaussians as defined in equation 1.15 and the normalization factors are included in his pre-exponential coefficients of the linear combinations. Moreover, these are the full normalization constants over all the spherical polar coordinates, as you can check using the procedures developed in Chapter 3, since it is customary to use full normalization for the Gaussian sets as in equation 1.15.

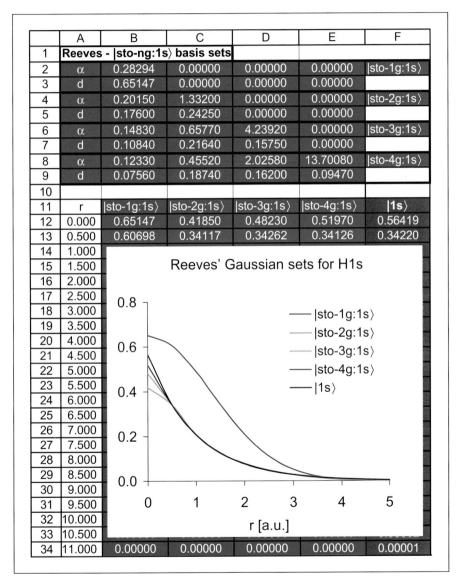

	A	B	C	D	E	F
1	Reeves - |sto-ng:1s⟩ basis sets					
2	α	0.28294	0.00000	0.00000	0.00000	|sto-1g:1s⟩
3	d	0.65147	0.00000	0.00000	0.00000	
4	α	0.20150	1.33200	0.00000	0.00000	|sto-2g:1s⟩
5	d	0.17600	0.24250	0.00000	0.00000	
6	α	0.14830	0.65770	4.23920	0.00000	|sto-3g:1s⟩
7	d	0.10840	0.21640	0.15750	0.00000	
8	α	0.12330	0.45520	2.02580	13.70080	|sto-4g:1s⟩
9	d	0.07560	0.18740	0.16200	0.09470	
10						
11	r	|sto-1g:1s⟩	|sto-2g:1s⟩	|sto-3g:1s⟩	|sto-4g:1s⟩	|1s⟩
12	0.000	0.65147	0.41850	0.48230	0.51970	0.56419
13	0.500	0.60698	0.34117	0.34262	0.34126	0.34220
14	1.000					
15	1.500					
16	2.000					
17	2.500					
18	3.000					
19	3.500					
20	4.000					
21	4.500					
22	5.000					
23	5.500					
24	6.000					
25	6.500					
26	7.000					
27	7.500					
28	8.000					
29	8.500					
30	9.000					
31	9.500					
32	10.000					
33	10.500					
34	11.000	0.00000	0.00000	0.00000	0.00000	0.00001

Figure 1.15a   Comparisons of the variations of the Reeves |sto-ng:1s⟩ basis sets with exact variation for the hydrogen |1s⟩ radial function. Note, the substantial disagreement for the simple single Gaussian function and the increasing agreement with the numbers of Gaussian primitives used.

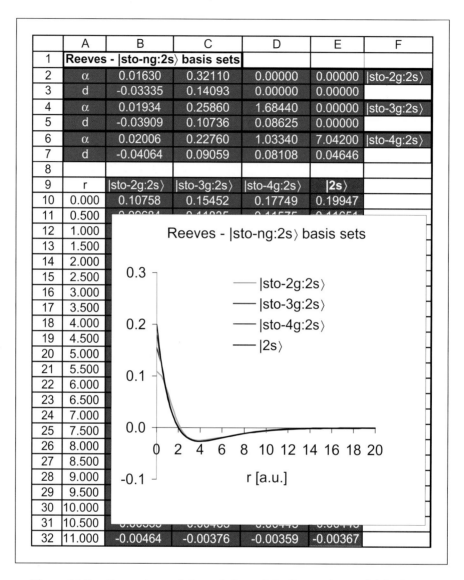

The spreadsheet/figure contains the following data:

	A	B	C	D	E	F
1	Reeves - |sto-ng:2s⟩ basis sets					
2	α	0.01630	0.32110	0.00000	0.00000	|sto-2g:2s⟩
3	d	-0.03335	0.14093	0.00000	0.00000	
4	α	0.01934	0.25860	1.68440	0.00000	|sto-3g:2s⟩
5	d	-0.03909	0.10736	0.08625	0.00000	
6	α	0.02006	0.22760	1.03340	7.04200	|sto-4g:2s⟩
7	d	-0.04064	0.09059	0.08108	0.04646	
8						
9	r	|sto-2g:2s⟩	|sto-3g:2s⟩	|sto-4g:2s⟩	|2s⟩	
10	0.000	0.10758	0.15452	0.17749	0.19947	
11	0.500					
12	1.000					
13	1.500					
14	2.000					
15	2.500					
16	3.000					
17	3.500					
18	4.000					
19	4.500					
20	5.000					
21	5.500					
22	6.000					
23	6.500					
24	7.000					
25	7.500					
26	8.000					
27	8.500					
28	9.000					
29	9.500					
30	10.000					
31	10.500					
32	11.000	-0.00464	-0.00376	-0.00359	-0.00367	

Chart title: Reeves - \|sto-ng:2s⟩ basis sets, with legend \|sto-2g:2s⟩, \|sto-3g:2s⟩, \|sto-4g:2s⟩, \|2s⟩; y-axis from -0.1 to 0.3, x-axis r [a.u.] from 0 to 20.

Figure 1.15b   Comparisons of the variations of the Reeves \|sto-ng:2s⟩ basis sets with exact variation for the hydrogen \|2s⟩ radial function. Note, Reeves did not recommend an \|sto-1g:2s⟩ function and that again, there is increasing agreement with the numbers of Gaussian primitives used.

Exercise 1.8.   Comparisons of the Reeves' \|sto-ng⟩ Gaussian sets with the exact hydrogen functions — variations with radial distance.

1.   Enter the appropriate Reeves parameters in the first few rows of appropriate cells of different worksheets in a new spreadsheet, named fig1-15.xls. provide for three worksheets and name them '1s'!, '2s'! and '2p'!.

	A	B	C	D	E	F
1	Reeves - |sto-ng:2p⟩ basis sets					
2	α	0.04527	0.00000	0.00000	0.00000	|sto-1g:2p⟩
3	d	0.02976	0.00000	0.00000	0.00000	
4	α	0.03240	0.13930	0.00000	0.00000	|sto-2g:2p⟩
5	d	0.01539	0.03952	0.00000	0.00000	
6	α	0.02465	0.07970	0.33630	0.00000	|sto-3g:2p⟩
7	d	0.00803	0.02870	0.03369	0.00000	
8	α	0.02018	0.05572	0.17430	0.73420	|sto-4g:2p⟩
9	d	0.00452	0.02042	0.02944	0.02544	
10						
11	r	|sto-1g:1s⟩	|sto-2g:1s⟩	|sto-3g:1s⟩	|sto-4g:1s⟩	|2p⟩
12	0.000	0.00000	0.00000	0.00000	0.00000	0.00000
13	0.500	0.01471	0.02672	0.03354	0.03700	0.03884
14	1.000					
15	1.500					
16	2.000					
17	2.500					
18	3.000					
19	3.500					
20	4.000					
21	4.500					
22	5.000					
23	5.500					
24	6.000					
25	6.500					
26	7.000					
27	7.500					
28	8.000					
29	8.500					
30	9.000					
31	9.500					
32	10.000					
33	10.500					
34	11.000	0.00137	0.00336	0.00450	0.00459	0.00448

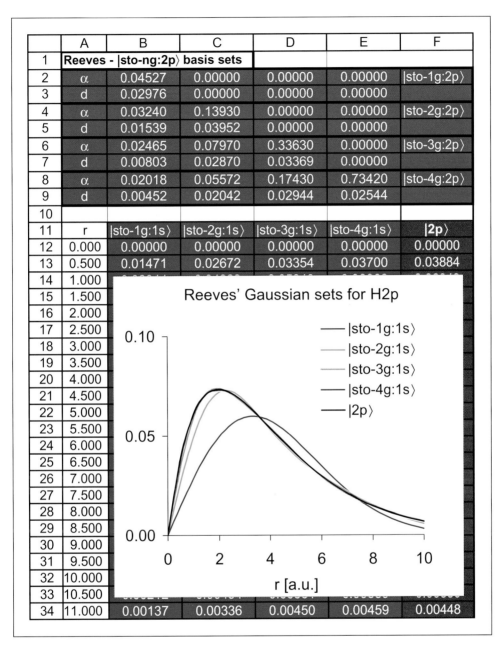

Figure 1.15c Comparisons of the variations of the Reeves |sto-ng:1s⟩ basis sets with exact variation for the hydrogen |1s⟩ radial function. Note, the substantial disagreement for the simple single Gaussian function and the increasing agreement with the numbers of Gaussian primitives used. But, note, too, that the |sto-2g:2p⟩ set does not return a good fit to the variation with radial distance of the exact hydrogen |2p⟩ function. The radial distance, in this example, is assumed to coincide with distance along the z-axis.

2. Choose a suitable radial mesh and generate the corresponding meshes of values for the hydrogenic 1s, 2s and 2p radial functions. Thus, for example, we find

'1s'!$B12 = $B$3***EXP**(-$B$2***POWER**($A12,2))
+$C$3***EXP**(-$C$2***POWER**($A12,2))
+$D$3***EXP**(-$D$2***POWER**($A12,2))
+$E$3***EXP**(-$E$2***POWER**($A12,2))

and, similarly, for the other active cells of columns B, C, D and E of the worksheet, '1s'!, [note the naming system], using the FILL sequence as necessary.

3. Generate the variation of the $|1s\rangle$ orbital with radial distance in the corresponding cells of column F. Reeves' normalization is over all the spherical polar coordinates and so it is necessary to use both normalization terms from Table 1.1. Thus, for example,

'1s'!$F12 = **SQRT**(1/**PI**( ))***EXP**(-$A12)

and, similarly, for the other cells of column F.

4. Repeat steps 2 and 3 for the cases of the $|2s\rangle$ Gaussian sets. Since the representation of the exact $|2s\rangle$ function is made using '1s' Gaussians, the only differences in '2s'! are the values of the parameters to define the primitive Gaussians and the formulae in column F, for the $|2s\rangle$ functions in hydrogen. For example, we find

'2s'!$E10 = **SQRT**(1/(32***PI**( )))*(2-$A10)***EXP**(-$A10/2)

since only three basis sets are listed in Table 1.5 for this case. Again, note the normalization constant used.

5. For the $|2p\rangle$ case, there is the substantial difference that the $|sto-nG:2p\rangle$ Gaussians include the radial factor for 2p-like functions. Thus, the formulae entries in columns C to F take the form, for example,

'2p'!$B12 = $A12*($B$3***EXP**(-$B$2***POWER**($A12,2))
+$C$3***EXP**(-$C$2***POWER**($A12,2))
+$D$3***EXP**(-$D$2***POWER**($A12,2))
+$E$3***EXP**(-$E$2***POWER**($A12,2)))

while the typical cell entry in column F takes the form

'2p'!$F12 = $A12***SQRT**(1/(32***PI**( )))***EXP**(-$A12/2)

6. Finally, follow the *wizard* instructions to construct the chart in Figure 1.15a, b and c.

As you can see, Reeves obtained good matches to the analytical radial functions of Table 1.1, based on the criterion that good orbital energies had been calculated, except in the cases of the smallest basis sets. As one might expect the match becomes better, in each case, the greater the number of primitive Gaussians in the linear combinations. However, this consideration is not the essential point and it is appropriate perhaps to remember the apocryphal principle, attributed to the great mathematician, Cauchy;

Give me four parameters and I will draw you an elephant. Give me five and I will make it wag its tail!

The Pople group chose in early studies, for example (33), to design Gaussian basis sets to match Slater functions. By 1984, with the publication of the Huzinaga *et al.* comprehensive listings of Gaussian basis sets (41), it was becoming standard to match Gaussian basis sets to the numerical radial functions output by such atomic structure programs as the Herman–Skillman one described earlier. In that compilation, too, data could be checked for overall quality using the additional requirement that atomic structure calculations based on the particular basis sets conform to the elusive Virial Theorem (16, 42). As the requirement that the mean kinetic, $\langle T \rangle$, and potential, $\langle V \rangle$, energies of the electrons satisfy the identity

$$-\frac{2\langle T \rangle}{\langle V \rangle} = 1 \qquad\qquad 1.20$$

A particular advantage of the Pople approach (25, 32, 33, 41) to the construction of Gaussian basis sets is the straightforward manner in which Slater's rules carry over to Gaussian theory. The interrelation (25) between a Gaussian exponent, α, and the Slater exponent, ζ, for a particular orbital in an atom, follows from the identity

$$e^{-(\zeta r)} = e^{-(\sqrt{\alpha}r)^2} \qquad\qquad 1.21$$

Therefore, to use a Gaussian set to approximate an atomic orbital, for which the Slater exponent is ζ, each Gaussian exponent must be scaled in the form

$$\alpha' = \alpha(\zeta = 1)\zeta^2 \qquad\qquad 1.22$$

Of course, all these exponents, potentially, are *Cauchy Principle* parameters for variation subject to whatever optimizing condition is applied!

Table 1.6 lists the Hehre, Stewart and Pople $|$sto-3g$\rangle$ Gaussian sets (33) for hydrogenic 1s, 2s and 2p orbitals, the minimum basis sets for the atoms of the first short period when scaled with suitable choices for the Slater exponents.

These Gaussian sets are very different to the linear combinations determined by Reeves, in that these basis sets were determined to match the equivalent Slater functions [see Section 2.6 page 70]. Like the Reeves' sets, though, because of the need, at the time, to calculate integrals as simply as possible, the Gaussian sets in the table for the 2s hydrogen orbital are linear combinations of 1s Gaussian primitives. If you read the Hehre, Stewart and Pople paper (33) you will see that their normalization criterion is over all the coordinates. Thus, in comparisons with Slater functions, we need to be careful to apply a consistent normalization condition. In this discussion, the choice is to normalize only over the radial coordinate and so to apply equation 1.9, with equation 1.15 multiplied by $(1/4\pi)^{1/2}$.

Table 1.6   **The Gaussian sets[20] proposed by Hehre, Stewart and Pople (33) for the representation of the Slater functions for the hydrogen 1s, 2s and 2p atomic orbitals. Note, again, that only one 'α' parameter is given for the 2s and 2p orbitals. To apply these basis sets to other atoms, they are scaled following equations 1.24 and 1.25. But note that the $|$sto-3g:2s$\rangle$ sets are for linear combinations of 1s Gaussians.**

$|\phi\rangle$	α_{1s}	d_{1s}	α_{2s}	d_{2s}	d_{2p}
$|$sto-2g$\rangle$	0.151623	0.678914	0.0974545	0.963782	0.612820
	0.851819	0.430129	0.384244	0.0494718	0.511541
$|$sto-3g$\rangle$	0.109818	0.444635	0.0751386	0.700115	0.391957
	0.405771	0.535328	0.231031	0.399513	0.607684
	2.22766	0.154329	0.994203	−0.0999672	0.155916
$|$sto-4g$\rangle$	0.080187	0.291626	0.0628104	0.497767	0.246313
	0.265204	0.532846	0.163541	0.558855	0.583575
	0.954620	0.260141	0.502989	0.00002968	0.286379
	5.21686	0.0567523	2.32350	−0.0622071	0.0436843
$|$sto-5g$\rangle$	0.0744527	0.193572	0.0544949	0.346121	0.156828
	0.197572	0.482570	0.127920	0.612290	0.510240
	0.578648	0.331816	0.329060	0.128997	0.373598
	2.017173	0.113541	1.03250	−0.0653275	0.107558
	11.3056	0.0221406	5.03629	−0.0294086	0.0125561
$|$sto-6g$\rangle$	0.0651095	0.130334	0.0485690	0.240706	0.1010708
	0.158088	0.416492	0.105960	0.595117	0.425860
	0.407099	0.370563	0.243977	0.250242	0.418036
	1.18506	0.168538	0.634142	−0.0337854	0.173897
	4.23592	0.0493615	2.04036	−0.0469917	0.0376794
	23.31030	0.00916360	10.3087	−0.0132528	0.00375970

Exercise 1.9.   Comparisons of the Hehre, Stewart and Pople, $|$sto-ng$\rangle$, Gaussian sets with the equivalent Slater functions — variations with radial distance.

1.   It is best, in this exercise, to use different spreadsheets for each Slater function and different workbooks for each basis set. So on a new worksheet, in a new spreadsheet, enter the primary data for the particular Slater orbital and basis set as in Figure 1.16, for the example of the $|$sto-6g:1s$\rangle$ basis set of Table 1.6.
2.   Work out the Slater exponents for the 1s, 2s and 2p orbitals and enter these in cell \$B\$5 of the $|$sto-6g$\rangle$ worksheets in each spreadsheet.
3.   Lay out a suitable radial mesh in column A of the '6g!' worksheet in each spreadsheet and establish links to the equivalent cells in the other columns A of each worksheet.

---

[20]You will be surprised at the sensitivity of the Gaussian parameters in the determination of results. Be warned that small errors in their transcription are significant. Since in real calculations large numbers of integrals over the Gaussians have to be calculated!

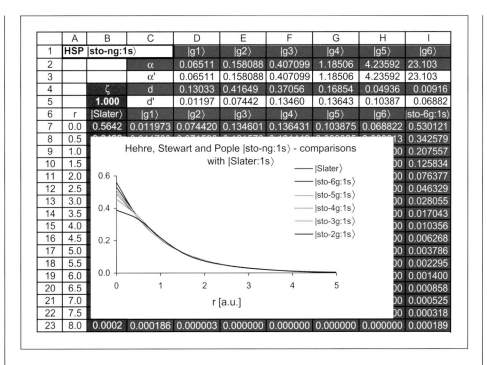

	A	B	C	D	E	F	G	H	I
1	HSP	|sto-ng:1s⟩		|g1⟩	|g2⟩	|g3⟩	|g4⟩	|g5⟩	|g6⟩
2			α	0.06511	0.158088	0.407099	1.18506	4.23592	23.103
3			α'	0.06511	0.158088	0.407099	1.18506	4.23592	23.103
4		ζ	d	0.13033	0.41649	0.37056	0.16854	0.04936	0.00916
5		1.000	d'	0.01197	0.07442	0.13460	0.13643	0.10387	0.06882
6	r	|Slater⟩	|g1⟩	|g2⟩	|g4⟩	|g5⟩	|g6⟩		sto-6g:1s⟩
7	0.0	0.5642	0.011973	0.074420	0.134601	0.136431	0.103875	0.068822	0.530121
8	0.5							13	0.342579
9	1.0							00	0.207557
10	1.5							00	0.125834
11	2.0							00	0.076377
12	2.5							00	0.046329
13	3.0							00	0.028055
14	3.5							00	0.017043
15	4.0							00	0.010356
16	4.5							00	0.006268
17	5.0							00	0.003786
18	5.5							00	0.002295
19	6.0							00	0.001400
20	6.5							00	0.000858
21	7.0							00	0.000525
22	7.5							00	0.000318
23	8.0	0.0002	0.000186	0.000003	0.000000	0.000000	0.000000	0.000000	0.000189

Figure 1.16   The Hehre, Stewart and Pople |sto-ng⟩ approximations to hydrogen 1s.

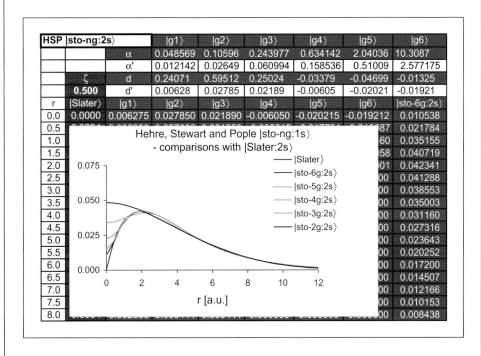

HSP	|sto-ng:2s⟩		|g1⟩	|g2⟩	|g3⟩	|g4⟩	|g5⟩	|g6⟩
		α	0.048569	0.10596	0.243977	0.634142	2.04036	10.3087
		α'	0.012142	0.02649	0.060994	0.158536	0.51009	2.577175
	ζ	d	0.24071	0.59512	0.25024	-0.03379	-0.04699	-0.01325
	0.500	d'	0.00628	0.02785	0.02189	-0.00605	-0.02021	-0.01921
r	|Slater⟩	|g1⟩	|g2⟩	|g3⟩	|g4⟩	|g5⟩	|g6⟩	sto-6g:2s⟩
0.0	0.0000	0.006275	0.027850	0.021890	-0.006050	-0.020215	-0.019212	0.010538
0.5							87	0.021784
1.0							60	0.035155
1.5							58	0.040719
2.0							01	0.042341
2.5							00	0.041288
3.0							00	0.038553
3.5							00	0.035003
4.0							00	0.031160
4.5							00	0.027316
5.0							00	0.023643
5.5							00	0.020252
6.0							00	0.017200
6.5							00	0.014507
7.0							00	0.012166
7.5							00	0.010153
8.0							00	0.008438

Figure 1.17   The Hehre, Stewart and Pople basis sets for Slater H_{2s} using 1s Gaussians.

4. Scale the Gaussian exponents following equation 1.22 **and** use these scaled values to calculate the normalizing pre-exponential factors of equation 1.14. Thus, for example, we find in figure1-15.xls

$$\text{‘6g!’}\$D3 = D2^*\textbf{POWER}(\$B\$5,2)$$

and,

$$\text{‘6g!’}\$D5 = D4^*(\textbf{POWER}(4^*D3,5/4)/\textbf{POWER}(2^*\textbf{PI}(\,),3/4))$$

but, in the case of the |sto-ng:2p⟩ Gaussian sets, because of equation 1.14,

$$\text{‘6g!’}\$D5 = D4^*(\textbf{POWER}(4^*D3,5/4)/\textbf{POWER}(2^*\textbf{PI}(\,),3/4))$$

and, similarly, throughout row 3 of each of the worksheets.

5. Enter the appropriate formulae for the Gaussian primitives in the active cells of columns.
6. Follow the *wizard* instructions to construct the charts in the Figures 1.16, 1.17 and 1.18, which are based on the spreadsheets fig1-16.xls, fig1-17.xls and fig1-18.xls.

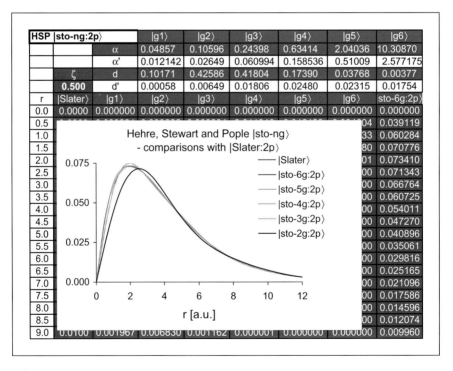

HSP	sto-ng:2p⟩		|g1⟩	|g2⟩	|g3⟩	|g4⟩	|g5⟩	|g6⟩
		α	0.04857	0.10596	0.24398	0.63414	2.04036	10.30870
		α'	0.012142	0.02649	0.060994	0.158536	0.51009	2.577175
	ζ	d	0.10171	0.42586	0.41804	0.17390	0.03768	0.00377
	0.500	d'	0.00058	0.00649	0.01806	0.02480	0.02315	0.01754

r	|Slater⟩	|g1⟩	|g2⟩	|g3⟩	|g4⟩	|g5⟩	|g6⟩	sto-6g:2p⟩
0.0	0.0000	0.000000	0.000000	0.000000	0.000000	0.000000	0.000000	0.000000
0.5							04	0.039119
1.0							33	0.060284
1.5							80	0.070776
2.0							01	0.073410
2.5							00	0.071343
3.0							00	0.066764
3.5							00	0.060725
4.0							00	0.054011
4.5							00	0.047270
5.0							00	0.040896
5.5							00	0.035061
6.0							00	0.029816
6.5							00	0.025165
7.0							00	0.021096
7.5							00	0.017586
8.0							00	0.014596
8.5							00	0.012074
9.0	0.0100	0.001967	0.006830	0.001162	0.000001	0.000000	0.000000	0.009960

Chart: Hehre, Stewart and Pople |sto-ng⟩ - comparisons with |Slater:2p⟩, with legend: —|Slater⟩, —|sto-6g:2p⟩, —|sto-5g:2p⟩, ·····|sto-4g:2p⟩, ······|sto-3g:2p⟩, —|sto-2g:2p⟩. Vertical axis 0.000 to 0.075; horizontal axis r [a.u.] from 0 to 12.

Figure 1.18   The Hehre, Stewart and Pople Gaussian basis sets to match the Slater hydrogen 2p orbital: comparisons of the variations with radial distance along the z-axis. Note, the normalization constant, $N = (128\alpha^5/\pi^3)^{1/4}$, follows from equation 1.14, when the angular factor is included.

1.7 SCALING FACTORS

Clearly, we need to be able to construct Gaussian approximations to the atomic orbitals of atoms other than hydrogen. For applications in molecular orbital theory, we need, in addition, to make allowance for the evident fact that an atom in a molecule is in a different environment to the isolated atom in free space, since any other atoms in a molecule give rise to extra potentials disturbing the free space electron distribution for that atom.

Since the Slater exponents lead to the Slater orbitals for any atom, we can expect that the application of equation 1.22 is the key to the generation of modern Gaussian basis sets. It is not too much of an extrapolation to develop this approach in the direction that different 'effective' Slater exponents might be appropriate in the molecular environment.

Consider the example of the boron atom and the making of comparisons of the Herman–Skillman numerical radial functions for 1s, 2s and 2p with the Slater functions and possible Gaussian basis set approximations.

Exercise 1.10a.   Construction of Figure 1.19: comparison of the numerical radial functions for boron 1s based on Slater and the Hehre, Stewart and Pople |sto-3g⟩ approximations to these Slater functions.

There are three distinct parts to exercises 1.10a, 1.10b and 1.10c.

In the first part, comparisons are made between the Slater functions and the numerical outputs of the Herman–Skillman program. In the second, the comparisons are between the Slater functions and their Gaussian representations. Finally, the role of the Slater exponent is emphasized by the restoration of the hydrogenic results using the hydrogenic Slater exponents.

Remember, the Herman–Skillman program outputs the 'numerical' radial function data, $rR(r)$ and so it is convenient to construct the Slater and |sto-3g⟩ approximations in this form. Remember, too, that there is a difference in the common practice with regard to the inclusion of normalization factors. The Herman–Skillman data are normalized only over the radial coordinate and this is usual, too, for Slater functions, but it is standard to normalize Gaussian functions over all the spherical polar coordinates.

1. Run the Herman–Skillman program for the boron atom using the atomic data in Table 1.3. Then transfer the output mesh and $rR(r)$ data for each atomic orbital to a new worksheet in a new spreadsheet. Paste the data, as text, into columns A and B starting from cell $A$9 as in Figure 1.19. Parse the text data into individual columns of r and $rR(r)$ values.
2. Fill in the basic parameters for the Gaussian primitives in cells $C$3 to $E$4 and, then, the primitive quantum number in cell $F$3, with the Slater exponent for the 1s orbital in boron in cell $G$3. Since the Slater function is to be included in the chart displaying the variations with radial distance, repeat these entries in cells $G$7 and $G$8.

3. The normalization constant of equation 1.14 involves the *double factorial*. So enter the formula

$$\$G\$4 = \textbf{FACTDOUBLE}(2*\$F\$3\text{-}1)$$

to calculate this component in the denominator of equation 1.14. Note this is the **FACTDOUBLE( )** intrinsic function in the '*Engineering Function* library' in EXCEL. So, if this library is not present on your basic version, add it under the TOOLS menu using the EXCEL CD.

4. Convert the hydrogen |sto-3g:H-1s⟩ parameters for the boron atom, with, for example, so, the entries in cells

$$\$C\$5 = \$C\$3*\textbf{POWER}(\$G\$3,2)$$

$$\$C\$6 = \textbf{SQRT}(\textbf{POWER}(2,(2*\$F\$3+1.5)))/$$
$$(\textbf{SQRT}(\textbf{PI}())*\$G\$4))*\textbf{POWER}(C\$5,((2*\$F\$3+1)/4))$$

$$\$C\$7 = C\$4*C\$6$$

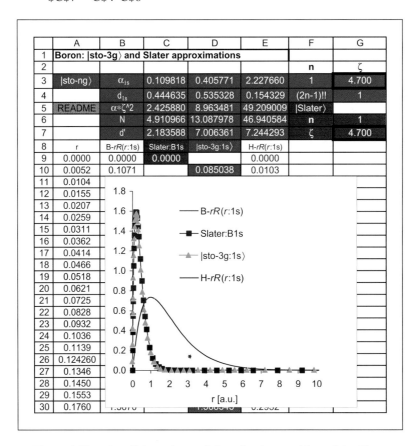

	A	B	C	D	E	F	G
1	Boron: |sto-3g⟩ and Slater approximations						
2						n	ζ
3	|sto-ng⟩	α_{1s}	0.109818	0.405771	2.227660	1	4.700
4		d_{1s}	0.444635	0.535328	0.154329	(2n-1)!!	1
5	README	$\alpha*\zeta^{\wedge}2$	2.425880	8.963481	49.209009	|Slater⟩	
6		N	4.910966	13.087978	46.940584	n	1
7		d'	2.183588	7.006361	7.244293	ζ	4.700
8	r	B-rR(r:1s)	Slater:B1s	|sto-3g:1s⟩	H-rR(r:1s)		
9	0.0000	0.0000	0.0000		0.0000		
10	0.0052	0.1071		0.085038	0.0103		
11	0.0104						
12	0.0155						
13	0.0207						
14	0.0259						
15	0.0311						
16	0.0362						
17	0.0414						
18	0.0466						
19	0.0518						
20	0.0621						
21	0.0725						
22	0.0828						
23	0.0932						
24	0.1036						
25	0.1139						
26	0.124260						
27	0.1346						
28	0.1450						
29	0.1553						
30	0.1760						

Figure 1.19a   Detail from the worksheet for the matching of the Slater and |sto-3g⟩ approximations to the 1s orbital for boron as the output of the Herman–Skillman program. Note, the deletion of entries in the calculated |Slater:nl⟩ and |sto-3g:nl⟩ to make clearer the graphs in the chart in this figure and Figure 1.19b for the match to the hydrogen function.

	A	B	C	D	E	F	G
1	Boron: |sto-3g⟩ and Slater approximations						
2						n	ζ
3	|sto-ng⟩	α_{1s}	0.109818	0.405771	2.227660	1	1.000
4		d_{1s}	0.444635	0.535328	0.154329	(2n-1)!!	1
5	README	α*ζ^2	0.109818	0.405771	2.227660	|Slater⟩	
6		N	0.481970	1.284475	4.606825	n	1
7		d'	0.214301	0.687616	0.710967	ζ	1.000
8	r	B-rR(r:1s)	Slater:B1s	|sto-3g:1s⟩	H-rR(r:1s)		
9	0.0000	0.0000	0.0000		0.0000		
10	0.0052	0.1071		0.008351	0.0103		
11	0.0104						
12	0.0155						
13	0.0207						
14	0.0259						
15	0.0311						
16	0.0362						
17	0.0414						
18	0.0466						
19	0.0518						
20	0.0621						
21	0.0725						
22	0.0828						
23	0.0932						
24	0.1036						
25	0.1139						
26	0.124260						
27	0.1346						
28	0.1450						
29	0.1553						
30	0.1760						

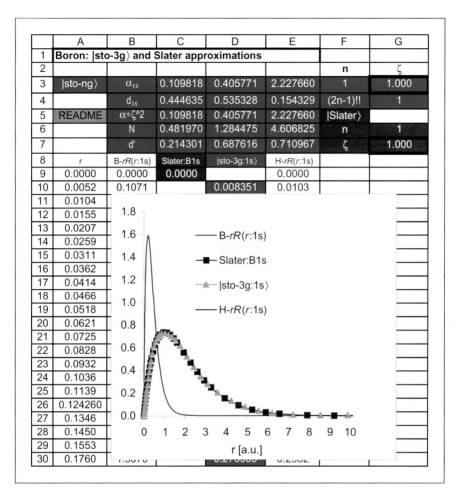

Figure 1.19b Restoration of the match to the hydrogen 1s orbital simply by changing the entries for the Slater exponents in cells $G$3 and $G$7.

with the second formula the complete normalization constant of equation 1.14. Fill these relationships across the equivalent cells in columns D and E.

5. Worksheet, '1s'! in fig1-19.xls is completed by generating the Slater, |sto-3g⟩ and |H-1s⟩ function values over the radial mesh of column A in the form $rR(r)$, with, for example,

$$\$C9 = \mathbf{POWER}((2^*G\$7),(G\$6+0.5))/\mathbf{SQRT}(\mathbf{FACT}(2^*G\$6))^*$$
$$\mathbf{EXP}(-G\$7^*\$A9)^*\mathbf{POWER}(\$A9,(G\$6-0.999999999999999))^*\$A9$$

$$\$D10 = (\$C\$7^*\mathbf{EXP}(-\$C\$5^*\$A10^2)$$
$$+\$D\$7^*\mathbf{EXP}(-\$D\$5^*\$A10^2)$$
$$+\$E\$7^*\mathbf{EXP}(-\$E\$5^*\$A10^2))^*\$A10$$

and

$$\$E\$9 = \$A9^*2^*\mathbf{EXP}(-\$A9)$$

and then filling these formulae down the active cells of these columns for the radial mesh in column A. Note, the device, **POWER** ($A9,(G$6-0.9999999999 99999)), used to avoid a non-defined error in the **POWER** function!

For ease of reading of the chart in '1s'!figure1-19.xls, alternate cells of columns C and D are left blank [delete after filling down!] so that the CONDITIONAL FORMATTING option in the FORMAT *dialogue* box can be set to particular patterns for entries greater than zero values as in Figures 1.19a and b.

6. Apply the chart *wizard* to generate the chart shown.
7. The great generality of the Slater and Gaussian basis sets approach to the matching of the numerical radial function data is emphasized in Figure 1.19b, which is generated in figure1-19a.xls, simply by changing the Slater exponent to the value for the hydrogen 1s orbital. Note, how the spreadsheet responds to this simple change of input. The chart in Figure 1.19b follows when the entries in cells $G$3 and $G$7 are set to 1.00.

Exercise 1.10b.   Construction of Figure 1.20: comparison of the numerical radial functions for boron 2s based on Slater functions and the Hehre, Stewart and Pople |sto-3g) approximations to these Slater functions.

1. Make a copy of figure1-18.xls and rename the copied file to be figure1-20.xls. This establishes the general form of the spreadsheet and the chart, which changes, as before, when the appropriate data and formulae for the 2s orbital are applied. The 2s boron radial function is represented using 1s Gaussian primitives, so while the Slater exponent in cells $G$3 and $G$7 is the 1.3 expected from Slater rules, there are new parameters for the 1s Gaussians in cells $C$3 to $E$4. These are specific, Table 1.6, for the 2s hydrogen function.
2. Run the Herman–Skillman program and transfer the output data for the numerical 2s radial function in boron to column B of the new spreadsheet.
3. Change the labels for the column headers.
4. The chart display will be incorrect, in that the graph for the variation of the boron 2s orbital turns out to be the variations found for the Slater and |sto-3g) approximations times -1. This is fixed easily by

   (i) generating the function $-rR(r)$ in column F, for example, with, for example, $F9 = -$B9 and filling this change down the active cells of column F [*remember you can use the sequence CTL/SHIFT/ and the appropriate keyboard arrow to select already occupied cells in rows or columns*];
   (ii) selecting the graph of for the numerical output of the Herman–Skillman program on the chart and changing the column address from B to F.

	A	B	C	D	E	F	G
1	Boron: |sto-3g⟩ and Slater approximations						
2						n	ζ
3	|sto-ng⟩	α_{1s}	0.075139	0.231031	0.994203	1	1.300
4		d_{1s}	0.700115	0.399513	-0.099967	(2n-1)!!	1
5		$\alpha{*}\zeta^2$	0.12698	0.39044	1.6802	|Slater⟩	
6		N	0.53744	1.24791	3.7285	n	2
7		d'	0.37627	0.49856	-0.3727	ζ	1.300
8	r	rR(r:2s)	Slater:B2s	|sto-3g:1s⟩	H-rR(r:2s)	-rR(r:2s)	
9	0.0000	0.0000	0.0000		0.0000	0.0000	
10	0.0052	0.0237		0.002600	-0.0036	-0.0237	
11	0.0104						
12	0.0155						
13	0.0207						
14	0.0259						
15	0.0311						
16	0.0362						
17	0.0414						
18	0.0466						
19	0.0518						
20	0.0621						
21	0.0725						
22	0.0828						
23	0.0932						
24	0.1036						
25	0.1139						
26	0.124260						
27	0.1346						
28	0.1450						
29	0.1553						
30	0.1760						

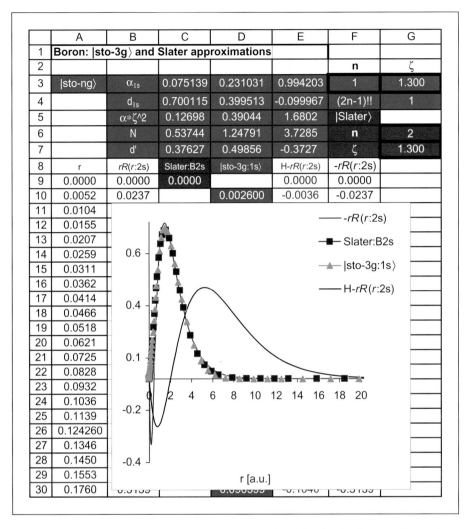

Figure 1.20 The matches to the numerical boron 2s orbital for the Slater and |sto-3g⟩ Hehre, Stewart and Pople basis of Table 1.6. Good matches are obtained in the valence region, but since both approximations are nodeless there is no match near the nucleus.

5. Generate the hydrogen 2s function in column E, with, for example,

$$\$E\$9 = \textbf{-SQRT}(1/8)^{*}(2\text{-}\$A9)^{*}\textbf{EXP}(\text{-}0.5^{*}\$A9)^{*}\$A9$$

which expression allows for the required change of sign and propagate this formula down column E.

6. As before, attempt to restore the hydrogen match, by substituting the Slater exponent value for the 2s orbital in hydrogen in cells $G$3 and $G$7. This leads to Figure 1.21.

	A	B	C	D	E	F	G
1	Boron: |sto-3g⟩ and Slater approximations						
2	Restoration of the match to hydrogen 2s					n	ζ
3	|sto-ng⟩	α_{1s}	0.075139	0.231031	0.994203	1	0.500
4		d_{1s}	0.700115	0.399513	-0.099967	(2n-1)!!	1
5		$\alpha*\zeta^2$	0.01878	0.05776	0.2486	|Slater⟩	
6		N	0.12819	0.29766	0.8894	n	2
7		d'	0.08975	0.11892	-0.0889	ζ	0.500
8	r	$rR(r{:}2s)$	Slater:B2s	|sto-3g:1s⟩	H-$rR(r{:}2s)$	-$rR(r{:}2s)$	
9	0.0000	0.0000	0.0000		0.0000	0.0000	
10	0.0052	0.0237		0.000620	-0.0036	-0.0237	
11	0.0104						
12	0.0155						
13	0.0207						
14	0.0259						
15	0.0311						
16	0.0362						
17	0.0414						
18	0.0466						
19	0.0518						
20	0.0621						
21	0.0725						
22	0.0828						
23	0.0932						
24	0.1036						
25	0.1139						
26	0.124260						
27	0.1346						
28	0.1450						
29	0.1553						
30	0.1760						

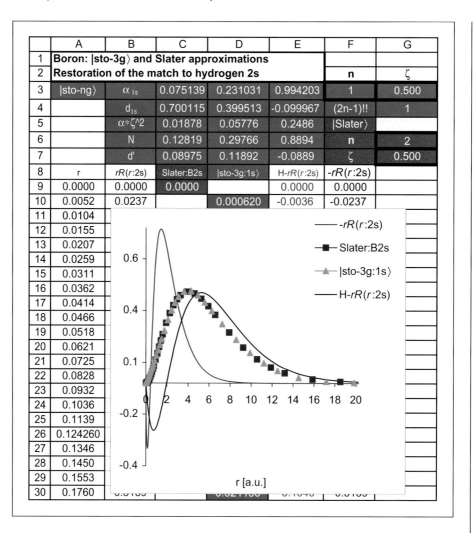

Figure 1.21 Attempt to restore a match to the hydrogen 2s orbital using the Slater and \|sto-3g⟩ approximations of Figure 1.20. Only the general feature of the valence region behaviour of the exact hydrogen function is reproduced.

The results, displayed in Figures 1.20 and 1.21, emphasize the great distinction between the approximations for the 1s and 2s orbitals. Since there is no attempt to reproduce the nodal behaviour of the numerical functions, both the Slater and \|sto-ng⟩ basis sets are deficient, although as you can see, they agree with each other. The matches can be improved by relaxing the manner in which the Slater exponents are determined, but more importantly, as you will see in later chapters, by re-imposing the fundamental requirement that the atomic functions should be orthonormal.

On the basis of the foregoing, it might be expected that a good match to the 2p orbital should be returned, since this is a nodeless function and is orthogonal to the 1s function. This is investigated in Exercise 1.10c.

Exercise 1.10c.   Construction of Figure 1.22: comparison of the numerical radial functions for boron 2p based on Slater functions and the Hehre, Stewart and Pople |sto-3g⟩ approximations to these Slater functions.

1. Again, make a copy of figure of the basic design. Rename the copy fig1-22.xls.
2. Enter the appropriate exponents and coefficients for the |sto-3g:2p⟩ function from the data in Table 1.6 and adjust the value of the Slater exponent in cells $G$3 and $G$7. Note, that both principal quantum numbers need to be set to the value 2.00, in this exercise and that this leads to a new value for the (2n-1)!! term in cell $G$4.
3. Run the Herman–Skillman program and transfer the output for the 2p orbital in boron to column B of the new spreadsheet.
4. While the Slater function formula entries in column C require only re-labelling in cell $C$8, the |sto-ng⟩ formula entries have to be modified to take account of the different functional form, since $n = 2$, required from equation 1.13. So, enter,

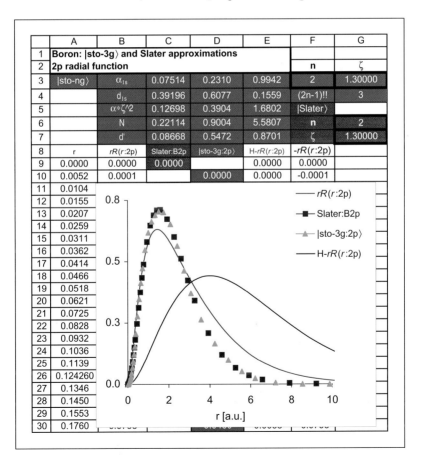

	A	B	C	D	E	F	G
1	Boron: |sto-3g⟩ and Slater approximations						
2	2p radial function					n	ζ
3	|sto-ng⟩	α_{1s}	0.07514	0.2310	0.9942	2	1.30000
4		d_{1s}	0.39196	0.6077	0.1559	(2n-1)!!	3
5		$\alpha*\zeta^2$	0.12698	0.3904	1.6802	|Slater⟩	
6		N	0.22114	0.9004	5.5807	n	2
7		d'	0.08668	0.5472	0.8701	ζ	1.30000
8	r	rR(r:2p)	Slater:B2p	|sto-3g:2p⟩	H-rR(r:2p)	-rR(r:2p)	
9	0.0000	0.0000	0.0000		0.0000	0.0000	
10	0.0052	0.0001		0.0000	0.0000	-0.0001	
11	0.0104						
12	0.0155						
13	0.0207						
14	0.0259						
15	0.0311						
16	0.0362						
17	0.0414						
18	0.0466						
19	0.0518						
20	0.0621						
21	0.0725						
22	0.0828						
23	0.0932						
24	0.1036						
25	0.1139						
26	0.124260						
27	0.1346						
28	0.1450						
29	0.1553						
30	0.1760						

Figure 1.22   Comparisons of the matches of the numerical boron radial 2p function with the Slater and |sto-3g⟩ approximations. Poor results are obtained.

$$\$D10 = (\$A10^*(\$C\$7^*EXP(-\$C\$5^*\$A10\wedge2)$$
$$+\$D\$7^*EXP(-\$D\$5^*\$A10\wedge2)$$
$$+\$E\$7^*EXP(-\$E\$5^*\$A10\wedge2)))^*\$A10$$

in cell $D10, for example, propagate and format this entry down the active cells of column D.

5.   Generate the hydrogen function in column E, with for example,

$$\$E9 = \mathbf{SQRT}(1/24)^*\$A9\wedge2^*\mathbf{EXP}(-\$A9/2)$$

As you can see in Figure 1.22, the results are not very good. The matches to the boron 2p function for either the Slater or |sto-3g⟩ approximate functions are much poorer that we have encountered up to this point. In contrast, the result is good when we compare the approximations with the hydrogen 2p function as in Figure 1.23.

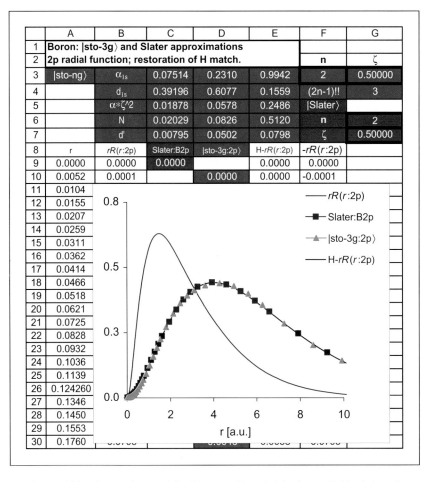

Figure 1.23   Comparisons of the Slater and |sto-3g⟩ basis set, Table 1.6, variations with radial distance and comparison with the exact hydrogen 2p function behaviour.

There are two considerations in the light of these comparisons. In the first place, we do not know the accuracy of the Herman–Skillman result. Secondly, it may be possible to improve the fit to the boron data by suitable choice of a modified Slater exponent or the use of a larger linear combination of primitive Gaussian functions.

The Herman–Skillman program (4) was written in the early 1960s and the original version of the code followed the Slater prescription for the electron–electron term. In 1972, Schwarz (43) suggested that it might be more accurate for calculations requiring atomic potentials constructed from atomic wave functions, to allow for different values in one of the parameters in the Slater exchange term. This parameter is the 'α' factor of the *Multiple Scattering Xalpha* program, a novel *semi-empirical* molecular orbital method (44) of that period. Schwarz suggest two possible values for this parameter on the basis of his atomic structure calculations. For the condition that the calculations should conform to the Virial Theorem, equation 1.20, to the greatest possible extent, Schwarz proposed that α_B should be 0.76756. For the condition that the calculations should be the best Hartree–Fock results possible[21] Schwarz reported the best value for boron to be 0.76531.

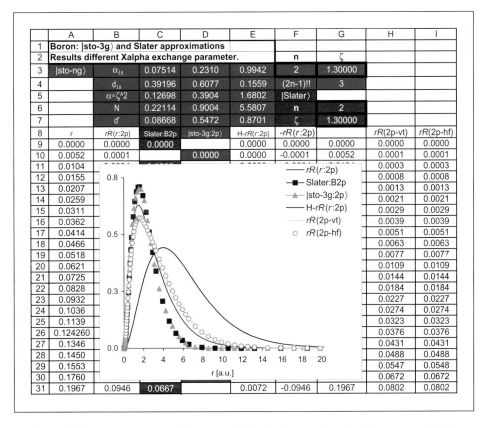

	A	B	C	D	E	F	G	H	I
1	Boron: |sto-3g⟩ and Slater approximations								
2	Results different Xalpha exchange parameter.					n	ζ		
3	|sto-ng⟩	α_{1s}	0.07514	0.2310	0.9942	2	1.30000		
4		d_{1s}	0.39196	0.6077	0.1559	(2n-1)!!	3		
5		$\alpha*\zeta^2$	0.12698	0.3904	1.6802	|Slater⟩			
6		N	0.22114	0.9004	5.5807	n	2		
7		d'	0.08668	0.5472	0.8701	ζ	1.30000		
8	r	rR(r:2p)	Slater:B2p	|sto-3g:2p⟩	H-rR(r:2p)	-rR(r:2p)		rR(2p-vt)	rR(2p-hf)
9	0.0000	0.0000	0.0000		0.0000	0.0000	0.0000	0.0000	0.0000
10	0.0052	0.0001		0.0000	0.0000	-0.0001	0.0052	0.0001	0.0001
11	0.0104							0.0003	0.0003
12	0.0155							0.0008	0.0008
13	0.0207							0.0013	0.0013
14	0.0259							0.0021	0.0021
15	0.0311							0.0029	0.0029
16	0.0362							0.0039	0.0039
17	0.0414							0.0051	0.0051
18	0.0466							0.0063	0.0063
19	0.0518							0.0077	0.0077
20	0.0621							0.0109	0.0109
21	0.0725							0.0144	0.0144
22	0.0828							0.0184	0.0184
23	0.0932							0.0227	0.0227
24	0.1036							0.0274	0.0274
25	0.1139							0.0323	0.0323
26	0.124260							0.0376	0.0376
27	0.1346							0.0431	0.0431
28	0.1450							0.0488	0.0488
29	0.1553							0.0547	0.0548
30	0.1760							0.0672	0.0672
31	0.1967	0.0946	0.0667		0.0072	-0.0946	0.1967	0.0802	0.0802

Figure 1.24 Comparisons of the numerical radial wave function for boron with the Slater and \|sto-3g⟩ approximations. The extra boron data were obtained by running the Herman–Skillman program using the Schwarz values for the Xα exchange term.

[21] See Chapters 4 and 5.

The results of this modification to the Herman–Skillman calculation are shown in Figure 1.24, constructed, using the outputs from the program, with the different possible choices for the α exchange parameter. As you can see in the diagram, two extra columns of Herman–Skillman output were processed to add the extra $r(Rr:2p)$ behaviour of the 2p numerical function.

The new results in Figure 1.24 show, at least, that it is possible to generate different numerical radial wave functions for the boron 2p orbital [and, of course, any other orbital obtained by approximate solution of the Schrödinger equation]. But, for a light atom like boron, it is unlikely that this is the major reason for any discrepancy.

The second possible improvement that we might seek is to vary the Slater exponent so that a better match to the boron function is obtained. This is a matter best investigated after Chapter 3, when we have considered the role of orthogonality as a fundamental requirement in solution procedures for calculation of orbital energies and orbital functions. But, it is a good exercise, to demonstrate the versatility of the spreadsheet approach. A 'better' result is shown in Figure 1.25. It follows, simply, from setting the Slater exponents for the Slater function and the $|sto\text{-}3g\rangle$ set equal to 1.00 in cells \$G\$3 and \$G47.

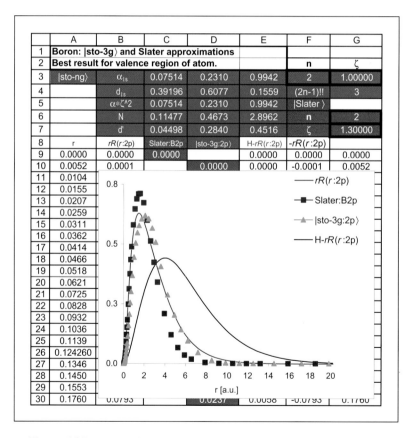

	A	B	C	D	E	F	G
1	Boron: $|sto\text{-}3g\rangle$ and Slater approximations						
2	Best result for valence region of atom.					n	ζ
3	$|sto\text{-}ng\rangle$	α_{1s}	0.07514	0.2310	0.9942	2	1.00000
4		d_{1s}	0.39196	0.6077	0.1559	(2n-1)!!	3
5		$\alpha*\zeta^2$	0.07514	0.2310	0.9942	$|Slater\rangle$	
6		N	0.11477	0.4673	2.8962	n	2
7		d'	0.04498	0.2840	0.4516	ζ	1.30000
8	r	$rR(r:2p)$	Slater:B2p	$|sto\text{-}3g:2p\rangle$	$H\text{-}rR(r:2p)$	$-rR(r:2p)$	
9	0.0000	0.0000	0.0000		0.0000	0.0000	0.0000
10	0.0052	0.0001		0.0000	0.0000	-0.0001	0.0052
11	0.0104						
12	0.0155						
13	0.0207						
14	0.0259						
15	0.0311						
16	0.0362						
17	0.0414						
18	0.0466						
19	0.0518						
20	0.0621						
21	0.0725						
22	0.0828						
23	0.0932						
24	0.1036						
25	0.1139						
26	0.124260						
27	0.1346						
28	0.1450						
29	0.1553						
30	0.1760	0.0793		0.0237	0.0058	-0.0793	0.1760

Figure 1.25   The choice that the Slater exponent should be that for one proton only is an interesting demonstration of the simplest model for the electronic structure of a many-electron atom, that each electron screens out one nuclear charge.

As you see there is quite a high degree of match in the outer regions of the atomic volume, but with some sacrifice of the match near the nucleus. This is a demonstration of the approximate validity of the simplest model for the electronic structure of the valence electron distribution in a many-electron atom. With the Slater exponent set to one atomic unit of charge, the fit to the numerical function is much better at large radii, as if the single 2p-electron experiences only the attraction of one nuclear proton, because of the effective screening of the other protonic charges by the complete sub shells $1s^2$ and $2s^2$.

It is instructive to make comparisons of the matches, which result when larger Gaussian basis sets are used to approximate the numerical 2p radial data. For the $|sto$-$6g\rangle$ of Table 1.6 the result is shown in Figure 1.26. As you see there is some improvement in the match, but the difficulty remains that it is not possible to obtain a uniform degree of match over the whole of the radial mesh range. What is required is the possibility of making matches independently over different ranges of the radial mesh and this, of course, is the origin of the *double-zeta* approach in Gaussian theory.

	A	B	C	D	E	F	G	H
1	Boron: |sto-6g⟩ match to			n	2	(2n-1)!!	Slater n	Slater ζ
2	2p numerical radial function.			ζ	1.30000	3	2	1
3	|sto-ng⟩	α_{1s}	0.04857	0.10596	0.24398	0.63414	2.04036	10.30870
4		d_{1s}	0.10107	0.42586	0.41804	0.17390	0.03768	0.00376
5		α*ζ^2	0.08208	0.1791	0.4123	1.0717	3.4482	17.4217
6		N	0.12817	0.3398	0.9639	3.1811	13.7081	103.8359
7		d'	0.01295	0.1447	0.4029	0.5532	0.5165	0.3904
8	r	rR(r:2p)	Slater:B2p	|sto-3g:2p⟩	H-rR(r:2p)	-rR(r:2p)		
9	0.0000	0.0000	0.0000		0.0000	0.0000		
10	0.0052							
11	0.0104							
12	0.0155							
13	0.0207							
14	0.0259							
15	0.0311							
16	0.0362							
17	0.0414							
18	0.0466							
19	0.0518							
20	0.0621							
21	0.0725							
22	0.0828							
23	0.0932							
24	0.1036							
25	0.1139							
26	0.124260							
27	0.1346							
28	0.1450							
29	0.1553							

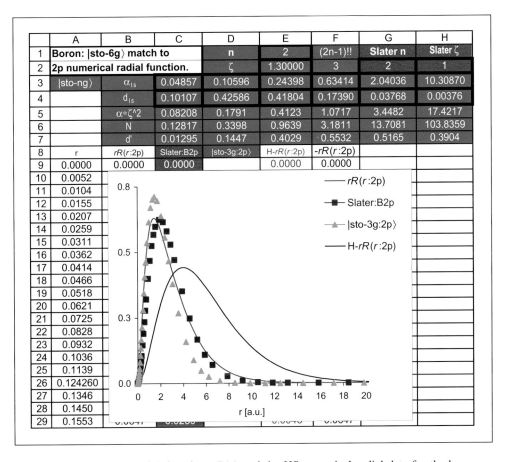

Figure 1.26   The $|sto$-$6g\rangle$ function $rR(r)$ and the HS numerical radial data for the boron 2p orbital.

Table 1.7   Huzinaga's |sto-ng:1s) basis sets for the 1s [Slater] orbital in hydrogen.

#	α_i	d_i	#	α_i	d_i
2	0.201527	0.82123	8	0.0525423	0.06412
	1.33248	0.27441		0.123655	0.35846
3	0.151374	0.64767		0.315278	0.42121
	0.681277	0.40789		0.886632	0.21210
	4.50038	0.07048		2.765179	0.06848
4	0.123317	0.50907		9.891184	0.01694
	0.453757	0.47449		43.93024	0.00322
	2.01330	0.13424		293.5708	0.00041
	13.3615	0.01906	9	0.0441606	0.03645
5	0.101309	0.37602		0.106151	0.29898
	0.321144	0.50822		0.250988	0.40433
	1.14680	0.20572		0.618330	0.25781
	5.05796	0.04575		1.714744	0.10769
	33.6444	0.00612		5.478296	0.03108
6	0.082217	0.24260		19.72537	0.00720
	0.224660	0.49221		87.39897	0.00138
	0.673320	0.29430		594.3123	0.00017
	2.34648	0.09280	10	0.0285649	0.00775
	10.2465	0.01938		0.0812406	0.20267
	68.1600	0.00255		0.190537	0.41300
7	0.060738	0.11220		0.463925	0.31252
	0.155858	0.44842		1.202518	0.14249
	0.436661	0.38487		3.379649	0.04899
	1.370498	0.15161		10.60720	0.01380
	4.970178	0.03939		38.65163	0.00318
	22.17427	0.00753		173.5822	0.00058
	148.2732	0.00097		1170.498	0.00007

Table 1.8   Huzinaga's |sto-ng:2p) basis sets for the Slater hydrogen 2p orbital.

#	α_i	d_i	#	α_i	d_i
2	0.032392	0.78541	5	0.017023	0.28504
	0.139276	0.32565		0.042163	0.52969
3	0.024684	0.57860		0.111912	0.27049
	0.079830	0.47406		0.346270	0.06550
	0.337072	0.09205		1.458369	0.00833
4	0.020185	0.41444	6	0.015442	0.21705
	0.055713	0.53151		0.035652	0.49334
	0.174211	0.18295		0.085676	0.32224
	0.7333825	0.02639		0.227763	0.10439
				0.710128	0.02055
				3.009711	0.00241

1.8   THE (4S/2S) BASIS SET, POLARIZATION AND SCALING FACTORS FOR MOLECULAR ENVIRONMENTS

Tables 1.7 and 1.8 list Huzinaga's (41,45) basis sets for the 1s and 2p Slater functions in hydrogen. These basis sets are interesting particularly because they were the first basis sets subjected to the *double-zeta* procedure of Slater theory. This procedure has the effect of reducing the number of terms for variation in calculations, but, more particularly, philosophically, provides for the better representation of details of the atomic radial functions, such as the cusp near the origin in s-functions, since the components subject to optimization separately, can be included in the linear combination.

Exercise 1.11.   **You should now test your understanding of this first chapter by writing spreadsheets to construct Figures 1.27 and 1.28, which compare Huzinaga's Gaussian sets and the corresponding Slater functions for the 1s and 2p orbitals in hydrogen.**

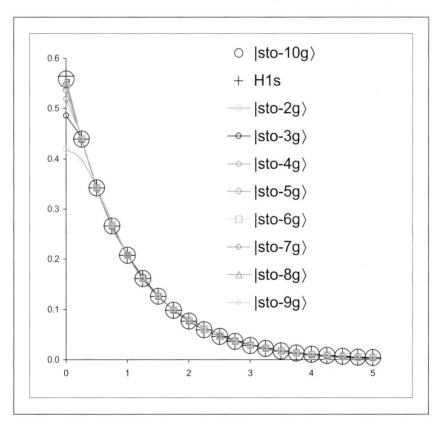

Figure 1.27   The matchings of Huzinaga's |sto-ng:1s⟩ basis sets to the 1s [Slater] function for hydrogen. Note, in particular, the match of the |sto-4g⟩ linear combination and the H 2s radial function for later comparison with the Dunning contraction of this basis set.

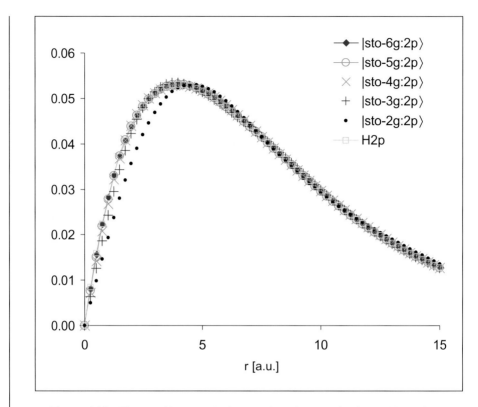

Figure 1.28 The matchings to Huzinaga's Gaussian sets for the 2p Slater function in hydrogen.

As you can see, from examination of the diagrams, very good overall agreement results.
While the basis sets are quite different to those that we have used earlier good agreement
with the exact 1s eigenvalues is obtained. Huzinaga pointed out that multiple minima
occur in the approximate eigenvalue calculations based on the use of linear combinations of Gaussians and so it is not surprising that *Cauchy's elephant* can be drawn in
many ways.

The (4–31) basis set for the 1s orbital in hydrogen is an early example of a split-
basis set[22]. From Huzinaga's original linear combination of four primitive Gaussians as
in Table 1.9,

$$\phi_{1s} = 0.50907^* g_{1s}(0.123317, \mathbf{r}) + 0.47449 g_{1s}(0.453757, \mathbf{r})$$

$$+ 0.13424^* g_{1s}(2.01330, \mathbf{r}) + 0.01906 g_{1s}(13.3615, \mathbf{r}) \hspace{2cm} 1.23$$

[22] Since the basis set can be applied to other atoms using a suitable ζ, the notation has a more general
meaning: 4–31 is accepted to mean a 4-term contraction of primitives for the core orbitals and then a split into
a single term and a 3-term contraction for the valence orbitals. For the minimal basis set of 1s, 2s and three
2p orbitals this is then a single 4-term linear combination for 1s and four pairs of the split basis set for the
valence s and p orbitals.

Table 1.9 Typical *double-zeta* quality Gaussian basis sets from the GAUSSIAN98 database as listed in the Environmental Molecular Sciences Laboratory Library of Basis functions, referred to in the text.

		α	d_s	d_p	d_d
$\lvert$sto-3-21g:H$\rangle$	c_1	0.183192000	1.00000000		
	c_2	5.44717800	0.156285000		
	c_2	0.824547000	0.904691000		
$\lvert$sto-6-31g*:Li$\rangle$	c_1	642.418920	0.0021426		
		96.7985150	0.0162089		
		22.0911210	0.0773156		
		6.2010703	0.245786		
		1.93511703	0.470189		
		0.63673580	0.3454708		
	c_2	2.3249184	−0.0350917	0.08941500	
	c_2	0.6324306	−0.1912328	0.1410095	
	c_2	0.0790534	1.0839878	0.9453637	
	c_3	0.0359620	1.0000000	1.0000000	
$\lvert$sto-6-31g**:Li$\rangle$	c_1	642.418920	0.0021426		
		96.7985150	0.0162089		
		22.0911210	0.0773156		
		6.2010703	0.245786		
		1.93511703	0.470189		
		0.63673580	0.3454708		
	c_2	2.3249184	−0.0350917	0.08941500	
	c_2	0.6324306	−0.1912328	0.1410095	
	c_2	0.0790534	1.0839878	0.9453637	
	c_3	0.0359620	1.0000000	1.0000000	
	c_4	0.0200000			1.0000000
$\lvert$sto-6-31g**:H$\rangle$	c_1	18.7311370	0.0334946		
		2.82539370	0.23472695		
		0.64012170	0.81375733		
	c_2	0.1612778	1.0000000		
	c_3	1.1000000		1.0000000	

determined in atomic structure calculations on hydrogen using different linear combinations of primitive Gaussians, Dunning (46) proposed the contraction to distinguish the most diffuse function [i.e. with smallest exponent] from the others to generate the *double-zeta* Gaussian basis, with

$$\phi_{1s}(1) = 1.0000 g_{1s}(0.123317, \mathbf{r}) \qquad\qquad 1.24$$

and

$$\phi_{1s}(2) = 0.817238 g_{1s}(0.453757, \mathbf{r}) + 0.231208 g_{1s}(2.01330, \mathbf{r})$$
$$+ 0.032828 g_{1s}(13.3615, \mathbf{r}) \qquad\qquad 1.25$$

This is a (4s)/[2s] contraction[23], in which the intrinsic norm of 1 of the primitive Gaussians is used for the first component of the contraction and then the remaining component of the original set of four primitives is renormalized to return the coefficients in the linear combination of equation 1.25.

Just as with Slater orbitals, it is common practice to use polarization functions as necessary and thus the adding of d-type components to the representation of ns and np orbitals and so on, limited only by the capacity of the computer used, are further extensions of basis set theory.

The level of agreement between the Huzinaga |sto-4g⟩ basis and the 1s orbital plots in Figure 1.27 is good and prompts the question as to why there is benefit in the use of the Dunning (4s/2s) function. The split-basis approach introduces flexibility in molecular calculations so that more effective modelling of molecular valence electron densities can be had. For the (42/2s) split, there are only two parameters per atom to change in calculations and so this is a considerable economy in the use of contracted functions, especially as the number of atomic orbitals being represented increases. Equations 1.24 and 1.25 are two copies of a Gaussian representation of the 1s orbital. Even for the atomic case, variation of the coefficients of the linear combination of these copies of the form

$$\phi = c_1\phi_{1s}(1) + c_2\phi_{1s}(2) \qquad\qquad 1.26$$

is more straightforward since this involves linear variation of the coefficients rather than the exponents of the Gaussian quadratic exponentials.

Formally, this example is similar to the Clementi *double-zeta* formulation within the Slater approximation. In order to use the (4s/2s) linear combination we need to know what values to use for the coefficient, c_1, in equation 1.26. To model the hydrogen 1s orbital, $\phi(r)$ needs to be normalized. Figures 1.29 and 1.30 show two conditions of the application of such a condition. In Figure 1.29, the normalized linear combination is a poor approximation to the 1s orbital. In Figure 1.30 close agreement with the 1s function is returned for the appropriate choice of coefficients in equation 1.26. This procedure of 'minimizing differences' is applied extensively in basis set theory.

As you see from an inspection of the second worksheet in this spreadsheet, we are beginning to make use of the subject matter of later chapters and, especially now, the need to work with orthonormal functions. This development in our use of basis functions becomes an essential pre-requisite when we consider how to use 1s hydrogenic Gaussians for other atoms, for other orbitals and in molecular applications.

There is one other detail about the contracted basis set notation. For hydrogen, there is only the valence 1s orbital and this is true too for helium. But for lithium, and any other atom in the Periodic Table, we divide the electron configuration into component core and valence orbitals and leave the core representation uncontracted. Thus, for lithium the 4-31 basis is the one linear combination of four primitive Gaussians of Table 1.8 scaled by the Slater exponent for 1s and then the two linear combinations of equations 1.21 and 1.22 scaled by the Slater exponent for lithium 2s and then the two linear combinations defined

---

[23]The common and modern practice is to use *segmentially* contracted basis sets with each primitive appearing only once in the overall linear combination of primitives. Thus, for example, different exponents would be applied for 2s and 2p functions in Table 1.7, the costs of the extra integrals to be calculated and stored being of less significance today with better integration routines and faster larger computers. Indeed, in 'direct scf' calculations it is normal to calculate all integrals as they are needed in each cycle of a self-consistent field calculation.

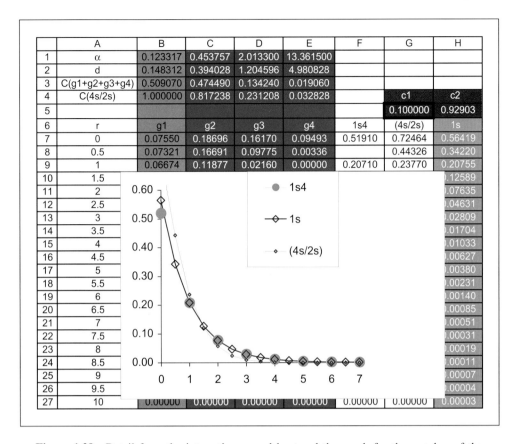

	A	B	C	D	E	F	G	H
1	α	0.123317	0.453757	2.013300	13.361500			
2	d	0.148312	0.394028	1.204596	4.980828			
3	C(g1+g2+g3+g4)	0.509070	0.474490	0.134240	0.019060			
4	C(4s/2s)	1.000000	0.817238	0.231208	0.032828		c1	c2
5							0.100000	0.92903
6	r	g1	g2	g3	g4	1s4	(4s/2s)	1s
7	0	0.07550	0.18696	0.16170	0.09493	0.51910	0.72464	0.56419
8	0.5	0.07321	0.16691	0.09775	0.00336		0.44326	0.34220
9	1	0.06674	0.11877	0.02160	0.00000	0.20710	0.23770	0.20755
10	1.5							0.12589
11	2							0.07635
12	2.5							0.04631
13	3							0.02809
14	3.5							0.01704
15	4							0.01033
16	4.5							0.00627
17	5							0.00380
18	5.5							0.00231
19	6							0.00140
20	6.5							0.00085
21	7							0.00051
22	7.5							0.00031
23	8							0.00019
24	8.5							0.00011
25	9							0.00007
26	9.5							0.00004
27	10	0.00000	0.00000	0.00000	0.00000	0.00000	0.00000	0.00003

Figure 1.29 Detail from the interactive spreadsheet and the graph for the matches of the Huzinaga and Dunning contracted bases with the 1s orbital function for hydrogen from Table 1.1. This diagram presents the results for the c_1 and c_2 starting choices of 0.1 and 0.929.

in equations. Thus, in the 4-31 *double-zeta* basis for lithium there are three functions to be varied in calculations, while in the 4-31 basis for fluorine there are nine functions (one for 1s, two as in equations 1.21 and 1.22 for each of 2s, $2p_x$, $2p_y$ and $2p_z$). Note, the comment in the legend to Figure 1.30, it is not that better agreement is reached that is important in these *split-basis* sets. Rather it is the economy of the representation in that with fewer variational parameters there is greater flexibility in molecular applications.

There are many split contracted basis-set schemes in the literature and in libraries available on the Web. Three examples are given in Table 1.10.

The $|sto\text{-}3\text{-}21g\rangle$ is a straightforward splitting of the minimal $|sto\text{-}3g\rangle$ basis set. The remaining two examples introduce the concept of polarization into basis set technology. In the presence of another nucleus, for example, in a molecular environment, the electron distribution associated with an electron is distorted in the direction of the other attractive potential field, the other nucleus. We describe this, in chemistry, as polarization and an obvious way to model polarization is to add to the basic linear combinations extra components of 'p' and 'd' characters to create an asymmetric electron distribution, familiar as hybridization in bonding theory.

	A	B	C	D	E	F	G	H
1	α	0.123317	0.453757	2.013300	13.361500			
2	d	0.148312	0.394028	1.204596	4.980828			
3	C(g1+g2+g3+g4)	0.509070	0.474490	0.134240	0.019060			
4	C(4s/2s)	1.000000	0.817238	0.231208	0.032828		c1	c2
5							0.509305	0.58037
6	r	g1	g2	g3	g4	1s4	(4s/2s)	1s
7	0	0.07550	0.18696	0.16170	0.09493	0.51910	0.51896	0.56419
8	0.5	0.07321	0.16691	0.09775	0.00336		0.34116	0.34220
9	1	0.06674	0.11877	0.02160	0.00000	0.20710	0.20708	0.20755
10	1.5							.12589
11	2							.07635
12	2.5							.04631
13	3							.02809
14	3.5							.01704
15	4							.01033
16	4.5							.00627
17	5							.00380
18	5.5							.00231
19	6							.00140
20	6.5							.00085
21	7							.00051
22	7.5							.00031
23	8							.00019
24	8.5							.00011
25	9							.00007
26	9.5							.00004
27	10	0.00000	0.00000	0.00000	0.00000	0.00000	0.00000	.00003

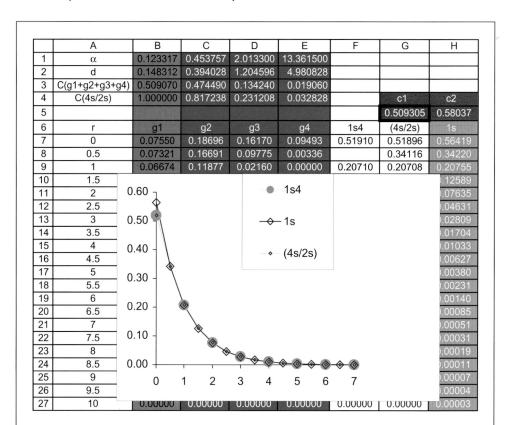

Figure 1.30   The 'best' result for Figure 1.29 with c_1 set equal to 0.509305 and, therefore, c_2 equal to 0.58037. This 'best' result was obtained using the 'least-squares' calculation of equation 3.6. Compare the 'best' pre-exponential factors, obtained by multiplying the coefficients in $G$5 and $H$5 with the entries in cells $B$4 to $E$4. As you might expect these 'best' results, $d_1 = 0.509305$, $d_2 = 0.474301$, $d_2 = 0.134186$ and $d_4 = 0.019052$, are close to the Huzinaga |sto-4g⟩ values in Table 1.8.

The examples shown in the table list the primitive Gaussians and the splitting schemes for the case of the lithium atom with added 'p' character in the form of an sp-hybrid and then dsp-hybrid character. Note the symbolism used in the labelling |6-31g⟩, which identifies the core linear combination to be comprised of six primitive Gaussians, while the valence orbital representation, |6-31g*⟩, is a contraction to two linear combinations of three and one primitives. Then, the |6-31g**⟩ basis includes the extra polarization effect of one added 'd' Gaussian. In basis set theory, to provide for the individual symmetry characters of the radial functions being modelled it is customary to define *six* d functions, the normal set of five in atomic orbital theory and then an additional s-function as $x^2 + y^2 + z^2$. Overall, the 6-31 contracted basis, for the atoms of the first short period, is the contraction (11s4p1d/4s), that is to say, there are only four coefficients to be varied in any calculation. Calculations at the |6-31g**⟩ level are the norm in modern molecular orbital theory.

It is common practice, too, to modify all basis sets to provide appropriate descriptions of the molecular environment using 'molecular' Slater exponents. Since we cannot expect

Table 1.10   A standard set of Slater ζ exponents for the STO-3G basis sets for molecular environments (39,47).

Atom	$\zeta[1s]$	$\zeta[2s/sp]$
H	1.24	—
He	[2.0925]	—
Li	2.69	0.75
Be	3.68	1.10
B	4.68	1.45
C	5.67	1.72
N	6.67	1.95
O	7.66	2.25
F	8.65	2.55

the electron distribution around an atom in a molecule to be the same as about the atom isolated in free space, it is reasonable to expect that approximations to the 'atomic orbitals' should be appropriate to this different environment. For the standard $|\text{sto-3g}\rangle$ basis set suitable 'molecular' Slater exponents are listed in Table 1.10. For other basis sets different scaling factors have been established and, for example, with the (4-31) basis and other *double-zeta* sets it is common to use a core exponent and a different valence exponent.

1.9   GAUSSIAN-LOBE AND OTHER GAUSSIAN BASIS SETS

Because of the dominance of the $|\text{sto-ng}\rangle$ contracted and polarized basis sets in modern applications, such as GAUSSIAN98, Gaussian-lobe (48) and Floating-Spherical Gaussian orbitals are little used. Their demise has been encouraged, too, by the development of the software technology to perform integrations involving other than 1s Gaussian primitives. The notion that Gaussian s-type functions, appropriately sited and combined about any centre, as *Gaussian lobes*, can approximate to atomic orbitals of higher angular momentum is appealing because then all the simplifications involving integrations with 1s Gaussians apply to all of a basis set.

The 'chemist's' perspective of the Lewis two-electron bond, also, has been utilized in basis set theory. In 1967, Frost (49) proposed the use of 'subminimal' basis sets of single Gaussian functions sited between the atomic positions in a molecular structure.

In an *extended* basis set, we double each valence orbital representation. A delocalized Gaussian, i.e. with smaller α exponent, of the same quantum number, for example, 1s and 1s' orbitals for hydrogen and 1s, 2s, 2s', 3 × 2p, and 3 × 2p' orbitals for carbon, nitrogen, and oxygen, facilitate better modelling of molecular electron density in the valence region because with suitable choices for the coefficients optimal modelling can be achieved. If, as well, we double the core orbitals, this extended *split-valence* set becomes a genuine *double-zeta* basis.

Then, for *polarized* basis sets, we add to each valence orbital basis functions of higher angular quantum number, for example: p orbitals for hydrogen and d orbitals for the carbon, nitrogen, and oxygen atoms. This permits small displacements of the centre of the electronic charge distributions away from the atomic positions (charge polarization) and is the chemist's model of hybridization. Finally, basis sets can be improved further with additional functions for the valence or core regions (to give *triple-zeta* and larger basis sets) and, indeed, by many other fine details to model particular properties being calculated, for example the effects of correlation (50) for which *Atomic Natural Orbitals* (51) and *Correlation Consistent Basis Sets* (52) are appropriate.

Nowadays, |sto-ng) basis sets have been proposed for most atoms of the elements of the Periodic Table. The William Wiley Environmental Molecular Sciences Laboratory maintains a basis set service at *http://www.emsl.pnl.gov:2080/forms/basisform.html*, which database includes performance information for each basis set in comparisons with HFS atomic structure calculations.

2

Numerical Integration

In this chapter, you will learn how to carry out numerical integration on a spreadsheet. The chapter begins with some revision of the two standard simple methods for numerical integration of a tabulated function. Then, this methodology is applied to the calculations of normalization and overlap integrals involving the hydrogen 1s orbital.

In quantum chemistry, every property that can be measured requires the evaluation of integrals over atomic orbitals, represented approximately as expansion over basis sets these in the form of basis sets. In *LCAO-MO* bonding theory, the basic integrals are the overlap, normalization, kinetic energy and potential energy integrals, which involve the components of whatever basis set is chosen. Within the Gaussian approximation, all these integrals can be transformed into analytical forms, as you will see in later chapters. However, numerical integration was a very important part of the early fabric of atomic and small molecule theory. In atomic theory, the Herman–Skillman program (4) is a good example of the application of this procedure. Moreover, it is still a very useful learning exercise to make familiar the mystery of the solution process of Schrödinger's equation.

Thus, in this chapter you will learn:

1. how to perform numerical integration on a spreadsheet, using Simpson's rule;
2. how to calculate a normalization constant using numerical integration;
3. how to calculate the angular normalization constants of Table 1.1 using the spreadsheet method;
4. how to calculate an overlap integral between two 1s atomic orbitals in H_2;
5. how to calculate the same integral using the $|$sto-3g$\rangle$ basis to approximate the hydrogen 1s orbitals;
6. how to determine the 'best' linear combinations of Gaussian functions for the modelling of Slater functions.

2.1 NUMERICAL INTEGRATION

The trapezium rule and Simpson's rule are the two standard procedures for numerical integration (53). In the trapezium-rule procedure, the integral, as the area under the curve of the integrand with respect to its argument, is divided into rectangular sections and the 'bits' which are left over, then, are approximated to triangles, formed by joining vertices of neighbouring rectangles, as shown in Figure 2.1. For 'h' identical intervals between

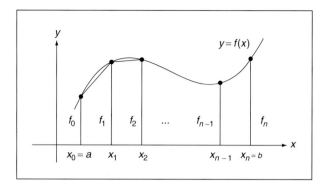

Figure 2.1   Division of $\int_a^b f(x)\,dx$ into contributions from individual polygons defined by the choice of mesh for $\Delta x = (x_2 - x_i) = dx$. [From *The Chemist's Maths Book, Erich Steiner, OUP 1996* © E. Steiner].

the limits of the integral and individual trapezoidal areas, the interval size multiplied by the function values in the intervals, $f_0, f_1, f_2, \ldots f_1, \ldots f_{n-1}, f_n$, the integral is given by

$$\int_a^b = \frac{b-a}{h}\left[\frac{f_0 + f_n}{2} + \sum_{i=1}^{n-1} f_i\right] \qquad 2.1$$

In this, the simplest approximation to an integral, for small intervals, the error reduces by a factor of four for every halving of the interval 'h'.

In the Simpson's rule procedure, a similar division of the area, defined between the bounds of the integration and the curve of the function is made. However, for Simpson's rule, it is necessary to choose an even number of intervals. Then, in the same notation as for the Trapezium rule procedure,

$$\int_a^b = \frac{b-a}{3h}\left[f_0 + f_n + 4\sum_{even} f_{even} + 2\sum_{odd} f_{odd}\right] \qquad 2.2$$

The error in a Simpson's rule integration is less than the corresponding trapezoidal rule value. Moreover, the Simpson's rule error reduces by a factor of 16 when the interval size is halved. This is not a particularly significant consideration with modern spreadsheets. On a modern spreadsheet, there are many thousands of rows of cells upon which to build a suitably large integration mesh[1] and so the exact shapes of the polygons in Figure 2.1 need not be important.

As you see in equation 2.2, with Simpson's rule, there is a need to provide for the multiplication of the odd and even integration components by different factors. The inclusion of a counter in the design of the integration spreadsheet provides a generally useful way to introduce such a condition into a calculation.

In the present application, such a device can serve, too, to facilitate a convergence check on the numerically calculated value of an integral. This extension of the basic design is introduced as an exercise, rather than a main component of the general spreadsheet. Convergence problems are not significant at the level of accuracy of interest in this book,

[1] In EXCEL97 and EXCEL2000 there are 65538 rows and 255 columns on the standard worksheet!

because fine integration meshes can be applied in equation 2.2, but it is proper to remain alert to the possibility that convergence errors can be important.

The spreadsheet, fig2-2.xls, establishes our general tool for definite integrals based on the application of Simpson's rule and the basic design features occur in most of the other spreadsheets throughout this book.

2.2   APPLICATION OF SIMPSON'S RULE TO CALCULATE A NORMALIZATION INTEGRAL

For the example of the 1s radial function in hydrogen, from Table 1.1,

$$|1s\rangle = e^{(-Zr/a_0)} \tag{2.3}$$

the defining condition for this normalization only over the radial coordinate follows from Figure 1.6 and is

$$\langle 1s|1s\rangle = N^2 \int_{r/a_0=0}^{r/a_0=\infty} e^{(-2Zr/a_0)} r^2 \, dr = 1 \tag{2.4}$$

We can check the normalization constant, N, given in Table 1 as

$$N = 2(Z/a_0)^{3/2} \tag{2.5}$$

by applying Simpson's rule as follows,

Exercise 2.1.   Simpson's rule applied to the calculation of the normalization constant of the H-1s radial orbital.

1.   Enter the column headings as marked in Figure 2.2 on a new spreadsheet, fig2-2.xls.
2.   Enter 0 in cell $A$2. To distinguish the entries in the column from the row and column headings in Figure 2.2, format the column, using Ctl/spacebar to select all the cells in the whole column. With the whole column selected, set the text colour to a personal choice, here black, choose *centred* and *integer* formats for all entries using the appropriate toolbar icons.
3.   Select $A$2². Choose the EDIT/FILL/SERIES command, select *column*, then *step value* $= 1$ and *stop value* $= 3200^3$.
4.   Enter 0.0 in $B$2. Select $B$2. Choose EDIT/FILL/SERIES. Set the *step value* $= 0.003125$ and the *stop value* $= 10$ a.u. These choices generate the 3200 radial grid for the integration. Format the cells to display numbers to four decimal places [*select cell $B$2 and then the whole column using the CTL/SPACEBAR sequence, use the TOOLBAR icon to set the number of decimal places*].
5.   Generate the projections of the 1s radial function on the radial mesh points in column C. For example, the first projection is the result

$$\$C2 = \textbf{EXP(-\$F\$4*\$B2/2)}$$

---

[2] Selections are cancelled by pressing the ESC [*Escape*] key!
[3] This limit to the counter is chosen to provide for a reasonable number of integration results in the convergence test: 3200 is $2^7 5^2$.

	A	B	C	D	E	F
1		r	$|1s\rangle$		Z =	1.00000
2	0	0.0000	1.00000	1	a_0 =	1.00000
3	1	0.0031	0.99688	4	n =)	1
4	2	0.0063	0.99377	2	ρ =)	2
5	3	0.0094	0.99067	4	mesh	Integral
6	4	0.0125	0.98758	2	3200	0.2499999
7	5	0.0156	0.98450	4	N	2.00000
8	6	0.0188	0.98142	2		
9	7	0.0219	0.97836	4		
10	8	0.0250	0.97531	2		
11	9	0.0281	0.97227	4		
12	10	0.0313	0.96923	2		
13	11	0.0344	0.96621	4		

Figure 2.2 The Simpson's rule integration procedure, on an EXCEL spreadsheet, to determine the normalization constant, N, for the 1s radial function in hydrogen defined in Table 1.1. The constant N in cell $F$7 is the inverse of the square root of the value of the integral in cell $F$6.

6. To propagate this formula down the active cells of column C, we need to select a large number of cells down this column of the spreadsheet. The simplest procedure is to press F8 on the keyboard and then use the EDIT/GOTO command to select from the first selected cell to the last one needed. So, in the present case, enter C3202 in the GOTO *dialogue box* of this command. Use the EDIT/FILL/DOWN command to propagate the formula values in column C. The keyboard sequence CTL/D enables direct access to the FILL/DOWN [over a selected range in a spreadsheet column] command in the EDIT *dialogue box*.

7. To provide for the alternating multiplication factors of 2 and 4 in equation 2.2, the spreadsheet design needs to include a conditional multiplier for the construction of the Simpson's rule integration as a sum over the projected values on the mesh points. **MOD( )** is one of the functions available in the library of the 'function wizard' in EXCEL, which is activated by the toolbar icon f_x. **MOD( )** returns the remainder in a division. The presence or absence of such a remainder, then, permits a logical operator such as the **IF( )** operator to be triggered as TRUE or FALSE. For example, for TRUE the cell entry can be set to 2, for FALSE the entry can be set to 4 and this procedure enables the various multiplications in equation 2.2. Thus, enter the formula

$$= \mathbf{IF}(\mathbf{MOD}(A3,2) = 0,2,4)$$

in cell $D3 and propagate this down column D to cell $D3201.

8. Enter 1 in cells $D$2 and $D$3202 to allow for the initial and final projections in equation 2.2.

9. Select column D and use the CONDITIONAL FORMAT option in the FORMAT *drop down* box to colour the cells in the column of value 4 as displayed in Figure 2.2.
10. Finally, complete the summation of equation 2.2, using the important array multiplication facility **SUMPRODUCT( )** from the function library, with the entry

$F$6 = ($B$3-$B$2)/3*$**SUMPRODUCT**(B$2:$B$3202,
$B$2:$B$3202,$D$2:$D$3202,$C$2:C$3202,$C$2:$C$3202)

Note, the inclusion of the r^2 term in the calculation of the integral, which is required by equation 1.9 because of the construction in Figure 1.6. The array multiplication in the **SUMPRODUCT( )** expression is the sum of all the individual products of the values of each equivalent cell in the columns involved in the calculation. Thus, the active cells of column B, the radial array, are included *twice* and this is true, too, for the cells of column C, since we require the product of the function, while the Simpson's rule multipliers, column D, occur only once. The integration is completed by multiplying by the grid size parameter, ($B$3-$B$2)/3.

11. Finally, calculate the normalization constant, N, as the inverse square root of the values of the integral, with

$$F$7 = **SQRT**(1/$F$6)$$

For the hydrogen 1s radial function the normalization constant is exactly 2.0. As you see, in Figure 2.2, the normalization constant has been calculated to be 2.00000 to five decimal places for the integration mesh of 3200 radial points in the range $0 \le r \le 10.00$. You should change the integration range and redo this calculation to determine the extent of the tail of the wave function.

It is instructive, too, to examine the convergence of the integration for different radial meshes. This is a good exercise to test your knowledge of spreadsheet design. A typical result is shown in Figure 2.3, based on the design changes given in Exercise 2.2.

	A	B	C	D	E	F	G	H	I	J	K	L
1		r	|1s⟩							Z =	1.00000	
2	0	0.0000	1.00000	1	1	1	1	1	1	a₀ =	1.00000	
3	1	0.0031	0.99688	4	0	0	0	0	0	n =⟩	1	
4	2	0.0063	0.99377	2	4	0	0	0	0	ρ =⟩	2	
5	3	0.0094	0.99067	4	0	0	0	0	0	mesh	Simpson's Rule Integral	N
6	4	0.0125	0.98758	2	2	4	0	0	0	3200	0.249999885912917	2.00000
7	5	0.0156	0.98450	4	0	0	0	0	0	1600	0.249999885793574	2.00000
8	6	0.0188	0.98142	2	4	0	0	0	0	800	0.249999886889649	2.00000
9	7	0.0219	0.97836	4	0	0	0	0	0	400	0.249999910419478	2.00000
10	8	0.0250	0.97531	2	2	2	4	0	0	200	0.250000297702832	2.00000
11	9	0.0281	0.97227	4	0	0	0	0	0	100	0.250006441023491	1.99997

Figure 2.3   Convergence tests on the Simpson's rule integration procedure to evaluate equation 2.4 on an EXCEL spreadsheet.

**Exercise 2.2.   Convergence tests on the Simpson's rule procedure for the cal-
culation of the normalization constant of the H-1s radial orbital.**

This exercise is an extension of the previous one and so the first step is to make a copy of fig2-2.xls. Then as you see in Figure 2.3, to provide for different integration meshes we need extra columns of Simpson's rule multiplier entries. The column D entries are as before, but for the other columns we need to allow cell entries equal to 0, as well. Thus, for example, the formula in cell $E3 is

$$\$E3 = \mathbf{IF(MOD}(\$A3,4) = 0,2,\mathbf{IF(MOD}(\$A3,2) = 0,4,0))$$

When, propagated down column E, the mesh size is doubled and there are only 1600 integration grid points, because of the actions of the second logical condition. You repeat this procedure in columns F to I but using the divisors 4, 8, 16 and 32 in the second **MOD()** function to provide for integration grids of 800, 400, 200, 100 and 50 points.

This result, demonstrated in Figure 2.3, confirms that we should have confidence in the integration procedure and that we can expect results equally as good as those which would follow from the application of the more formal procedures, used, for example, by Hartree (54). It seems that the capacity to use large numbers of mesh points in the spreadsheet integration procedure renders insignificant, the matter of the generally poor representation of the polygon remainders by the quadrature of the Simpson's rule procedure, when the function approximated is an exponential.

2.3   CALCULATIONS OF NORMALIZATION CONSTANTS OVER THE ANGULAR COORDINATES

It is appropriate, too, to test the utility of the spreadsheet integration procedures for the determination of the orthonormality relationships among the angular parts of the atomic orbital functions. These are the integrations of equation 1.8, over the unit sphere, of the form

$$\langle Y_{lm}(\theta, \phi) \mid Y_{l'm'}(\theta, \phi) \rangle = \int_0^{2\pi} \int_0^{\pi} Y_{lm}^*(\theta, \phi) Y_{lm}(\theta, \phi) \sin(\theta) \, d\theta \, d\phi \qquad 2.6$$

For the case of 1s atomic orbitals, Table 1.1, we expect the integrations to return the value 4π. The integration over the azimuthal angle is trivial and amounts to 2π. The integration over the polar angle is simply the sums of the values of $\sin(\theta)$ in the range of θ from 0 to π radians. This, as a spreadsheet procedure is set out in Figure 2.4 and it is evident that good convergence is achieved even for relatively crude sampling of the integration area defined by the sine function over the limits 0 to π.

Since this example is a straightforward modification of the basic spreadsheet, the design details are left again as an exercise. They are evident in the file, fig2-4.xls, on the CDROM, and it is good practice to demonstrate the correctness of the other normalization constants given for the angular functions in Table 1.1 using this spreadsheet.

#	θ	Sin(θ)							mesh	Simpson's Rule Integral
0	0.0000	0.00000	1	1	1	1	1	1	720	2.000000000004030
1	0.2500	0.00436	4	0	0	0	0		360	2.000000000064440
2	0.5000	0.00873	2	4	0	0	0		180	2.000000001031060
3	0.7500	0.01309	4	0	0	0	0		90	2.000000016498710
4	1.0000	0.01745	2	2	4	0	0		45	1.999187620887650
5	1.2500	0.02181	4	0	0	0	0			
6	1.5000	0.02618	2	4	0	0	0			
7	1.7500	0.03054	4	0	0	0	0			
8	2.0000	0.03490	2	2	2	4	0			

Figure 2.4 Simpson's rule-based numerical integration of the 1s atomic orbital, over θ and ϕ converging to the expected value of 2.0000000, at least, for most meshes.

2.4 NUMERICAL INTEGRATION IN A CYLINDRICAL VOLUME: DIATOMIC AND LINEAR MOLECULAR GEOMETRIES

There will be a need, in Chapter 6, to carry out numerical integrations by extending a planar array into a cylindrical volume. For linear molecular structures, it is appropriate to construct a planar array and then integrate each result, in the plane, over the circumference of the circle, with radius defined by the position of the array point with respect to the molecular axis. The construction is set out in Figure 2.5 and the example of a grid for the space of the diatom, H_2, is sketched in Figure 2.6.

The details of the integration procedure are set out in Figure 2.7. Depending on whether the molecular structure is symmetrical across the origin on the z-axis, e.g., A_2 and AB_2-type

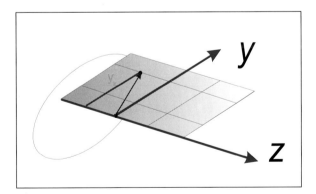

Figure 2.5 The two-dimensional mesh for numerical integrations in the MO theory of diatomic molecules. Any point on the 2D mesh in the yz-plane is transformed into an integration component in 3D by weighting each value by $2\pi y$ as marked for the y-value identified by the pink vector.

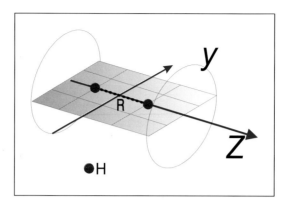

Figure 2.6   The H_2 molecular geometry on the integration array.

	A	B	C	D	E	F
3			Atom 1	Atom 2		
4		Z =⟩	1	1	R	
5		Positions	1	-1	2	
6		N =⟩	0.564189584	0.564189584		
7						
8		Integral =⟩	0.58644122			
9		Exact =⟩	0.58645289			
10	1	2	3	4	5	6
11		1	4	2	4	2
12	y/z	-7.500	-7.425	-7.350	-7.275	-7.200
13	7.500	5.84279E-10	2.59608E-09	1.44108E-09	3.19798E-09	1.77320E-09
14	7.425	6.49638E-10	2.88801E-09	1.60398E-09	3.56139E-09	1.97576E-09
15	7.350	7.21926E-10	3.21108E-09	1.78437E-09	3.96403E-09	2.20032E-09
16	7.275	8.01825E-10	3.56838E-09	1.98398E-09	4.40985E-09	2.44911E-09
17	7.200	8.90082E-10	3.96328E-09	2.20473E-09	4.90318E-09	2.72458E-09
18	7.125	9.87505E-10	4.39945E-09	2.44870E-09	5.44872E-09	3.02939E-09
19	7.050	1.09497E-09	4.88089E-09	2.71814E-09	6.05161E-09	3.36643E-09
20	6.975	1.21345E-09	5.41194E-09	3.01553E-09	6.71741E-09	3.73888E-09
21	6.900	1.34396E-09	5.99730E-09	3.34354E-09	7.45220E-09	4.15018E-09

	GR	GS	GT	GU	GV	GW	GX
10	200	201	202				
11	2	4	1				
12	7.350	7.425	7.500				
13	1.44108E-09	2.59608E-09	5.84279E-10	1	6.346E-05	Integral =	1.84236
14	1.60398E-09	2.88801E-09	6.49638E-10	4	2.902E-04		
15	1.78437E-09	3.21108E-09	7.21926E-10	2	1.659E-04		
16	1.98398E-09	3.56838E-09	8.01825E-10	4	3.793E-04		
17	2.20473E-09	3.96328E-09	8.90082E-10	2	2.168E-04		
18	2.44870E-09	4.39945E-09	9.87505E-10	4	4.954E-04		
19	2.71814E-09	4.88089E-09	1.09497E-09	2	2.830E-04		
20	3.01553E-09	5.41194E-09	1.21345E-09	4	6.465E-04		
21	3.34354E-09	5.99730E-09	1.34396E-09	2	3.691E-04		

Figure 2.7   The first few and last few columns of the spreadsheet applied to the calculation of the overlap integral in H_2 using the approximate molecular orbital of equation 2.7.

structures, the array plane can be chosen to embrace all the atomic positions or just the one half of them, with the resulting integration being over the whole of the integration cylinder or just one-half of it. In fact, it is good practice to use this extra flexibility, where possible, to estimate the convergence of any one integral, since the convergence test, applied in fig2-3.xls, is not simple to include in the case of the cylindrical procedure.

The number of columns[4] on the typical spreadsheet is a limiting factor in the setting of mesh size along the z-axis of Figure 2.5 for the simplest design of one block of data per row. The best approximate integration follows from the use of the finest mesh and the most appropriate procedure consistent with the size of PC available. Therefore, it is worth making a little effort to generate a reliable integration spreadsheet for the cylindrical procedure. Since a few columns are required for the final calculations over and above the values of the integrand on the mesh, it is necessary to exclude the last few columns on the extreme right of the spreadsheet. In addition, for the Simpson's rule procedure, an even number of mesh points is needed, which further condition fixes the finest mesh along the z-axis to intervals of 0.05 atomic units and a maximum range $-12.4 \leq z \leq +12.4$ for the integration using the doubling device. The advantage of such a choice is that some estimation of convergence can be had by some increases or decreases in mesh size. Moreover, when the doubling procedure cannot be used, for example, when the integrations for diatomic species are required, the precision of particular results can be judged from these observations.

The calculation of the overlap integral between two hydrogen 1s atomic orbitals, required, for example, in the *LCAO*-MO theory of the dihydrogen molecule, is a good example of this numerical procedure.

The standard simple approximation to the $1\sigma_g{}^+$ molecular orbital of H_2 is the linear combination

$$1\sigma_g{}^+ = 1s(r - R/2) + 1s(r + R/2) \qquad 2.7$$

with R the bond length.

The overlap integral, S, in this case, is easy to calculate exactly (55) and is given by

$$S = e^{-R}(1 + R + R^2/3) \qquad 2.8$$

Exercise 2.3.   Calculation of the overlap integral between two 1s atomic orbitals using the cylindrical procedure.

The Simpson's rule spreadsheet, fig2-7.xls, for the calculation is constructed as follows,

1. Enter the titles and basic data in the first few rows of the spreadsheet as in Figure 2.7.
2. Note that the bond length, R, cell $E$5, is a primary parameter and that the entries in cells $C$5 and $D$5 are linked as $+/- (R/2)$.
3. The normalization constants, N, in cells $C$6 and $D$6 are entered as given in Table 1.1, but note the alternative power symbol '^' in the equations, for

---

[4]Remember, there are 255 columns, on the standard EXCEL spreadsheet.

example, the formula

$$\$C\$6 = \$C\$4^1.5^*\textbf{SQRT}(1/\textbf{PI}())$$

Note, too, that normalization is required over all the coordinates.

4. The entry in cell $\$C\$9$ is the exact result for the overlap integral given by equation 2.7

$$\$C\$9 = \textbf{EXP}(-\$E\$5)^*(1+\$E\$5+\$E\$5^2/3)$$

5. The entries in pink in row 10 are the counters used to obtain the Simpson's rule multipliers of row 11.
6. Then, in row 12 and column A, the yz-mesh is established using the EDIT/FILL SERIES command. In this example, the mesh is set out over the cells bound by column GT and row 113, from -7.5 to $+7.5$ along the z-axis and for a maximum radius on the y-axis of 7.5, with mesh points at intervals of 0.075 a.u.
7. The first element of the overlap integral is generated in cell B13 with the formula entry,

$$B13 = B\$11^*\textbf{EXP}(-\$C\$4^*(\textbf{SQRT}(\textbf{SUMSQ}$$
$$(\$A13,(B\$12-\$C\$5)))))^*\textbf{EXP}(-\$D\$4^*(\textbf{SQRT}$$
$$(\textbf{SUMSQ}(\$A13,(B\$12-\$D\$5)))))$$

which, then, is propagated down the active cells of column A and, then again, in turn across the active cells in the 114 rows of the x-mesh, to complete the integral array in column GT. Note, that this expression includes the multiplication factors of row 11.

8. Rather than use another counter to establish the Simpson's rule multipliers in column GU, required for the second integration, it is simpler to copy these from row 11 and then, with corrections for the first and last entries, use the EDIT /PASTE SPECIAL/TRANSPOSE option.
9. In column GV the first sets of Simpson's rule integrations are carried out over the z-mesh for each point on the x-mesh. These integrals are transformed to values over the circles defined in Figure 2.5 and multiplied by the same factors to return the final integral in cell $\$GU\$13$. Thus, the formula entry in cell $\$GV\$13$ is

$$\$GV13 = 2^*\textbf{PI}()^*\$A13^*(\$C\$12-\$B\$12)/3^*$$
$$\textbf{SUM}(\$B13:\$GT13)^*\$A13^*\$GU13$$

Note, the multiplication by $2\pi r$ returning the value over the circle and the final multiplication by cell $\$GU\$13$ to set the second Simpson's rule integration over the x-coordinate. Remember, the first Simpson's rule multipliers for the z-coordinate integration are included in the individual expressions in each cell of each row.

10. Finally, obtain the overall integral as the sum formula in cell $\$GX\$12$ with

$$\$GX\$13 = \textbf{SUM}(\$GV13:\$GV113)^*(\$A\$13-\$A\$14)/3$$

and copy this result into cell $\$C\$8$.

The value for the overlap integral, at the experimental bond length, is equal to the exact result to the fifth-decimal place, which is sufficient for our later applications of this kind of spreadsheet.

Overlap is a very important consideration in the determination of bonding interactions. Chemists assess the strength of chemical bonds, largely based on overlap considerations; for example, there is the familiar maxim 'strong bonds result from significant overlap'. Figure 2.8 illustrates the variation of overlap integral for 1s Slater orbitals with bond length to simulate the different radial functions for atoms from hydrogen to beryllium.

The calculations of the overlap integrals for the different atoms identified by the different atomic numbers do not involve the generation of a distinct spreadsheet for each case. The entries in cells $B$10 to $I$13 are the overlap integrals for the effective atomic numbers, the Slater exponents, listed in cells $A$10 to $A$13 and the bond lengths in cells $B$9 to $I$9. The results in this 'two-variable table' follow from the table master entry

$$\$A\$9 = \$C\$6^*\$D\$6^*\$GX\$13 \qquad\qquad 2.9$$

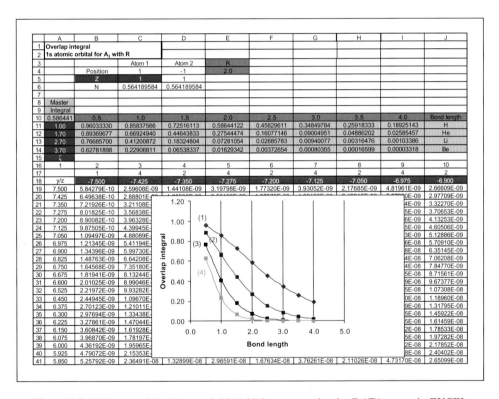

Figure 2.8 The use of the 'two-variable table' macro under the DATA menu in EXCEL to extend the calculations of Figure 2.6 to include 1s atomic orbitals, approximated to as Slater orbitals, from different atoms identified in cells $J$11 to $J$14[5].

[5]Note that, depending on the version of EXCEL you are running on your PC, you may activate the notice that the file fig2-8.xls contains MACROS, a warning used by 'virus checking programs'. In the present case and others on the CDROM this warning simply indicates that macros in the EXCEL software are enabled in the spreadsheets.

using the macro *wizard* to specify the inputs cells for the variables, i.e. the bond length in cell $D$4 and the Slater exponent in cell $B$5, which value also determines the atomic number in cell $C$5, because only two variables are permitted in the calculation.

Only hydrogen and lithium form stable neutral diatomic molecular species for the four atoms considered in Figure 2.8. The bond lengths for these diatomics are 1.4 a.u. in H_2 and 5.05 a.u. in Li_2 (56). Figure 2.8 illustrates the point that while the bond in dihydrogen is between the hydrogen 1s orbitals, it is only between the 2s orbitals in lithium, because there is little overlap of the 1s functions at the larger bond length in dilithium.

Exercise 2.4.   Use of the two-variable table to generate Figure 2.8.

1. Copy fig2-7.xls and rename the copy fig2-8.xls.
2. Rearrange the occupied cells in the first few rows of data, by selecting these cells and using the *drag and drop* facility by activating the *drag handle* in the top-left cell of the selection.
3. Insert rows 7 to 14.
4. Make sure that cell $A$9 contains the master integral formula

$$\$F\$9 = \$C\$6^*\$D\$6^*\$GX\$13$$

form cell $C$6 of fig2-7.xls.
5. Enter the bond length range of interest in cells B$9 to I$9 and the Slater exponents for the series H to Be in cells $A10 to $A13.
6. Select cells $A$9 to $I$13 and activate the TABLE wizard under the DATA box in the toolbar.
7. Enter cell $D$4 as the bond length variable and cell $B$5 as the Slater exponent variable in the appropriate cells of the wizard window.
8. Use the results to plot the variations in overlap integral with this effective atomic number and bond length using the CHART wizard [*note, that the individual graphs have been labelled using PAINT, in Figure 2.8*].

2.5   CALCULATION OF THE OVERLAP INTEGRAL BETWEEN 1S ORBITALS IN A GAUSSIAN BASIS

It is very instructive to recalculate the overlap integral between the two 1s atomic orbitals in a Gaussian basis, because we begin to see the great benefit of the Gaussian representation and the importance of the simplifications, which follow from equations 1.16 to 1.19. The calculation, too, is a good further exercise in the application of the TABLE macro facility in EXCEL. The *two-variable table* macro provides a very convenient procedure to calculate integrals over basis sets involving linear combinations of two or several components and, with suitable design details can be applied also to *double-zeta* Slater approximations to atomic radial functions. This calculation is left as an exercise for the reader.

Exercise 2.5.   Calculation of the overlap integral using a Gaussian basis.

To establish the general integration procedure for 1s Gaussians the simplifications of equation 1.16 to 1.19 are introduced into the spreadsheet calculation. The spreadsheet is fig2.9.xls on the CDROM and is constructed on a copy of fig2-8.xls

1. Enter the basic |sto-3g⟩ parameters, the location vectors, R, and the Slater exponents for hydrogen in cells $C$6 to $H$7, including the various labels as appropriate.
2. Provide for the scaling required to model other atoms or the molecular environment, following Table 1.11, using equation 1.22 in row 8, for example,

$$\$C\$8 = \textbf{POWER}(\$D\$4,2)$$

3. Apply the normalization factor for the 1s Gaussian as in equation 1.15, so that in row 9, for example, we find

$$\$C\$9 = C\$7*\textbf{POWER}(2*\$C\$8/\textbf{PI}(\,),0.75)$$

The next instructions introduce the general simplification of 1s Gaussian products of equations 1.16 to 1.19.
4. Enter labels for all the possible Gaussian products as in row 10.
5. Provide for the pre-exponential term of equation 1.16 in row 11, with for example,

$$\$C\$11 = \textbf{EXP}(-C\$8*C\$8/(C\$8+C\$8)*\textbf{ABS}((\$E\$4-\$E\$4)\hat{}2))$$

6. Form the required product of the coefficients needed in the Gaussians products in row 12 again, for example

$$\$C\$12 = C\$9*C\$9$$

7. Now, establish the primary row and column of the *two-variable table*. Repeat the labels for the individual Gaussian products in row 15 and then enter the summed exponents for each of these in the next row, for example

$$\$C\$16 = (C\$8+C\$8)$$

and the Gaussian product location vector R_p of equation 1.18 as column entries in column B, with, for example,

$$\$B\$17 = (C\$8*\$E\$4+C\$8*\$E\$4)/(C\$8+C\$8)$$

8. Enter dummy variables for the TABLE in cells $I$4 and $J$4. As long as the choice of these does not cause irregular behaviour in the quadratic exponential form of the Gaussian, the values chosen do not matter. Here the choice is 1.00 and 0.00, respectively.
9. Generate the cylindrical integration array as in the previous spreadsheet from $A$41 to $GT$143, based on the formula for the primitive Gaussian formed with the dummy product sum and location vector of cells $I$4 and $J$4, for

example, in cell $B$43

$$\$B\$43 = \textbf{EXP}(-\$I\$4*\textbf{SUMSQ}(\$A43,(B\$42-\$J\$4)))*B\$41$$

and note the inclusion of the Simpson's rule multiplier.

10.   Enter the master formula for the *two-variable table* in cell $B$16, the formula

$$\$B\$16 = \textbf{SUM}(\$GV43:\$GV143)*(\$A\$43-\$A\$44)/3$$

11.   Select the table area from cell $B$16 to cell $W$37 and use the table wizard to identify cells $I$4 and $J$4 as the variables in the calculation.

12.   Render invisible [select the font colour 'white'] all table entries other than the diagonal elements and copy these to the equivalent cells in row 13.

13.   Calculate the total contributions to the normalization and overlap integrals as the products, for example,

$$\langle 1|1 \rangle = \$C\$14 = C\$11*C\$12*C\$13$$

14.   Collect these components into appropriate sums in cell $J$7, $J$8 and $J$9 to finish the calculation of the normalization and overlap integrals.

As you see, in Figure 2.9, the same level of accuracy has been achieved in the calculation of the overlap integral as was found using the simple cylindrical procedure on the actual 1s orbital. Later, we return to this spreadsheet design as the basic template for the more complicated calculations in later chapters.

2.6   DESIGNING GAUSSIAN BASIS SETS TO MODEL SLATER ORBITALS

The integration 'on a spreadsheet' procedure developed in this chapter can be applied very successfully to the testing and proving of different Gaussian sets for modelling Slater orbitals.

In their seminal 1969 paper (33) on the design of Gaussian basis sets to model Slater functions, Hehre, Stewart and Pople proposed that basis sets be determined using the least-squares minimum conditions,

$$\varepsilon_{1s} = \int (\phi_{1s-\text{Slater}} - \phi_{1s-\text{gaussian}})^2 \, d\tau \qquad 2.10$$

and

$$\varepsilon_{2s} + \varepsilon_{2p} = \int (\phi_{2s-\text{Slater}} - \phi_{2s-\text{gaussian}})^2 \, d\tau + \int (\phi_{2p-\text{Slater}} - \phi_{2p-\text{gaussian}})^2 \, d\tau \qquad 2.11$$

The minima found for these integrals for the choices of Gaussian exponents and coefficients, listed in Table 1.6, are given in Table 2.1.

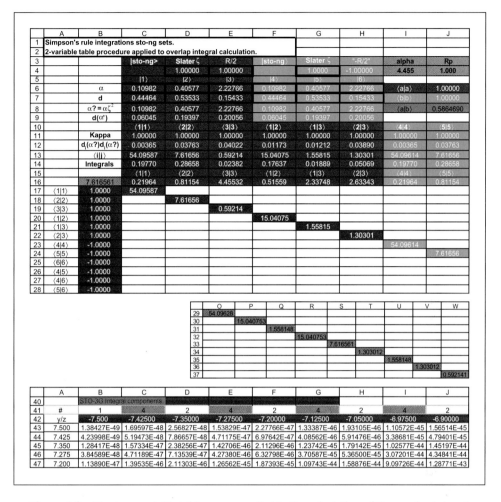

Figure 2.9 The *two-variable table* macro applied to the calculation of the overlap integral between two |sto-3g:1s⟩ Gaussian sets for hydrogen. Note, that the Slater exponent is set equal to 1.00, in this example, rather than scaled to 1.24 [Table 1.11] as is customary in molecular calculations. *Note, the 'blanking out' of the off-diagonal entries in the table, by choosing the white font colour.*

Table 2.1   The Hehre, Stewart and Pople values (33) for the integrals of equations 2.9 and 2.10.

|sto-ng⟩	ε_{1s}	ε_{2s}	ε_{2p}
|sto-2g⟩	$3.16 \ 10^{-3}$	$5.6 \ 10^{-3}$	$3.48 \ 10^{-3}$
|sto-3g⟩	$3.31 \ 10^{-4}$	$6.42 \ 10^{-4}$	$3.60 \ 10^{-4}$
|sto-4g⟩	$4.38 \ 10^{-5}$	$7.95 \ 10^{-5}$	$4.82 \ 10^{-5}$
|sto-5g⟩	$6.88 \ 10^{-6}$	$1.17 \ 10^{-5}$	$7.28 \ 10^{-6}$
|sto-6g⟩	$1.24 \ 10^{-6}$	$2.01 \ 10^{-6}$	$1.22 \ 10^{-6}$

It is straightforward and instructive to repeat the Pople, Hehre and Stewart calculations using the spreadsheet method.

Exercise 2.6a.   Demonstration that the results in Table 2.1 are returned by the spreadsheet calculation using the example |sto-6g:2s⟩.

1. Provide for up to six primitive Gaussian functions to model a Slater function, by entering either zero values or the exponents and coefficients listed in Table 1.6 in cells \$D\$5 to \$I\$6. In Figure 2.10, the cell values correspond to the |sto-6g:2s⟩ basis set.
2. Provide for scaling following equation 1.22 in cells \$D\$7 to \$I\$7 and for the product of the coefficients of the expansion and the normalization constants of the individual Gaussians following equation 1.14, with for example

$$\$D\$7 = \$D\$5*\$F\$4^2$$

and

$$\$D\$8 = \$D\$6*\mathbf{POWER}(2*\$D\$7/\mathbf{PI}(),0.75)$$

with the appropriate Slater exponent in cell \$F\$4 and the principal quantum number in \$H\$4.

	A	C	D	E	F	G	H	I	J	K	L
1	Hehre, Stewart and Pople |sto-ng⟩										
2	Evaluation of ⟨φ−φ⟩² for ns										
3											
4			|sto-ng⟩	Slater ζ	0.5	n =⟩⟩	2				
5	α		0.0485690	0.1059600	0.2439770	0.6341420	2.0403600	10.3087000			
6	d		0.2407060	0.5951170	0.250242	-0.0337854	-0.0469917	-0.0132528			
7	αζ²		0.01214	0.02649	0.06099	0.15854	0.51009	2.57718			
8	d(αζ²)		0.00628	0.02785	0.02189	-0.00605	-0.02021	-0.01921			
9			1/N²[sto-ng]	1.0000	⟨φ−φ⟩²	2.01E-06	1/N²[Slater]	1.0000			
10	r	#	|g1⟩	|g2⟩	|g3⟩	|g4⟩	|g5⟩	|g6⟩	|sto-ng:2s⟩	|Slater⟩	⟨φ−φ⟩
11	0.00	1	0.00628	0.02785	0.02189	-0.00605	-0.02021	-0.01921	0.01054	0.00000	-0.0105
12	0.01	4	0.00628	0.02785	0.02189	-0.00605	-0.02021	-0.01921	0.01054	0.00057	-0.0100
13	0.02	2	0.00628	0.02785	0.02189	-0.00605	-0.02021	-0.01919	0.01056	0.00114	-0.0094
14	0.03	4	0.00628	0.02785	0.02189	-0.00605	-0.02021	-0.01917	0.01059	0.00170	-0.0089
15	0.04	2	0.00627	0.02785	0.02189	-0.00605	-0.02020	-0.01913	0.01063	0.00226	-0.0084

Figure 2.10   Calculation of the least-squares integral for the matching of the |sto-6g⟩, Hehre, Stewart and Pople, Gaussian basis to the Slater 2s function for the case of the hydrogen atom. The least-squares result cell, \$G\$9, is as listed in Table 2.1.

3. Form the projections of the normalized primitive Gaussian functions on the radial mesh in columns \$D\$11 to \$I\$3011[6] in the usual manner. Enter the basic formula in cell \$D\$11, with

$$\$D\$11 = \$D\$8*\mathbf{EXP}(-\$D\$7*\$A11^2)$$

---

[6]There is no significance in this choice of the number of mesh points. It is just marginally easier to work with 3000, rather than 3200, since there is no need to check for good convergence.

for the 1s Gaussian representation of the 2s Slater function. Note and remember to use absolute referencing of the rows of transformed input exponents and coefficient and to fix the column of the radial array.

4. Propagate this formula down column D. Select cells $D$11 to $D$3011 and apply the EDIT/FILL/DOWN sequence in the EDIT drop down box or using the CTL/D keyboard stroke.

5. Hold the SHIFT key down and extend the selection to the cells of columns E to I using the EDIT/FILL/RIGHT sequence or the CTL/R keyboard stroke.

6. Add these individual contributions to form the projections of the overall linear combination on the radial mesh using the **SUM( )** function in cell $J$11 and propagate this result down the column.

7. Form the projections of the Slater function to be matched in column K, with, for example, the cell entry

$$\$K\$11 = \textbf{POWER}(2^*\$F\$4, (\$H\$4+0.5))^*(1/\textbf{SQRT}(\textbf{FACT}(2^*\$H\$4))^* \\ \$A11\char`^(\$H\$4-1)^*\textbf{EXP}(-\$F\$4^*\$A11))/\textbf{SQRT}(4^*\textbf{PI}()))$$

and propagate this down the cells of column K.

8. Subtract the projections of the Slater and |sto-ng) functions in column L using the select and fill-down procedures as before.

9. Finally, calculate the least-squares integral in cell $G$9 as

$$\$G\$9 = 4^*\textbf{PI}()^*(\textbf{SUMPRODUCT}(\$A\$11:\$A\$3011,\$A\$11:\$A\$3011, \\ \$C\$11:\$C\$3011,\$L\$11:\$L\$3011,\$L\$11:\$L\$3011))^* \\ (\$A\$12-\$A\$11)/3$$

There are two immediate instructive applications of this spreadsheet, fig2-9.xls.

We can check straightforwardly that our understanding of the details about a basis set is correct. For example, consider the parameters given in Table 1.6 for the Gaussian modelling of the Slater functions for hydrogen 2s and 2p orbitals. The 2s Gaussian sets are linear combinations of 1s Gaussian primitives but scaled, as required, for applications to particular atoms. Thus, in Exercise 2.5, while the linear combinations are based on 1s primitive functions, the Slater exponent is set at 0.5, the value for the 2s Slater exponent for hydrogen. On the other hand, the 2p Gaussian set is based on the Slater function for the 2p radial function. Again, the Slater exponent is 0.5, the principal quantum number is 2, but the normalization factor is the pre-exponential factor in the defining equation, from equation 1.14, for the primitive Gaussians

$$g_{2p}(\alpha, r, \theta, \phi) = \left(\frac{128\alpha^5}{\pi^3}\right)^{1/4} r \, \exp(-\alpha r^2)\cos(\theta) \qquad\qquad 2.12[7]$$

in the Gaussian sets for the $2p_z$ orbital of hydrogen.

---

[7]This is $(3/(4^*\pi))^{0.5*}[2^{5.5}/(3^*\pi)]^{0.5}\alpha^{5/4}$ and so normalizes the Gaussian function over all the coordinates, see equation 1.16.

Exercise 2.6b.   Calculation of all the results in Table 2.1.

For the $|$sto-ng:s$\rangle$ Gaussian sets, the only modification required is the entry of the data in the cells of rows 5 and 6, from \$D\$5 to \$I\$6, with zero values as appropriate. For the $|$sto-ng:2p$\rangle$ Gaussian sets the spreadsheet must be modified as in Figure 2.11 to include the normalization factor and the extra multiplication by r in equation 2.13. Thus, on a copy of fig2-10.xls, renamed to be fig2-11.xls, make the following changes

1. Enter the formula

$$= C\$6^*POWER(2,1.75)^*POWER(C\$7,5/4)^*(1/POWER(PI(\,),0.75))$$

in cell \$C\$8.

	A	B	C	D	E	F	G	H	I	J	K
1	Hehre, Stewart and Pople STO-nG										
2	Evaluation of $\langle\phi-\phi\rangle^2$ for 2p										
3											
4			STO-3G	Slater ζ	0.5000	n =$\rangle\rangle$		2			
5	α		0.0751386	0.2310310	0.9942030	0.0000000	0.0000000	0.0000000			
6	d		0.3919570	0.6076840	0.1559160	0.0000000	0.0000000	0.0000000			
7	$\alpha' = \alpha\zeta^2$		0.01878	0.05776	0.24855	0.00000	0.00000	0.00000			
8	d(α')		0.00389	0.02453	0.03900	0.00000	0.00000	0.00000			
9	r	#	1/N^2	1.00000	$\langle\phi-\phi\rangle^2$	3.60E-04	1/N^2	1.0000			
10			$|$g1$\rangle$	$|$g2$\rangle$	$|$g3$\rangle$	$|$g4$\rangle$	$|$g5$\rangle$	$|$g6$\rangle$	$|$STO-Ng$\rangle$	$|$Slater$\rangle$	$\langle\phi-\phi\rangle$
11	0.0000	1	0.00389	0.02453	0.03900	0.00000	0.00000	0.00000	0.00000	0.00000	0.00000
12	0.0025	4	0.00389	0.02453	0.03900	0.00000	0.00000	0.00000	0.00017	0.00025	0.00008
13	0.0050	2	0.00389	0.02453	0.03900	0.00000	0.00000	0.00000	0.00034	0.00050	0.00016
14	0.0075	4	0.00389	0.02453	0.03900	0.00000	0.00000	0.00000	0.00051	0.00075	0.00024

Figure 2.11   The match between the $|$sto-3g:2p$\rangle$ and the Slater-2p functions for the minimization of the least squares integral of \$F\$9 value of Table 2.1.

2. Use the COPY/RIGHT procedure to propagate this normalization along the active cells of row 8.
3. Include the extra multiplication factor, the radial distance of equation 2.13, in column I, the summed projections of the Gaussian primitives weighted by the coefficients of the linear combination, with for example the entry

$$\$I\$11 = \$A11^*SUM(\$C11:\$H11)$$

and propagate this down the active cells of the column.
4. Calculate the difference function of cell \$F\$9 as before. For the $|$sto-3g:2p$\rangle$ basis set of Table 1.6, the exact result of Table 2.1 is reproduced and, as you see, in Figure 2.11, there is good agreement with the Slater function.

Exercise 2.7.   Use of the chart facility to reproduce Figure 1.17.

It is more convenient to construct charts in fig2-10.xls and fig2-11.xls, displaying the degrees of fit of the $|$sto-ng$\rangle$ linear combinations to the Slater functions, on a coarser radial mesh. This exercise provides, too, for the reproduction of the comparisons in Figure 1.16 to 1.18.

1. So, for example, in fig2-10.xls, layout a coarse radial array in column M in intervals of 0.5 a.u. from $r = 0$ to $r = 20$.
2. Project the normalized $|\text{sto-ng}\rangle$ function on the mesh, with, for example, the entry

$$\$N11 = (1/\textbf{SQRT}(\$E\$9))^{*}(D\$8^{*}\textbf{EXP}(-D\$7^{*}\$M11^{\wedge}2)$$
$$+E\$8^{*}\textbf{EXP}(-E\$7^{*}\$M11^{\wedge}2)+F\$8^{*}\textbf{EXP}(-F\$7^{*}\$M11^{\wedge}2)$$
$$+G\$8^{*}\textbf{EXP}(-G\$7^{*}\$M11^{\wedge}2)+H\$8^{*}\textbf{EXP}(-H\$7^{*}\$M11^{\wedge}2)$$
$$+I\$8^{*}\textbf{EXP}(-I\$7^{*}\$M11^{\wedge}2))$$

so that up to a linear combination of six primitives can be used.

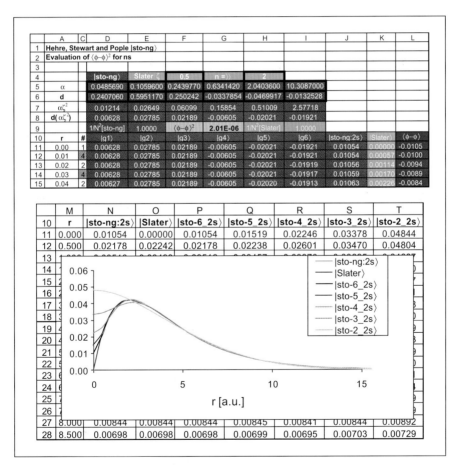

Figure 2.12   Reproduction of Figure 1.16 on the spreadsheet for the checking of basis set modelling of Slater functions. Particular *least-squares* answers, cell $\$G\$9$, are returned for the appropriate input in cells $\$D\$5$ to $\$I\$6$; compare Table 2.1.

3. Form the appropriate Slater function projection in column O, with, for example, the entry

$$\$O\$11 = \textbf{POWER}(2^*\$F\$4,(\$H\$4+0.5))^*(1/\textbf{SQRT}(\textbf{FACT}(2^*\$H\$4)))^*$$

[remember that normalization for Slater orbitals normally is over only the radial coordinate].

4. For each |sto-ng⟩ basis projected in this manner, select the projection and then use the EDIT/PASTE SPECIAL/VALUES sequence to enter a permanent record of this projection in adjacent columns.

5. Use the *INSERT CHART* macro to construct the chart in Figure 2.12 and the *ADD DATA* facility for extra projections after selecting the chart.

The second application of this spreadsheet is to the improving and even the design of basis sets. The procedure is illustrated in Figure 2.13. All that is required is the

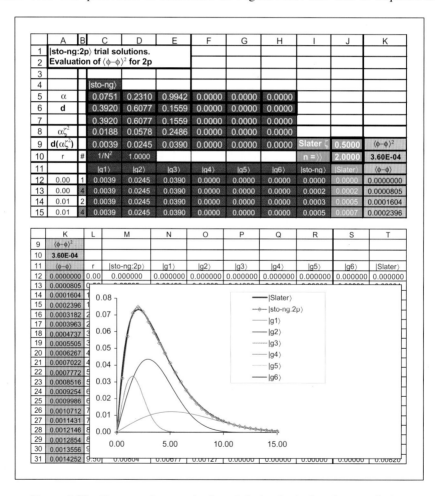

Figure 2.13 The procedure to check and design basis functions applied to the Slater 2p function.

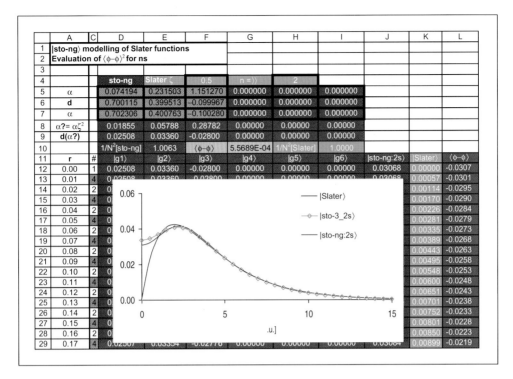

	A	C	D	E	F	G	H	I	J	K	L
1	|sto-ng⟩ modelling of Slater functions										
2	Evaluation of ⟨φ−φ⟩² for ns										
3											
4			sto-ng	Slater ζ	0.5	n =⟩⟩	2				
5	α		0.074194	0.231503	1.151270	0.000000	0.000000	0.000000			
6	d		0.700115	0.399513	-0.099967	0.000000	0.000000	0.000000			
7	α		0.702306	0.400763	-0.100280	0.000000	0.000000	0.000000			
8	α?= αζ²		0.01855	0.05788	0.28782	0.00000	0.00000	0.00000			
9	d(α?)		0.02508	0.03360	-0.02800	0.00000	0.00000	0.00000			
10			1/N²[sto-ng]	1.0063	⟨φ−φ⟩	5.5689E-04	1/N²[Slater]	1.0000			
11	r	#	|g1⟩	|g2⟩	|g3⟩	|g4⟩	|g5⟩	|g6⟩	|sto-ng:2s⟩	|Slater⟩	⟨φ−φ⟩
12	0.00	1	0.02508	0.03360	-0.02800	0.00000	0.00000	0.00000	0.03068	0.00000	-0.0307
13	0.01	4	0.02508	0.03360	-0.02800	0.00000	0.00000	0.00000	0.03068	0.00057	-0.0301
14	0.02	2								0.00114	-0.0295
15	0.03	4								0.00170	-0.0290
16	0.04	2								0.00226	-0.0284
17	0.05	4								0.00281	-0.0279
18	0.06	2								0.00335	-0.0273
19	0.07	4								0.00389	-0.0268
20	0.08	2								0.00443	-0.0263
21	0.09	4								0.00495	-0.0258
22	0.10	2								0.00548	-0.0253
23	0.11	4								0.00600	-0.0248
24	0.12	2								0.00651	-0.0243
25	0.13	4								0.00701	-0.0238
26	0.14	2								0.00752	-0.0233
27	0.15	4								0.00801	-0.0228
28	0.16	2								0.00850	-0.0223
29	0.17	4	0.02507	0.03354	-0.02776	0.00000	0.00000	0.00000	0.03084	0.00899	-0.0219

Figure 2.14   Matching of |sto-3g:2s⟩ Hehre, Stewart and Pople basis to the Slater 2s in hydrogen. If the criterion that a match be dependent on the least-square integral minimum, then the modified data in rows 5 and 6 could be claimed to be superior. In fact, all that is happening is a cancellation effect of no value in the matching process!

addition of the individual primitive functions variations with radial distance in the chart and the renormalization of the coefficients of the linear combination to provide a direct comparison with the data in Table 1.6. Note, the renormalization step introduced in row 7, so that the basis set parameters are defined exactly as in Table 1.6 and the original Hehre, Stewart and Pople paper (33), with, for example, $D$7 = $D$6/**SQRT** (1/$E$10).

Exercise 2.8.   Application of the general spreadsheet to design new basis functions.

1. Make a copy of fig2-10.xls or fig2-11.xls depending on the application intended and modify the chart, the Slater function, the |sto-ng⟩ function and the individual primitive Gaussians, variations with radial distance as in Figure 2.13.
2. In attempting to find a rough match of any chosen |sto-ng⟩ set, initially use the immediate visual effect of any individual changes to the exponents in row 5 or the coefficients in row 7.

3. For finer control and to introduce two extra useful macros in EXCEL, use the
    GOALSEEK and SOLVER[8] wizards in the TOOLS drop-down menu to activate
    internal minimization routines on the least-squares integral in cell $F$9 [*SOLVER
    is not a standard item loaded in a typical installation of the EXCEL program, so
    you may have to enable it using the Add-In option in the TOOLS menu*].

Figure 2.14 illustrates the application of the spreadsheet checking and improvement
scheme to the original |sto-3g:2s) data of Table 1.6. It illustrates the important point that
one has to be very careful with this kind of analysis. The criterion of equations 2.10
and 2.11, that the two representations exhibit a small difference integral over the radial
mesh, does not discriminate against cancellation events like the one shown. In fact, in
the example displayed, the match is less good, even though the least-squares integral is
less, because of the positive and negative difference effects exhibited in different ranges
of the radial mesh.

This matter is returned to later, at the end of Chapter 3, when matching possibilities
to the actual radial functions, rather than Slater approximations to these functions, is
discussed.

Exercise 2.9.

Increase the number of primitive Gaussians functions that can be used in fig2-9.xls
and further minimize the least-squares integral of cell $F$9. Start the analysis by
identifying the major regions of disagreement between the |sto-ng) and the Slater
bases, then, choose the exponents of the added Gaussians to have most effect in
these regions. Use the added coefficients in row 6 to improve the overall match.

[8]The SOLVER macro in EXCEL and other spreadsheet programs was developed by Frontline Systems.
Much useful information is available at *http://www.frontsys.com/index.html*.

3

Orthonormality

Orthonormality is an intrinsic property of the eigenfunctions of Schrödinger's equation. Therefore, before the corresponding approximations to the eigenvalues of a Schrödinger equation can be calculated, we arrive at a point in the calculation, when it is necessary to impose the orthonormality requirement. In modern programs, this step invariably occurs as a concerted transformation (42,47) of the Hall–Roothaan (57,58) matrix form of the Schrödinger equation within the *LCAO-MO* theory, returning approximations to the eigenvalues and the orthonormal eigenfunctions for the basis set applied in the calculation. However, it is instructive to consider orthonormalization independently first of all, before proceeding to the concerted procedure described in Sections 3.6 and 3.7.

So, for the most part, this chapter is devoted to the application of the integration techniques, developed in Chapter 2, to render the approximate functions, the basis sets of modern molecular orbital theory, mutually orthonormal. You will learn:

1. how to construct mutually orthonormal Slater orbitals and improve the modelling of the hydrogenic orbitals;
2. how to construct mutually orthonormal Gaussian orbitals and, again, improve the modelling of the hydrogenic orbitals;
3. how to determine the 'best' linear combinations of approximate functions for the modelling of radial atomic orbitals in many-electron atoms;
4. how to use split-basis sets;
5. how to model using only hydrogen 1s basis sets;
6. then in Sections 3.6 and 3.7, the concerted procedures of modern computer code are described and you will learn how to apply canonical and symmetric orthogonalizations for the rendering of Slater and Gaussian basis sets mutually orthonormal as part of the matrix eigenvalue problem.

In spherical polar coordinates orthonormality between two atomic orbitals, $|i\rangle$ and $|j\rangle$, requires that

$$\langle i|i\rangle = \langle j|j\rangle = \int_{r=0}^{r=\infty} \int_{\theta=0}^{\theta=\pi} \int_{\phi=0}^{\phi=2\pi} \phi_i(r,\theta,\phi)\phi_i(r,\theta,\phi)r^2 \, dr \, \sin(\theta) \, d\theta \, d\phi$$

$$= \int_{r=0}^{r=\infty} \int_{\theta=0}^{\theta=\pi} \int_{\phi=0}^{\phi=2\pi} \phi_j(r,\theta,\phi)\phi_j(r,\theta,\phi)r^2 \, dr \, \sin(\theta) \, d\theta \, d\phi$$

$$= 1 \hspace{8cm} 3.1$$

which, demonstrates the normality of the functions and that

$$\langle i|j\rangle = \langle j|i\rangle = \int_{r=0}^{r=\infty} \int_{\theta=0}^{\theta=\pi} \int_{\phi=0}^{\phi=2\pi} \phi_i(r,\theta,\phi)\phi_j(r,\theta,\phi)r^2 \, dr \, \sin(\theta) \, d\theta \, d\phi$$

$$= 0 \qquad\qquad 3.2$$

because of the mutual orthogonality of eigenfunctions of any Schrödinger equation i.e. $\langle i|j\rangle = \delta_{ij}$.

The significance of the orthonormality requirement becomes evident in the algebra leading to the general expression for the energy of any solution to Schrödinger's equation, be that solution exact or approximate.

The symbolic form of the equation is

$$H\varphi = E\varphi \qquad\qquad 3.3$$

and, in the standard solution procedure, we use the orthonormality property of the set $\{\varphi\}$ of eigenfunctions, left-multiply by φ, assuming all the eigenfunctions are real, and integrate over the whole of the atomic volume,

$$\int \varphi H\varphi \, d\tau = E \int \varphi\varphi \, d\tau \qquad\qquad 3.4$$

or

$$E = \frac{\int \varphi H\varphi \, d\tau}{\int \varphi\varphi \, d\tau} \qquad\qquad 3.5$$

Clearly, in terms of equation 3.5, the energy returned in any calculation depends on the value of the denominator in the equation and, if this integral is not of unit magnitude, which is the normalization condition, the calculated result is not correct.

Equally, if $|\varphi\rangle$ is not orthogonal to all solutions, except the exact eigenfunction being modelled, an average result only will be obtained for the energy. This average is over the mixture of the exact solutions to be found in the approximation[1]. For example, if $|\varphi\rangle$ is a weighted mixture of two eigenfunctions, $|\phi_1\rangle$ and $|\phi_2\rangle$, with

$$|\varphi\rangle = |\phi_1\rangle + \lambda|\phi_2\rangle \qquad\qquad 3.6$$

with

$$E = \frac{\langle\phi_1 + \lambda\phi_2|H|\phi + \lambda\phi_2\rangle}{\langle\phi_1 + \lambda\phi_2|\phi_1 + \lambda\phi_2\rangle} \qquad\qquad 3.7[2]$$

Even in the simpler case, with $|\varphi\rangle$ normalized, equation 3.7 does not return the appropriate approximation to the eigenvalue, rather the average

$$E = \langle\phi_1|H|\phi_1\rangle + \lambda^2\langle\phi_2|H|\phi_2\rangle \qquad\qquad 3.8$$

Mutual orthonormality is also a key requirement in our interpretation of the physical information about the electron distribution implicit in one-electron atomic or molecular

---

[1]If $|\varphi\rangle$ is not an eigenfunction, it can be written as a linear combination of any suitable basis sets [axes] of the algebra and so, for example, the set of exact eigenfunctions for the Schrödinger equation [see, for example, references (53)).

[2]Note, for later reference, that equation 3.7 is the basis of the *LCAO-MO* approximation, when suitably extended as a linear combination of atomic functions (42,47,53).

orbitals. We require the orbitals to be normalized, since the total probability, as the square of the orbital for the particular energy of the electronic state, must be one over the atomic or molecular space. Equally, orbitals are required to be mutually orthogonal, because otherwise the independent electron distributions corresponding to the different 'stationary states' are not distinct.

Orthonormality is not an inherent property of Slater-type orbitals nor the 'best fit' Gaussian basis sets to these orbitals. This deficiency is compounded in all the kinds of linear combinations we might use to approximate atomic and molecular orbitals. The taking of linear combinations of normalized functions, in general, destroys the normalization. So before considering the concerted transformations of Sections 3.6 to 3.7, it is appropriate to investigate the orthonormality, with respect to the various basis functions that we encountered in Chapter 1.

3.1   ORTHONORMALITY IN SLATER ORBITAL AND BASIS SET THEORY

In the previous chapter, the normalization integrals for the exact hydrogenic 1s atomic orbitals were determined as test cases for the various numerical integration procedures developed as spreadsheet operations and then the overlap integral between hydrogenic 1s orbitals, as exact functions and $|$sto-3g$\rangle$ approximations, was calculated for the case of H_2.

In this chapter, we apply this approach to the checking and establishing of orthonormality for Slater and Gaussian basis sets. Reasonable convergence of results have been achieved using modified versions of the spreadsheets described in Chapter 2, but you should, at least in a few cases, test for the convergence of the integrals of the most contracted [*largest exponential exponent*] and least contracted [*smallest exponential exponent*] components in some of the linear combinations used, on your own computer.

Care is needed if data are copied, especially from the early literature on basis sets, either linear combinations of Slater or Gaussian functions, since it is dangerous to assume that a particular normalization or orthogonalization condition has been imposed. The normalization constants in the expressions for the basic functions may not be included in the pre-exponential coefficients and individual preferences certainly determined whether normalization constants were defined over all the integration coordinates, or simply the radial coordinate, in the modelling of the radial wave functions.

Thus, keep in mind the starting definitions of Chapter 1. The Slater function approximation to the radial component of an atomic orbital is defined by

$$R(r:n\zeta) = (2\zeta)^{n^*+1/2}[(2n^*)!]^{-1/2}r^{n^*-1}\exp(-\zeta r) \qquad\qquad 1.12$$

and the pre-exponential factor is the normalization constant over the radial coordinate.

In advanced Slater theory, more than one Slater function is taken in a linear combination to generate the 'best' approximation to particular atomic orbitals and we have seen that this 'best' standard could be based on the 'degree of fit' to the numerical radial functions or the linear combinations that returned the variation principle 'best' eigenvalue. In such cases, these coefficients are undetermined until the 'best' eigenvalues have been calculated and the overall requirement of normalization is imposed. This is a general problem, which leads us to the theory of the *self-consistent field* (57,58,61,62, 42,47,53) developed by Hartree in his early calculations (1) and to Chapter 5.

With the advent of the Gaussian approximation, there were, in the beginning, many investigations of how best to use Gaussian linear combinations in atomic and molecular orbital theory. The options were, to devise a Gaussian or linear combinations of Gaussians to model the numerical atomic orbitals, to model Slater radial functions with different Gaussians or to attempt to obtain the exact eigenvalues for the hydrogen orbitals. The first option is the general standard applied in modern calculations and so we find basis sets as linear combinations of 1s, 2p and 3d Gaussians, determined to match the numerical data from good atomic structure calculations, separately or all combined together, scaled appropriately for application to model molecular electronic structure.

Thus, in the simplest applications to molecular theory, we expect to encounter linear combinations of *primitive* 1s Gaussians of the form of equation 1.15

$$g(r{:}\alpha) = (2\alpha^3/\pi)^{0.75} \exp(-\alpha r^2) \qquad\qquad 1.15$$

in which the pre-exponential factor includes the normalization constant $(1/(4\pi))^{1/2}$.

However, we should not forget the general normalization constant defined in equation 1.14

$$N_g = \left[\frac{2^{2n+3/2}}{(2n-1)!!\pi^{1/2}} \right]^{1/2} [\alpha]^{(2n+1)/4} \qquad\qquad 1.14$$

and take care, if necessary, to multiply in the appropriate term for normalization over the angular coordinates.

In practice, too, with Gaussian basis sets, it is common to write the complete functional forms in Cartesian coordinates for the angular components of the wave function with

$$g(\alpha|lmn) = (\alpha^3/\pi)^{0.75} \exp(-\alpha r^2/a_0^3)x^l y^m z^n \qquad\qquad 3.9$$

arrived at, with the taking of sums and differences of the parent spherical harmonics and identified as the Cartesian Gaussians[3].

3.2   ORTHONORMALITY AND SLATER ORBITALS

The pre-exponential factors in the equation 1.12 normalize the Slater approximations to the radial components of atomic orbitals. Normality is not an inherent property of linear combinations of Slater orbitals, for example, as in Table 1.3, and it is important to check any published coefficients to determine whether normalization is included. In addition, the Slater orbitals for a set of atomic orbitals in an atom are not mutually orthogonal. The results of atomic structure calculations using Slater orbitals, either as single functions or in linear combinations, as in *double-zeta* sets, of course, are mutually orthogonal, since this property of the eigenfunctions, is mirrored in the final linear combinations returned by the calculations for the eigenvalues.

---

[3] Again the EMSL database is worth an inspection. Note, one point of detail, for technical reasons it is simplest to use 'over complete' sets of Cartesian Gaussians to atomic orbitals of higher angular momenta. Thus, for example, six 'd' orbitals are defined, xy, xz, yz, $x^2 - y^2$, z^2 and $x^2 + y^2 + z^2$, which is of 's'-like character at the origin.

It is straightforward to design a spreadsheet to check the mutual orthonormality of Slater orbitals. We need simply to modify the designs used in Chapter 2 to take account of the form of the Slater function, equation 1.12. Fig3-1.xls is a spreadsheet designed to test orthogonality between two Slater functions for the lithium 1s and 2s orbitals. Normalized Slater functions can be used or not in this exercise, depending on the logical flags in cells $D$3 and $E$3.

Exercise 3.1. Orthonormality checks for Slater orbitals.

1. Enter the basic parameters in cells $D$4 to $E$8, i.e. the atomic number for lithium, cells $D$6 and $E$6, the principal quantum numbers for the atomic orbitals in cells $D$6 and $E$6 and the calculated Slater exponents in cells $D$7 and $E$7.

2. The entries in cells $D$8 to $E$9 provide for the option to check the normalization constants. The entries 1.00000 in cells $D$8 and $E$8 are dummy factors for the options 'no' or 'yes' in cells $D$3 and $E$3. For the choice 'yes' the condition

    $$= IF(\$D\$3=\text{"yes"},\$D\$8,\$D\$8^*\mathbf{POWER}(2^*\$D\$7,(\$D\$6+0.5))^* \\ \mathbf{SQRT}(1/\mathbf{FACT}(2^*\$D\$6)))$$

    in cell $D$9 returns the trivial pre-exponential factor 1.00000 of $D$8 and a normalization check is performed, while for the choice 'no' the normalization factor of equation 1.14 is applied.

3. Enter the correct formulae for the normalization constants for the 1s and 2s Slater functions in cells $H$5 and $H$7,

    $$\$H\$5 = D\$8^*\mathbf{POWER}(2^*D\$7,(\$D\$6+0.5))^*\mathbf{SQRT}(1/\mathbf{FACT}(2^*D\$6))$$

    and

    $$\$H\$7 = E\$8^*\mathbf{POWER}(2^*E\$7,(\$E\$6+0.5))^*\mathbf{SQRT}(1/\mathbf{FACT}(2^*E\$6))$$

4. Lay out the standard integration procedure from cell $A11 to cell $E$3012 for a radial mesh ranging from the origin to 20 a.u. in intervals of 20/3000.

5. Complete the spreadsheet design by using the **SUMPRODUCT** function to return the normalization integrals, as checks [*ncheck* = 'yes'] or for the calculation of the normalization constants and the overlap integral. These are the entries,

    $$\$F\$5 = \mathbf{SUMPRODUCT}(\$B\$12:\$B\$3012,\$C\$12:\$C\$3012, \\ \$C\$12:\$C\$3012,\$D\$12:\$D\$3012,\$D\$12:\$D\$3012)^* \\ (\$C\$13-\$C\$12)/3$$

    $$\$F\$7 = \mathbf{SUMPRODUCT}(\$B\$12:\$B\$3012,\$C\$12:\$C\$3012, \\ \$C\$12:\$C\$3012,\$E\$12:\$E\$3012,\$E\$12:\$E\$3012)^* \\ (\$C\$13-\$C\$12)/3$$

$$\$F\$9 = \textbf{SUMPRODUCT}(\$B\$12{:}\$B\$3012,\$C\$12{:}\$C\$3012,$$
$$\$C\$12{:}\$C\$3012,\$D\$12{:}\$D\$3012,\$E\$12{:}\$E\$3012)^*$$
$$(\$C\$13{-}\$C\$12)/3$$

with the normalization constants in cells $\$G\$5$ and $\$G\$7$ as inverse square roots, for example,

$$\$G\$5 = 1/\textbf{SQRT}(\$F\$5)$$

Figure 3.1 displays the spreadsheet result for the choice that the 2s Slater function normalization constant is checked. As you see, there is good agreement to seven figures between the calculated numerical results and those based on the exact relationship, equation 1.12, in cell $\$G\$5$. These results are based on the integration mesh interval of 20/3000. It is instructive to check the convergence, in this example, since it is straightforward to improve the result by generating a finer mesh.

A simple standard procedure to render functions mutually orthogonal is to apply the Schmidt orthogonal transformation (39,42,47,53). For a set of functions $\{|i\rangle\}$, the Schmidt orthogonalization transformation is to render each function, in turn, orthogonal to the preceding one in the series. The examples discussed in this chapter, concern only the rendering of two functions, $|1\rangle$ and $|2\rangle$, mutually orthogonal, so the Schmidt transformation

	A	B	C	D	E	F	G	H
1			Slater radial functions.					
2			Orthonormality checks.					
3			ncheck	no	yes			
4				$|1\rangle$	$|2\rangle$	$\langle 1|1\rangle$	N_1	
5			Z	3	3	1.000000028	1.00000	8.87311
6			n*	1	2	$\langle 2|2\rangle$	N_2	
7			ζ	2.70000	0.65000	6.4638970466	0.39333	0.39333
8			d	1.00000	1.00000	$\langle 1|2\rangle = \lambda$		
9			d'	8.87311	1.00000	0.16626		
10								
11			r	$|1\rangle$	$|2\rangle$			
12	0	1	0.0000	8.87311	0.00000			
13	1	4	0.0067	8.71482	0.00664			
14	2	2	0.0133	8.55936	0.01322			
15	3	4	0.0200	8.40666	0.01974			
16	4	2	0.0267	8.25670	0.02621			

Figure 3.1   The spreadsheet detail for orthonormality checking in this example of the Slater approximations to the lithium 1s and 2s radial orbitals.

	A	B	C	D	E	F	G	H
1			Slater radial functions.					
2			The orthogonal |2s-1s⟩					
3			ncheck	no	yes			
4				|1⟩	|2⟩	⟨1|1⟩	N_1	
5			Z	3	3	1.000000028	1.00000	8.87311
6			n*	1	2	⟨2|2⟩	N_2	
7			2.70000	0.65000	6.4638970466	0.39333	0.39333
8			d	1.00000	1.00000	⟨1|2⟩=λ	⟨2-λ1⟩²	
9			d'	8.87311	1.00000	0.42272	6.28521	
10								
11			r	|1⟩	|2⟩	|2-λ1⟩		
12	0	1	0.0000	8.87311	0.00000	3.75079		
13	1	4	0.0067	8.71482	0.00664	3.67725		
14	2	2	0.0133	8.55936	0.01322	3.60495		
15	3	4	0.0200	8.40666	0.01974	3.53388		
16	4	2	0.0267	8.25670	0.02621	3.46402		

Figure 3.2 Generation of the orthogonal Slater lithium 2s radial function by Schmidt transformation.

takes the form

$$|2 - \lambda 1\rangle = |2\rangle - \frac{\langle 1|2\rangle}{\langle 1|1\rangle}|1\rangle \qquad 3.10$$

with $\lambda = \langle 1|2\rangle/\langle 1|1\rangle$ and we should note, especially, the normalization correction $\langle 1|1\rangle$ in equation 3.10.

The result in cell $G$9 of Figure 3.1 demonstrates that the Slater 1s and 2s radial functions for lithium are not mutually orthogonal. Equation 3.10 can be applied straightforwardly to generate the appropriate linear combination, which is orthogonal to the 1s function. The result is shown in Figure 3.2.

Exercise 3.2. Construction of a lithium 2s function orthogonal to the 1s function.

1. Make a copy of fig3-1 and rename it fig3-2.xls.
2. Make the linear combination required by equation 3.10 in the cells of column F, with

$$\$F12 = -(\$E12-\$F\$9^*\$D12/\$F\$5)^4$$

[4]Note the taking of the negative of the linear combination. This is simply to conform to the convention that the function should be positive near the nucleus and, also, to make the match in Figure 3.3.

filled down the active cells of the column. $F$9 is the integral to test the degree of orthogonality between the two functions, $F$5 is the normalization integral for the |1s⟩ Slater function.

3.   Finally calculate the normalization integral for the orthogonal function, |2s-λ1s⟩, in the usual manner,

$$\$G\$9 = \mathbf{SUMPRODUCT}(\$B\$12{:}\$B\$3012,\$C\$12{:}\$C\$3012,$$
$$\$C\$12{:}\$C\$3012,\$F\$12{:}\$F\$3012,\$F\$12{:}\$F\$3012)^*$$
$$(\$C\$13{-}\$C\$12)/3$$

It is to be expected[5] that the orthonormal function formed by applying equation 3.10, should be a good fit, when normalized, to the numerical radial function obtained by direct solution of the Schrödinger equation for the lithium atom. This proposition is examined in fig3-3.xls, in which the chart, displayed in Figure 3.3, incorporates the Herman–Skillman

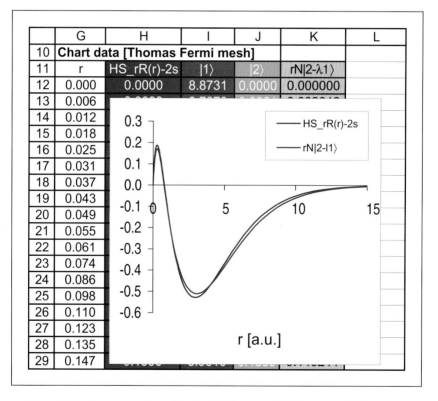

	G	H	I	J	K	L
10	Chart data [Thomas Fermi mesh]					
11	r	HS_rR(r)-2s	|1⟩	|2⟩	rN|2-λ1⟩	
12	0.000	0.0000	8.8731	0.0000	0.000000	
13	0.006					
14	0.012					
15	0.018					
16	0.025					
17	0.031					
18	0.037					
19	0.043					
20	0.049					
21	0.055					
22	0.061					
23	0.074					
24	0.086					
25	0.098					
26	0.110					
27	0.123					
28	0.135					
29	0.147					

Figure 3.3   Demonstration of the good fit between the Slater 2s radial approximation, when rendered orthogonal to the 1s function, and the numerical radial function as output of the Herman–Skillman program. The other details of the spreadsheet are as in Figures 3.1 and 3.2.

[5]This is an important point. Our expectation here depends on the fact that the first function is the exact 1s atomic orbital. Therefore, the second function, which is orthogonal has to be close to the 2s atomic orbital. This need not be true, in general, for the orthogonalization procedures described in Sections 3.6 to 3.8.

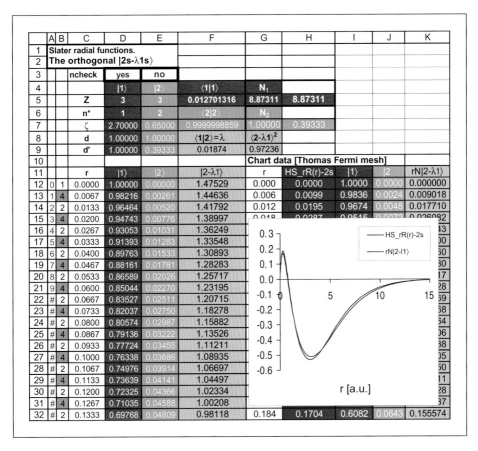

Figure 3.4 Demonstration that the internal normalization checks can be changed at will [ncheck = 'yes/no'] in the spreadsheet fig3-1.xls without affecting the match to the numerical radial function for the lithium 2s orbital, output by the Herman–Skillman program. But, note that, as we might expect, the results in $F\$9$ and $G\$9$ are different.

output for the lithium 2s numerical radial function. Since we chose to generate the negative of the orthogonal linear combination of the $|2s\rangle$ and $|1s\rangle$ Slater functions, the only extra detail in this extension of Figure 3.2 is the inclusion of the normalization constant as the factor $(1/\mathbf{SQRT}(G\$9))$ in the formulae entries in column K.

Finally, it is comforting to reassure oneself that all the procedures described are consistent. We can ignore the pre-exponential term in equation 1.18 and redo the calculation of the orthogonal lithium 2s function, simply by choosing the 'yes' option for both the 1s and 2s functions in the general spreadsheet fig3.1-to-3.3.xls. As you can see in Figure 3.4, while the normalization and overlap integrals are different to those appearing in Figure 3.2, the same overall degree of fit to the numerical 2s function results.

3.3   ORTHONORMALITY AND GAUSSIAN ORBITALS

Calculations with Gaussian basis sets, as approximations to the Slater functions for atomic radial functions, in general, involve more operations since for even a minimal

|sto-3g⟩ basis set, there are many more terms to be determined even to check the normal-ization. However, all the design details to calculate the integrals have been worked out for fig2-9.xls and so we need only provide a second worksheet for the generation and display of the orthonormal function formed by applying equation 3.10 to the case of 1s and 2s functions modelled by |sto-ng⟩ Gaussian sets.

Exercise 3.3.   Application of the *two-variable table* macro to calculate the inte-grals needed to render the Pople *et al.* |sto-3g:2s⟩ Gaussian basis orthonormal to the 1s function for the case of the lithium atom.

1. Make a copy of fig2.9.xls using the COPY/PASTE facility on the toolbar in EXPLORER. Rename this spreadsheet as fig3.5.xls.
2. Freeze the *automatic* calculation facility under the OPTIONS menu in the TOOLS dropdown box, to allow the necessary changes to be made without having to wait for calculations to be completed each time.
3. Enter the appropriate exponent and pre-exponential values from Table 1.6 for the 1s and 2s |sto-3g⟩ linear combinations.
4. Change the Slater exponent values in cells $D$3 and $G$3 to the 1s and 2s values 2.7 and 0.65.
5. Since this is an atomic problem, remove the redundant location vector entries and the cylindrical integration mesh.
6. Construct the standard Simpson's rule numerical integration procedure in cells $A$34 to $C$3034.
7. Provide for a *two-variable* table calculation in rows 10 to 31 of the spreadsheet, with the input row data, the sums of the exponents for the integrations over the primitive Gaussian products and the input column data, the products of the coefficients in the primitive Gaussian products. Make cells $H$4 and $I$4 the master cells for the macro.
8. Colour code the cells for the orthonormality calculation over the two Gaussian sets. Write the master formula in cell $A$10

$$\$A\$10 = \textbf{SUMPRODUCT}(\$A\$34:\$A\$3034,\$A\$34:\$A\$3034$$
$$\$C\$34:\$C\$3034,\$D\$34:\$D\$3034)^*(\$A\$35-\$A\$34)^*4^*\textbf{PI}(\ )/3$$

9. Collect the components of each normalization and overlap integration in cells $I$6 to $I$8, with

$$\$I\$6 = \$B\$11+\$C\$12+\$D\$13+2^*(\$E\$14+\$F\$15+\$G\$16)$$

$$\$I\$7 = \$H\$17+\$I\$18+\$J\$19+2^*(\$K\$20+\$L\$21+\$M\$22)$$

$$\$I\$8 = \$N\$23+\$O\$24+\$P\$25+\$Q\$26+$$
$$\$R\$27+\$S\$28+\$T\$29+\$U\$30+\$V\$31$$

10. Project the linear combination of the two starting |sto-3g⟩ Gaussian sets on the radial array in the form of equation 3.10.

11. Create a new worksheet using the INSERT menu, if the default in your version of EXCEL is one.

12. Use the rename facility to rename the first worksheet 'onorm' and the second 'orbitals'.

13. Construct the projections of the orthonormal linear combination on the radial mesh using the overlap integral, cell 'onorm!'$I$8, and the normalization integral of the 1s functions, 'onorm!'$I$6, with, for example,

$$\$D5 = -('onorm'!\$E\$8^*\mathbf{EXP}(-'onorm'!\$E\$7^*\$A5^2)+$$
$$'onorm'!\$F\$8^*\mathbf{EXP}('onorm'!\$F\$7^*\$A5^2)+$$
$$'onorm'!\$G\$8^*\mathbf{EXP}(-'onorm'!\$G\$7^*\$A5^2)-$$
$$('onorm'!\$I\$8/'onorm'!\$I\$6)^*$$
$$('onorm'!\$B\$8^*\mathbf{EXP}(-'onorm'!\$B\$7^*\$A5^2)+$$
$$'onorm'!\$C\$8^*\mathbf{EXP}(-'onorm'!\$C\$7^*\$A5^2)+$$
$$'onorm'!\$D\$8^*\mathbf{EXP}(-'onorm'!\$D\$7^*\$A5^2)))$$

14. Calculate the normalization integral for the orthogonal function, with

$$'orbitals'!\$E\$4 = 4^*\mathbf{PI}(\,)^*\mathbf{SUMPRODUCT}(\$A\$5:\$A\$3005,$$
$$\$A\$5:\$A\$3005,\$C\$5:\$C\$3005,\$D\$5:\$D\$3005,$$
$$\$D\$5:\$D\$3005)^*(\$A\$6-\$A\$5)/3$$

15. Open fig3.4.xls and copy the Thomas–Fermi mesh data for the Herman–Skillman lithium numerical 2s radial function.

16. Project the orthonormal linear combination of the Gaussian functions on the Thomas–Fermi mesh appropriate for lithium and use the CHART wizard to construct Figure 3.6.

The beauty of the representation and the diagrams, based on the spreadsheet for Figures 3.5 and 3.6, is that the only changes required for different atoms are the input values of the Slater exponents. Everything remains in register. Once a close fit to the Slater orbitals has been achieved, the fit is maintained for all applications. You should check the correctness of this claim for other atoms.

In addition, while we need to remember that the original purpose in the design of Gaussian sets was to approximate Slater orbitals, it is worth continuing with the analysis until the 'best' fits are achieved between the orthogonal Gaussian and Slater functions and the numerical function from the Herman–Skillman program. My result is set out in Figure 3.7 and provides another example of the application of the variation of the parameters of linear combinations in all this theory for atoms and molecules.

The Slater exponents are parameters in any calculations involving these Gaussians sets, which we can vary to achieve a particular result. In the present case, the desired result is a fit to the numerical radial data for lithium 2s. For the valence shell regions of the lithium 2s orbital the fit in Figure 3.7 is almost complete for the choices $\zeta_{1s} = 2.7$ and $\zeta_{2s} = 0.675$. However, that, of course, does not mean that we have discovered a universal set of Slater exponents for lithium. All we have done in this analysis is set the criterion for the choice of Slater exponents, the match to the output of the Herman–Skillman program

	A	B	C	D	E	F	G	H	I
1	Simpson's rule integrations of STO-3G								
2	2-variable Table calculations.								
3		|sto-3g⟩	Slater ζ	2.70000	|sto-3g⟩	Slater ζ	0.65000	alpha	pre-exp
4		|1a⟩	|2a⟩	|3a⟩	|1b⟩	|2b⟩	|3b⟩	1.00000	1.00000
5	exponent	0.10982	0.40577	2.22766	0.07514	0.23103	0.99420		
6	coefficient	0.44464	0.53533	0.15433	0.70012	0.39951	-0.09997	⟨a|a⟩	1.00000
7	alpha	0.80057	2.95807	16.23964	0.03175	0.09761	0.420051	⟨b|b⟩	1.00000
8	coeff	0.26820	0.86057	0.88979	0.03753	0.04972	-0.03717	⟨a|b⟩	0.17443
9									
10	5.56833	0.07193	0.74058	0.79174	0.23081	0.23865	0.76573	0.00141	0.00247
11	1.60115	0.19770							
12	5.91614		0.28658						
13	32.47928			0.02382					
14	3.75864				0.17637				
15	17.04021					0.01889			
16	19.19771						0.05069		
17	0.06349							0.49016	
18	0.19522								0.15961
19	0.84010								
20	0.12936								
21	0.45180								

	A	B	C	D
33	r		#	⟨1|1⟩
34	0.0000	1	1	1.00000
35	0.0100	2	4	0.99990
36	0.0200	3	2	0.99960
37	0.0300	4	4	0.99910
38	0.0400	5	2	0.99840
39	0.0500	6	4	0.99750
40	0.0600	7	2	0.99641

Figure 3.5 The forming of the orthonormal 2s radial wave function in lithium from the 1s and 2s |sto-3g⟩ Gaussian basis sets of Table 1.6, using the *two-variable table* procedure of fig2-9.xls. Figure 3.6 reproduces the chart demonstrating the match with the Herman–Skillman data.

and, as you may remember from Chapter 1, Table 1.11, the best choices for applications in atomic theory appear to be 2.69 and 0.75!

3.4 ORTHONORMALITY AND *DOUBLE-ZETA* SLATER ORBITALS

The study of orthonormality relationships in basis sets involving more than one Slater function is a very instructive exercise in the development of our understanding of the *LCAO-MO* theory and continues the discussion on the role of variation parameters. Because these are linear in action they lead to increased computational efficiency, since the integrals between the components need to be calculated only once, which, until recently, was a major consideration.

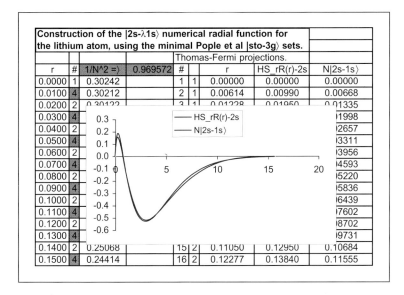

r	#	1/N^2 =⟩	0.969572	#		r	HS_rR(r)-2s	N|2s-1s⟩
0.0000	1	0.30242		1	1	0.00000	0.00000	0.00000
0.0100	4	0.30212		2	1	0.00614	0.00990	0.00668
0.0200	2	0.30122		3	1	0.01228	0.01950	0.01335
0.0300	4							1998
0.0400	2							2657
0.0500	4							3311
0.0600	2							3956
0.0700	4							4593
0.0800	2							5220
0.0900	4							5836
0.1000	2							6439
0.1100	4							7602
0.1200	2							8702
0.1300	4							9731
0.1400	2	0.25068		15	2	0.11050	0.12950	0.10684
0.1500	4	0.24414		16	2	0.12277	0.13840	0.11555

Construction of the |2s-λ1s⟩ numerical radial function for the lithium atom, using the minimal Pople et al |sto-3g⟩ sets. Thomas-Fermi projections.

(Chart legend: HS_rR(r)-2s, N|2s-1s⟩)

Figure 3.6 Completion of Exercise 3.3 with the generation of the orthonormal |sto-3g:2s⟩ function for lithium on the Thomas–Fermi mesh.

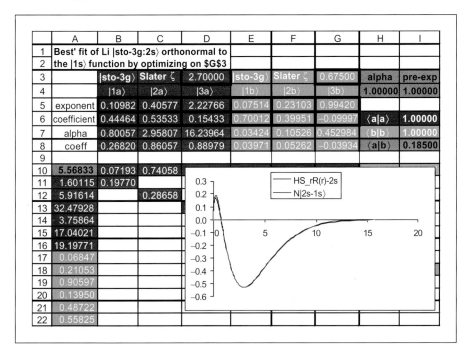

	A	B	C	D	E	F	G	H	I
1	Best' fit of Li |sto-3g:2s⟩ orthonormal to								
2	the |1s⟩ function by optimizing on $G$3								
3		|sto-3g⟩	Slater ζ	2.70000	|sto-3g⟩	Slater ζ	0.67500	alpha	pre-exp
4		|1a⟩	|2a⟩	|3a⟩	|1b⟩	|2b⟩	|3b⟩	1.00000	1.00000
5	exponent	0.10982	0.40577	2.22766	0.07514	0.23103	0.99420		
6	coefficient	0.44464	0.53533	0.15433	0.70012	0.39951	-0.09997	⟨a|a⟩	1.00000
7	alpha	0.80057	2.95807	16.23964	0.03424	0.10526	0.452984	⟨b|b⟩	1.00000
8	coeff	0.26820	0.86057	0.88979	0.03971	0.05262	-0.03934	⟨a|b⟩	0.18500
9									
10	5.56833	0.07193	0.74058						
11	1.60115	0.19770							
12	5.91614		0.28658						
13	32.47928								
14	3.75864								
15	17.04021								
16	19.19771								
17	0.06847								
18	0.21053								
19	0.90597								
20	0.13950								
21	0.48722								
22	0.55825								

(Chart legend: HS_rR(r)-2s, N|2s-1s⟩)

Figure 3.7 'Best' fit results to make a match with the numerical radial function for the lithium 2s orbital in the valence region of the atom by optimizing using the Slater 2s exponent in cell $G$3. A further improvement is to be expected for an appropriate change in the 1s Slater ζ, here unchanged at 2.7, in cell $D$3, especially in the K-shell region.

So, consider again the rendering of the lithium 2s Slater function orthogonal to the 1s radial function, but, in this case, let us use the *double-zeta* basis data set out in Table 1.3. As before, we need to take the appropriate linear combination of the 1s and 2s functions, but now with

$$|\text{Li}-1\text{s}\rangle = c_1[2(2.4331)^{3/2}]e^{(-2.4331r)} + c_2[2(4.5177)^{3/2}]e^{(-4.5177r)} \qquad 3.11$$

and

$$|\text{Li}-2\text{s}\rangle = c_3\left(\tfrac{4}{3}\right)^{0.5}(0.6714)^{2.5}re^{(-0.6714r)} + c_4\left(\tfrac{4}{3}\right)^{0.5}(1.9781)^{2.5}e^{(-1.9781r)} \qquad 3.12$$

Our problem, in the abstract, is that we have no constraints that we can apply to assign relative values to the coefficients in equations 3.11 and 3.12. Since atomic orbitals are required to be normalized, it seems sensible that both approximate functions should be normalized and then render the normalized 2s linear combination orthogonal to the normalized 1s linear combination. This means that for each linear combination forming the *split-basis* or *double-zeta* sets, we choose only coefficients linking the parts of the two approximate representations, so that

$$c_i^2\langle\phi_i|\phi_i\rangle + c_j^2\langle\phi_j|\phi_j\rangle + 2c_ic_j\langle\phi_i|\phi_j\rangle = 1 \qquad 3.13$$

which for a choice of c_i fixes the value of c_j to be

$$c_j = \frac{-2c_i\langle\phi_i|\phi_j\rangle - [(2c_i\langle\phi_i|\phi_j\rangle)^2 - 4\langle\phi_j|\phi_j\rangle(c_i^2\langle\phi_j|\phi_j\rangle))]^{1/2}}{2(c_i^2\langle\phi_j|\phi_j\rangle)} \qquad 3.14$$

Subject to the normality condition of equation 3.13, the *double-zeta* Clementi 2s function, equation 3.12 is rendered mutually orthonormal to the *double-zeta* Clementi 1s function using the spreadsheet of two worksheets set out in Figure 3.8. Thus, with the identifications, $|1\rangle = \phi_i$ and $|1'\rangle = \phi_j$ on the dz1s![6] and similarly on dz2s! for the 2s functions in equation 3.13, the worksheets are laid out as previously, with the following extra details now important for the development of the mutually orthonormal linear combinations.

Exercise 3.4.   Orthonormality in *double-zeta* basis sets — construction of the Lithium 2s radial orbital from the *double-zeta* Clementi basis.

1. Much of the design of the spreadsheet fig3-8.xls should be familiar. To allow for the orthonormalization stage it is useful to keep the '1s' and '2s' parts of the calculation separate on two different worksheets. So, in a new spreadsheet name two worksheets 'dz1s' and 'dz2s'.
2. Provide for linked Simpson's rule integrations on the worksheets, by entering the appropriate formulae in columns A to C of 'dz1s' and then linking through

---

[6]Remember, the symbol '!' identifies a particular worksheet on a multiple page spreadsheet.

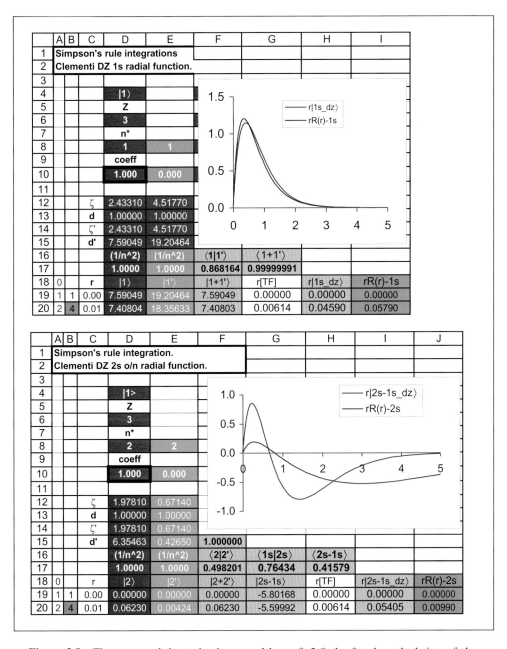

Figure 3.8 The two worksheets in the spreadsheet, fig3-8.xls, for the calculation of the orthonormal *double-zeta* Slater radial function of the lithium 2s orbital. The comparison graphs in this figure show the possible starting situation, with both c_1 coefficients equal 1 and the c_2 equal 0. In both cases, changing the value of c_i in cell \$C\$11 leads to the 'best fit' results shown in the next diagram.

to the corresponding cells in columns A to C of 'dz2s' with for example

$$\$A19 = \text{'dz1s'!}\$A19$$

$$\$B19 = \text{'dz1s'!}\$B19$$

$$\$C19 = \text{'dz1s'!}\$C19$$

and extending these relationships down the active cells in the second worksheet.

3. Enter the appropriate data from Table 1.3 into the input cells of both work-sheets as displayed in Figure 3.8.

4. For each exponent calculate the normalization pre-exponential factor in cells $\$D\$15$ and $\$E\$15$ with for normalization over 'r' only

$$\$D\$15 = \$D\$13^*\textbf{POWER}(2^*\$D\$14,(\$D\$8+0.5))^*\textbf{SQRT}(1/\textbf{FACT}(2^*\$D\$8))$$

and copy this formula into cell $\$E\$15$.

5. Project the Slater functions over the radial arrays on both worksheets and calculate normalization and overlap integrals in cells $\$D\$17$ to $\$F\$17$ on both worksheets, with

$$\$D\$17 = \textbf{SUMPRODUCT}(\$C\$19:\$C\$3019,\$C\$19:\$C\$3019,$$
$$\$B\$19:\$B\$3019,\$D\$19:\$D\$3019,\$D\$19:\$D\$3019)^*$$
$$(\$C\$20-\$C\$19)/3$$

$$\$E\$17 = \textbf{SUMPRODUCT}(\$C\$19:\$C\$3019,\$C\$19:\$C\$3019,$$
$$\$B\$19:\$B\$3019,\$E\$19:\$E\$3019,\$E\$19:\$E\$3019)^*$$
$$(\$C\$20-\$C\$19)/3$$

$$\$F\$17 = \textbf{SUMPRODUCT}(\$C\$19:\$C\$3019,\$C\$19:\$C\$3019,$$
$$\$B\$19:\$B\$3019,\$D\$19:\$D\$3019,\$E\$19:\$E\$3019)^*$$
$$(\$C\$20-\$C\$19)/3$$

6. Now, provide for the coefficients of the linear combinations, defined in equation 3.13 and 3.14 with

$$c_1 = \$D\$10 = [\text{enter a suitable choice between 0 and 1}]$$

$$c_2 = \$E\$10 = (-2^*\$F\$17^*\$D\$10+\textbf{SQRT}(\textbf{POWER}(2^*F17^*D10,2)$$
$$-4^*\$E\$17^*(\textbf{POWER}(D10^*\$D\$17,2)-1)))/(2^*\$E\$17)$$

on both worksheets.

7. With these coefficients defined, project the linear combination forming the *double-zeta* representation of the 1s and 2s Slater functions in column F of both worksheets, with, for example,

$$\$F19 = \$D\$10^*\$D\$15^*\$C19\hat{\ }(\$D\$8-1)^*\textbf{EXP}(-D\$14^*\$C19)+$$
$$\$E\$10^*\$E\$15^*\$C19\hat{\ }(\$E\$8-1)^*\textbf{EXP}(-E\$14^*\$C19)$$

8. Check the overall normalization in cell $\$G\$17$

$$\$G\$17 = \textbf{SUMPRODUCT}(\$C\$19:\$C\$3019,\$C\$19:\$C\$3019,$$
$$\$B\$19:\$B\$3019,\$F\$19:\$F\$3019,\$F\$19:\$F\$3019)^*$$
$$(\$C\$20-\$C\$19)/3$$

9. COPY and PASTE the Herman–Skillman data for lithium 1s from previous spreadsheets and provide for the comparison of the Herman–Skillman output numerical radial function and the *double-zeta* Slater function using columns G to I and the INSERT/CHART macro, in the usual manner on worksheet 'dz1s'!.

10. Complete the construction of the orthonormal 2s radial function, by calculating the overlap integral between the two *double-zeta* Slater functions and apply equation 3.10. So, using the projections of the 1s and 2s functions on the radial array calculate their overlap integral in $G$17 of 'dz2s'!, with

$$\langle 1s|2s \rangle = \$G\$17 = \textbf{SUMPRODUCT}(\$C\$19{:}\$C\$3019,\$C\$19{:}\$C\$3019,$$
$$\$B\$19{:}\$B\$3019,\$F\$19{:}\$F\$3019,$$
$$dz1s!\$F\$19{:}dz1s!\$F\$3019)^*(\$C\$20{-}\$C\$19)/3$$

11. Then, project the orthogonal 2s function on the radial array down column G in the form

$$\text{'dz2s'!}\$G19 = \$F19{-}\$G\$17^*dz1s!\$F19/dz1s!\$G\$17$$

and so on down the active cells of column G.

12. Calculate the normalization integral in cell $H$17 with

$$\text{'dz2s'!}\$H\$17 = \textbf{SUMPRODUCT}(\$C\$19{:}\$C\$3019,\$C\$19{:}\$C\$3019,$$
$$\$B\$19{:}\$B\$3019,\$G\$19{:}\$G\$3019,$$
$$\$G\$19{:}\$G\$3019)^*(\$C\$20{-}\$C\$19)/3$$

Finally, finish the design with the chart for the comparison of the *double-zeta* orthonormal 2s function and the numerical output of the Herman–Skillman program for the lithium 2s orbital.

Figure 3.8 displays a typical starting result in the determination of the *double-zeta* orthonormal 2s function, with only the first Slater component of each pair in Table 1.3 enabled, by choosing c_1 equal 1.0 in both cases, so that the c_2 are zero. While, as we might expect, there is fair agreement with the 1s function the 2s Slater orthonormal function does not compare at all with the Herman–Skillman numerical output. Figure 3.9 shows the 'best' result, the closest matching of the 1s and 2s functions visually appears to occur for the c_1 choices shown in the figure legend[7] and as you can see there is now good agreement for both matchings.

It is important, now, to change focus somewhat. We have concentrated on the matching of Herman–Skillman data and the orthonormal functions, which we can generate using Slater functions. This approach is defective for two reasons. Firstly, we have not determined how good are the Herman–Skillman data. Secondly, the whole objective in basis set theory is to provide functions, which lead to simpler integrations, from which can

---

[7]The lithium Herman–Skillman data are values only over the Thomas–Fermi mesh and so a *least-square integral* fit is not appropriate, because of the small number of terms in any attempted numerical integration.

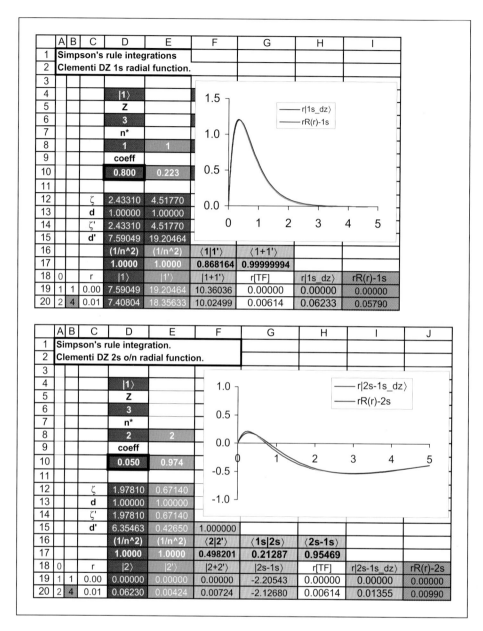

Figure 3.9   Best choices for the coefficients of the linear combinations forming the *double-zeta* representation of the 1s and 2s Slater orbitals for lithium in the formation of the orthonormal function.

be determined the 'best' energies of electrons in atoms and molecules and the corresponding distributions of electron density as probability distributions within the spirit of the Copenhagen Interpretation.

The detail of Clementi's original calculations (63) on atoms is no longer available, but it is possible to redo his calculations using the Roothaan–Hartree–Fock SCF program

Table 3.1   The coefficients of the Slater functions for the 1s and 2s radial functions in lithium after 16 cycles of Mitroy's scf program (64) to convergence. The 'best' fit overall coefficients based on the orthonormal functions are in the last two rows[8].

1s	α	2.43310	4.51770	0.67140	1.97810
	c_i	0.88193	0.13265	0.01611	−0.0033
2s	α	2.43310	4.51770	0.67140	1.97810
	c_i	−0.151579	−0.02297	1.04142	−0.0613
1s	c_i	0.88193	0.13347	0.00000	0.00000
2s	c_i	−0.14766	−0.02235	1.0200	−0.0414

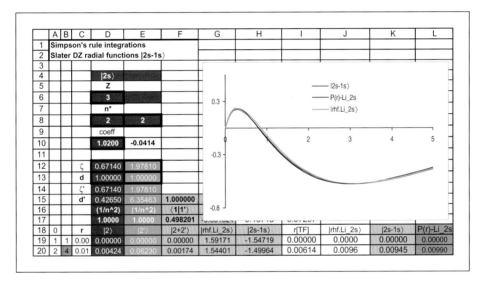

Figure 3.10 Comparisons of the spreadsheet calculated Slater DZ radial function as rR(r) with the same function as outputs from the Herman–Skillman program and Mitroy's RHF atomic program.

available and on public access at *http://lacebark.ntu.edu.au/j_mitroy/research/atomichf/ atomichf.htm* (64). When we make the 'best' basis set condition, the one, which returns the lowest, by the variation principle, energy, we find not only different coefficients for the linear combinations of the Slater functions, but also that all the Slater functions are involved in both approximations to the 1s and 2s radial functions. These data are presented in Table 3.1 and, then, Figure 3.10 compares these mutually orthonormal linear combinations with those, herein, based on the fit to the Herman–Skillman output and the Herman–Skillman output itself.

That the rendering of the 1s and 2s functions orthonormal leads to linear combinations ranging over all four of the Slater functions in the *double-zeta* basis for the two orbitals

---

[8]Note, I have started with the RHF result for the first coefficient and since the orthonormalization is to render the second linear combination orthogonal to the first, I have two coefficients only for the |1s⟩ result.

Table 3.2   Dunning's (4–31) and (4–211) contracted Gaussian sets for the 1s radial function in the hydrogen atom.

Huzinaga	(4s)	α	13.3615	2.01330	0.453757	0.123317
		d_i	0.01906	0.13424	0.47449	0.50907
Dunning[9]	(4s)/[2s]	d_i	0.032828	0.231208	0.817238	1.000000
	(4s)/[3s]	d_i	0.130844	0.921539	1.000000	1.000000

should not be surprising, since it is implicit in the orthonormalization procedure. We can extract from the spreadsheet the overall coefficients as the products of the entries in cells $E$10 and $F$10 of the worksheets with the normalization and overlap constants for the various components of the linear combinations. These are given as the last two rows of Table 3.1 and you can see that the two criteria, the best fit to the Herman–Skillman results and the coefficients, lead to the 'best' energy are not too different. This matter is dealt with in more detail in Chapter 4.

3.5   ORTHONORMALITY AND *SPLIT-BASIS* OR *DOUBLE-ZETA* GAUSSIAN BASIS SETS

The split-basis (46), outlined in Chapter 1, the separating of the Huzinaga |sto-4g⟩ basis set, Table 1.7, into the Gaussian, with the smallest exponent and so the most diffuse function, as one component and the other three primitives as the second component in the contraction, is a simple Gaussian equivalent of Clementi's *double-zeta* Slater functions. It provides a good example of the *double-zeta* approach applied in Gaussian basis set theory.

In the manner of equations 1.24 and 1.25, some other of these basis sets by Dunning are listed in Table 3.2. As you see, the Dunning procedure is to renormalize over the number of primitive Gaussians in each component of the split-basis. The following spreadsheet design details are particular to the case of the (4s)/[2s] and follow straightforwardly from those used for the *double-zeta* Slater functions in the previous section. This alternative to the procedure applied in fig3-5.xls is important for later calculations.

Exercise 3.5.   A general spreadsheet for orthonormality and split-basis calculations with Gaussian basis sets — the matching of hydrogen and lithium 1s and 2s radial orbitals.

1. Open a new spreadsheet and provide for two worksheets as in the case of the *double-zeta* design. Name the worksheets '1s' and '2s'. Then, build both worksheets, as follows, with suitable links for the common design features.

[9]Dunning renormalized the Huzinaga basis functions for a Slater exponent of 1.2, which is to be compared with the common choice [Chapter 6] for molecular environments. Since we renormalize after making the orthonormal linear combination this choice does not matter.

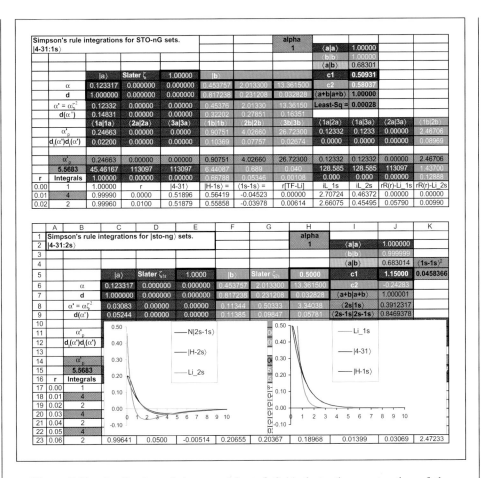

Figure 3.11 Application of the spreadsheet fig3-11.xls to the construction of the 'best' SOLVER based $|4-31:2s\rangle$ Dunning basis approximation to the 2s radial function in hydrogen. Minimize '2s'!\$K\$5 using c_1 for the *split-basis* to the Slater 2s.

2. Provide in cells \$D\$6 to \$E\$7 and \$F\$6 to \$H\$7 for up to two $|sto-3g\rangle$ inputs and, in the next two rows, provide for scaling and pre-exponential normalization factors in the usual manner. These data are input only on '1s' and linked through to '2s'.

3. Provide the usual Simpson's rule integration routine over the radial array in cells \$A\$16 to \$C\$3017.

4. Label the cells in row 10 of column C to W, as in Figure 3.11, to identify the possible integrals over the primitive Gaussians defined in rows 6 and 7.

5. Enter the possible sums and products of the exponents and pre-exponential factors, required in the integrals, in the active cells of rows 8 and 9, for example,

$$C\$11 = C\$8+C\$8$$

$$C\$12 = C\$9*C\$9$$

6. Use the α'_p copies of the data in row 8 as the variables in the master integral of cell $B$15 and the variable table results of the integrations over the Gaussian primitives determined for these entries as the changing value of the master input $H$2, with

$$\$B\$15 = 4^*PI()^*SUMPRODUCT(\$A\$17:\$A\$3017,\$A\$17:\$A\$3017,$$
$$\$B\$17:\$B\$3017,C\$17:C\$3017)^*(\$A\$18-\$A\$17)/3$$

7. Calculate the components of the normalization and overlap integrals in row 16, with, for example, the entry

$$C\$16 = C\$12^*C\$15$$

and so on in the other active cells of the row.

8. Divide the total integral over the primitive functions into parts corresponding to normalization of the first part of the split-basis, the second-part and the overlap component, with the formulae in cells $J$2 to $J$4 given by

$$\$J\$2 = SUM(C16:E16)+2^*SUM(I16:K16)$$

$$\$J\$3 = SUM(F16:H16)+2^*SUM(L16:N16)$$

$$\$J\$4 = SUM(O16:W16)$$

9. With the overlap integral of $J$4 calculated, define the normalization condition on the coefficients of the split-basis, as previously using equation 3.19 and 3.20,

$$c_1 = \$J\$5$$

$$c_2 = (-2^*\$J\$5^*J4+SQRT(POWER(2^*\$J\$5^*\$J\$4,2)$$
$$- 4^*\$J\$3^*(POWER(\$J\$5,2)^*\$J\$2 - 1)))/(2^*\$J\$3)$$

10. For a particular choice of the coefficient c_1 the $|4{-}31\rangle$ function over the radial array is defined in column E, with for example the formula entry in $E$18 of '1s'!

$$\$E\$18 = (\$J\$5^*(\$C\$9^*EXP(-\$C\$8^*\$A17\hat{}2)$$
$$+\$D\$9^*EXP(-\$D\$8^*\$A17\hat{}2)+\$E\$9^*EXP(-\$E\$8^*\$A17\hat{}2))$$
$$+\$J\$6^*(\$F\$9^*EXP(-\$F\$8^*\$A17\hat{}2)+\$G\$9^*EXP(-\$G\$8^*\$A17\hat{}2)$$
$$+\$H\$9^*EXP(-\$H\$8^*\$A17\hat{}2)))/SQRT(\$J\$7)$$

with $J$7 the overall normalization integral for the split-basis function.

11. It is useful to provide for the calculation of the *least-squares* integrals, compare Figure 2.10 and fig2-10.xls, between the hydrogenic and approximate functions, so, on both worksheets, include this procedure with, in '1s'! the projections of cell formulae exemplified by

$$|H\text{-}1s\rangle = \text{'1s'!}\$F\$18 = 1/SQRT(PI())^*EXP(-\$A17)$$

$$\langle r|1s\rangle - \langle r|4{-}31\rangle = \$E18-\$F18$$

and

$$\langle 1s-(4-31)\rangle^2 = \text{'1s'!\$J\$8} = 4^*\textbf{PI}()^*\textbf{SUMPRODUCT}(\$A\$17{:}\$A\$3017,$$
$$\$A\$17{:}\$A\$3017,\$B\$17{:}\$B\$3017,G\$18{:}G\$3018,$$
$$\$G\$18{:}\$G\$3018)^*(\$A\$18{-}\$A\$17)/3$$

and, similarly, on worksheet '2s'.

12. Now, on worksheet '1s'!, follow Figure 3.11 and enter the Herman–Skillman output with two empty columns of cells after the Thomas–Fermi mesh to allow for the recovery of the actual lithium 1s and 2s radial function over the radial array, as in Chapter 1, with, as before, the irregular behaviour at the origin avoided using the **INTERCEPT** function of the EXCEL program, for example for the 1s function

$$\$I\$18 = \textbf{INTERCEPT}(\$I\$19{:}\$I\$23,\$H\$19{:}\$H\$23)$$

13. The remainder of the design of the spreadsheet for the $|4-31\rangle$ basis is closely similar to the detail described for the *double-zeta* Slater basis and so is not described, further, in this exercise.

14. Note, one convenient feature, which is different is that the two comparison charts are shown on '2s' and the Slater exponents' values are determined by the values in cells $F$3 and $F$4 on '2s'.The 'best' matches, using SOLVER are shown in Figure 3.11.

The use of Gaussian sets, designed for 1s orbitals [the Slater orbital, for this case] to model the actual radial functions for different atoms as in Figure 3.10 and 3.11 is quite different from the application of different 1s Gaussian basis sets to model Slater functions for the higher atomic orbitals, which we have considered previously. In fact, this approach can be applied quite generally and it is worth examining how to apply this spreadsheet to other cases.

Exercise 3.6.   Application of the split-basis spreadsheet to model the 2s radial functions in atoms, using $|sto-ng:1s\rangle$ basis sets — hydrogen and lithium 2s radial orbitals.

Little modification of fig3-8.xls is required to test the modelling capacity of 1s designed basis sets for 2s and higher orbitals in hydrogen and by scaling in other atoms.

1. Make a copy of fig3-8.xls and rename the copied file as required by steps 2 and 3 below.

2. To generate Figure 3.12 and attempt to model the hydrogen 2s radial function using the $|sto-ng:1s\rangle$ Gaussian sets of Table 1.7, we need simply to set c_1 and c_2 equal to 1.0 on both spreadsheets. To avoid any singularities in the arithmetic set the values in unused input cells at 1.0 for exponents and 0.0 for coefficients.

3. Direct entry of values for c_2 would cause the deletion of the control exerted by equation 3.20 in cell $J$6, so, to avoid this write a conditional energy for this cell using the flag in $K$5 in the '2s' worksheet, with

$$\$J\$6 = IF(\$K\$5 = 1, \$K\$6, (-2^*\$J\$5^*\$J\$4 + SQRT(POWER$$
$$(2^*\$J\$5^*\$J\$4, 2) - 4^*\$J\$3^*(POWER(\$J\$5, 2)^*$$
$$\$J\$3 - 1)))/(2^*\$J\$3))$$

This device allows either direct entry of the value for the flag equal 1 and equation 3.20 for c_2 for the flag set to zero. Link the flag to the similar condition to set c_2 in '1s' with

$$\text{'1s'!}\$J\$6 = IF(\text{'2s'!}\$K\$5 = 1, \text{'2s'!}\$K\$6, (-2^*\$J\$5^*\$J\$4 +$$
$$SQRT(POWER(2^*\$J\$5^*\$J\$4, 2) - 4^*\$J\$3^*$$
$$(POWER(J5, 2)^*\$J\$3 - 1)))/(2^*\$J\$3))$$

4. For both investigations apply SOLVER to improve the matching of the 2s functions and note that the *least-squares* integrals for the 1s 'best' results agree with the Hehre, Stewart and Pople data in Table 2.1 for hydrogen as one would expect.

Figures 3.13 and 3.14 demonstrate that the modelling of atomic radial functions using Gaussian basis sets need not involve the step that the basis sets be proved against Slater orbitals and so connect to early work with Gaussian basis sets. This should not be

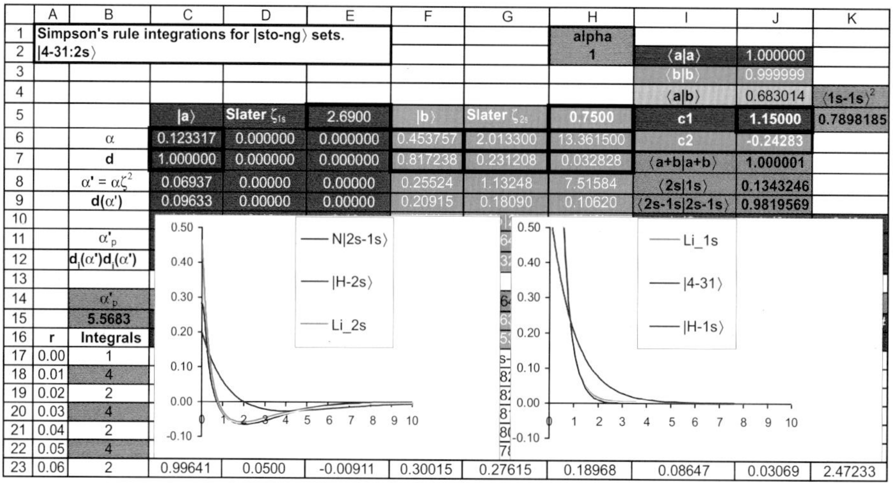

Figure 3.12 Demonstration of the 'scalability' of the split-basis results. In this diagram, the matches are to the output for the Herman–Skillman program for lithium 1s and 2s radial functions. Close agreement is obtained simply by changing the Slater exponents to the Table 1.3 values for the case of lithium, compare Figure 3.11.

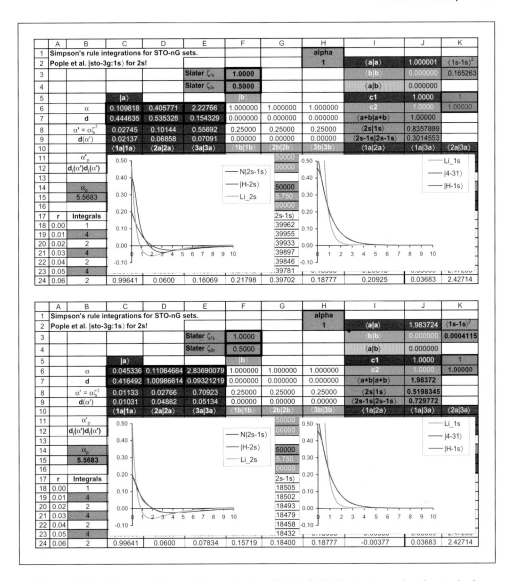

Figure 3.13a   Two conditions of the worksheet '2s'! in fig3-13.xls. Approximations to both the 1s and 2s orbitals in hydrogen are made using the same |sto-3g⟩ basis of Table 1.6. This leads to the indifferent agreement shown in the first diagram, which is improved dramatically when SOLVER is applied to minimize $K$3 with respect to the coefficients and exponents of the basis set in cells $C$6 to $E$7. The '1s'! worksheet is not shown. Note that the design of the spreadsheet is based on fig3-11.xls, but now with both coefficients c_1 and c_2 redundant and set to 1.0. Note, too, especially, the changed constants, $J$7, $J$8 and $J$9, on each of these spreadsheets.

surprising, since, ultimately, in any application to atoms, the Slater orbitals need to be rendered mutually orthonormal to model the actual atomic orbitals.

In Figure 3.13a the results of applying the orthogonalization procedure to the |sto-3g⟩ basis of Table 1.6 are presented. In the first diagram, poor agreement with the hydrogen

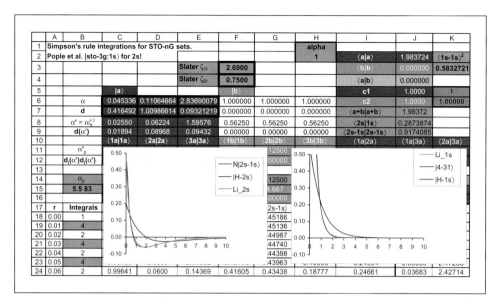

Figure 3.13b   Demonstration of the same 'scalability' property for the $|$sto-3g$\rangle$ 'best' basis set of Figure 3.13a. The substitution of the Slater exponents for lithium 1s and 2s leads to good agreement with the output of the Herman–Skillman programme.

2s function is returned using only the 1s basis and the 2s Slater exponent. But as you can see a dramatic improvement follows the application of SOLVER to the coefficients and exponents of the 1s basis. Finally, once the basis has been fitted to the 2s function, the same scalability is found as with the other example in Figures 3.11 and 3.12.

Figure 3.14a and b present the similar analysis for the example of the $|$sto-6g$\rangle$ basis set of Table 1.6. As you see the *least-squares* agreement now is much better for both 1s and 2s approximations, cell '2s'!$K$3 on all the spreadsheets, although it is recommended that you try to improve the $|$sto-6g$\rangle$ approximation further by suitable changes in the input parameters.

3.6   THE JACOBI TRANSFORMATION, DIAGONALIZATION OF A SYMMETRIC MATRIX AND CANONICAL ORTHOGONALIZATION

So far, in this chapter, basis sets have been rendered into mutually orthonormal forms using the Schmidt procedure. This approach has the pedagogical advantage that the Schmidt procedure is the most familiar method to render functions mutually orthogonal since it is a standard part of courses on Group Theory and Elementary Linear Algebra. However, the Schmidt procedure is rarely, if ever, applied in modern calculations, because it is more efficient to apply global orthonormalization to the expression for the approximate eigenvalues. These are the canonical and symmetric orthonormality procedures (53,47) found in the large modern computer programs. Fundamental to this approach is the need to diagonalize the overlap matrix of the approximate functions used. The Jacobi transformation (59) is a standard procedure for this purpose.

Consider, again, equations 3.1 and 3.2, but assume now that the two functions $|i\rangle$ and $|j\rangle$ while not mutually orthogonal are normalized in the form of equation 3.1.

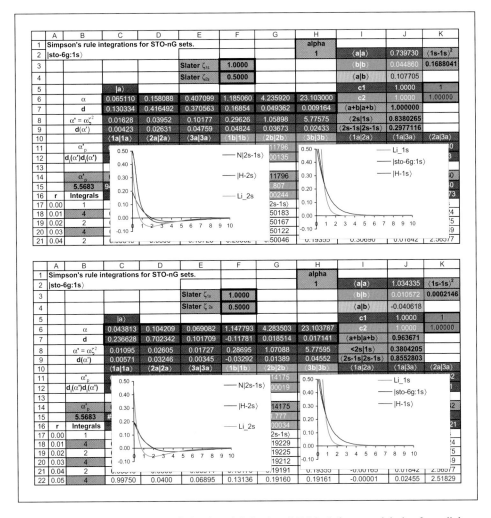

Figure 3.14a Application of the |sto-6g⟩ basis of Table 1.6 to model the 2s radial function in hydrogen. Note the good fit obtained, based on the *least-squares* integral $K$3, compared with the results for the minimal basis in the previous diagrams.

The orthonormality properties of the two functions can be represented by the symmetric matrix

$$\begin{pmatrix} \langle i|i\rangle & \langle i|j\rangle \\ \langle j|i\rangle & \langle j|j\rangle \end{pmatrix} = \begin{pmatrix} 1 & S \\ S & 1 \end{pmatrix}$$

3.15

The matrix is *symmetric* because it is equal to its transpose, which is generated by interchanging the rows and columns. The standard procedure to diagonalize any matrix, A, is to find a suitable similarity transform of the form $X^{-1}AX$, which renders the off-diagonal elements of A to zero values and leaves unchanged, of course, the trace of A. For real symmetric matrices, as in equation 3.15, the transformation matrix, X, is orthogonal since the eigenvectors of A are real and orthonormal. The *similarity transform*, then, is an orthogonal transform involving the eigenvector matrix and its transpose.

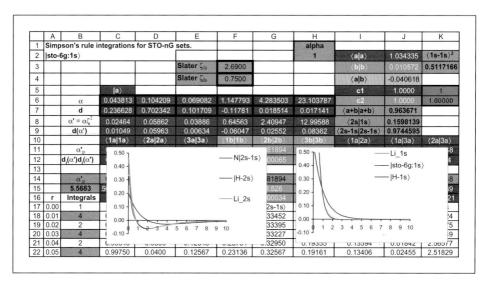

Figure 3.14b Application of the $|$sto-6g:1s$\rangle$ basis set of Table 1.6 to the construction of an orthonormal approximation to the lithium 2s orbital. These are the results of Figure 3.14a scaled to match the numerical lithium 2s orbital calculated using the Herman–Skillman program.

The Jacobi transform on an $n \times n$ symmetric matrix of overlap integrals, like that in 3.15, amongst linear combinations of atomic orbitals taken to form approximations to the molecular orbitals, for a particular molecule, is performed as a series of similarity transforms on the matrix, which have the effect of rendering more and more off-diagonal elements equal to zero, until only the diagonal elements, the eigenvalues remain. Since, on the plane, vector rotation is described in terms of the sines and cosines made with the coordinate axes [*the eigenvectors*] the transformation of the $n \times n$ matrix to the diagonal form in its eigenvectors is known, also, as rotation[10]. Each transforming matrix, X_{pq}, of each similarity transform is constructed in the form of a mostly diagonal matrix except for two rows and columns in which we enter cosines and sines of the angle needed to cause one rotation of the general basis of the original matrix about the diagonal, *viz*

$$X_{pq} = \begin{pmatrix} 1 & 0 & 0 & 0 & 0 & 0 & 0 \\ 0 & \cdots & 0 & 0 & 0 & \cdots & 0 \\ 0 & 0 & \cos(\theta) & \cdots & \sin(\theta) & 0 & 0 \\ 0 & 0 & \vdots & 1 & \vdots & 0 & 0 \\ 0 & 0 & -\sin(\theta) & \cdots & \cos(\theta) & 0 & 0 \\ 0 & \cdots & 0 & 0 & 0 & \cdots & 0 \\ 0 & 0 & 0 & 0 & 0 & 0 & 1 \end{pmatrix} \qquad 3.16$$

When the similarity transform is applied the matrix A is changed only in the relevant matrix elements in the rows and columns identified by the labels 'p' and 'q', with the

---

[10]Remember too that, in a rotation, the vector length remains unchanged since $\sin^2 + \cos^2 = 1$.

effect that in this section of the matrix, we find

$$
A'_{pq} =
\begin{pmatrix}
 & \cdots & a'_{1p} & \cdots & a'_{1q} & \cdots & \\
 & & \vdots & & \vdots & & \vdots \\
a'_{p1} & \cdots & a'_{pp} & \cdots & a'_{pq} & \cdots & a'_{pn} \\
 & & \vdots & & \vdots & & \vdots \\
a'_{q1} & \cdots & a'_{qp} & \cdots & a'_{qq} & \cdots & a'_{qn} \\
 & & \vdots & & \vdots & & \vdots \\
 & \cdots & a'_{np} & \cdots & a'_{nq} & \cdots &
\end{pmatrix}
\tag{3.17}
$$

for which the action of the similarity transform has returned the identities

$$
a'_{rp} = \cos(\theta)a_{rp} - \sin(\theta)a_{rq}
\tag{3.18}
$$

$$
a'_{rq} = \cos(\theta)a_{rq} + \sin(\theta)a_{rp}\, r \neq p, r \neq q
\tag{3.19}
$$

$$
a'_{rq} = \cos(\theta)a_{rq} + \sin(\theta)a_{rp}
\tag{3.20}
$$

$$
a'_{pp} = \cos^2(\theta)a_{pp} + \sin^2(\theta)a_{qq} - 2\sin(\theta)\cos(\theta)a_{pq}
\tag{3.21}
$$

$$
a'_{qq} = \sin^2(\theta)a_{pp} + \cos^2(\theta)a_{qq} + 2\sin(\theta)\cos(\theta)a_{pq}
\tag{3.22}
$$

$$
a'_{pq} = (\cos^2(\theta) - \sin^2(\theta))a_{pq} - \sin(\theta)\cos(\theta)(a_{pp} - a_{qq})
\tag{3.23}
$$

These identities provide for the rendering to zero of off-diagonal elements in sequence. For example, the requirement that a'_{pq} be zero[11] after the Jacobi rotation in equation 3.16 is the identity

$$
\tfrac{1}{2}(a_{pp} - a_{qq})\sin(2\theta) - a_{pq}\cos(2\theta) = 0
\tag{3.24}
$$

in which

$$
\theta_{pq} = \frac{1}{2}\tan^{-1}\left(\frac{2a_{pq}}{a_{pp} - a_{qq}}\right)
\tag{3.25}
$$

For the case of the matrix 3.15, above, equation 3.25 is simply

$$
\theta_{12} = \frac{1}{2}\tan^{-1}\left(\frac{2s}{1-1}\right) = \frac{\pi}{4}
\tag{3.26}
$$

the eigenvalues of the overlap matrix follow from equations 3.21 and 3.22, with

$$
a'_{11} = \cos^2(\theta_{12}) + \sin^2(\theta_{12}) - 2\sin(\theta_{12})\cos(\theta_{12})s
\tag{3.27}
$$

and

$$
a'_{22} = \sin^2(\theta_{12}) + \cos^2(\theta_{12}) + 2\sin(\theta_{12})\cos(\theta_{12})s
\tag{3.28}
$$

[11] And, perhaps, we should just remember, here, that this is the defining condition of orthogonality between two vectors described as in equation 3.15.

while the column eigenvector matrices over $|1\rangle$ and $|2\rangle$ are the linear combinations

$$(\cos(\theta_{12})\quad \sin(\theta_{12}))\begin{pmatrix} S_1 & 0 \\ 0 & S_2 \end{pmatrix}\begin{pmatrix} |1\rangle \\ |2\rangle \end{pmatrix}$$

and

$$(\sin(\theta_{12})\quad -\cos(\theta_{12}))\begin{pmatrix} S_1 & 0 \\ 0 & S_2 \end{pmatrix}\begin{pmatrix} |1\rangle \\ |2\rangle \end{pmatrix} \qquad 3.29$$

Equation 3.29 corresponds to canonical orthogonalization, which (47) as a plane rotation leads to orthogonalization by choosing one vector to bisect the angle between the originals and the second, of course, to be at right angles.

Exercise 3.7.   Application of the canonical and symmetric orthogonalization routines to the construction of $|$sto-3g$\rangle$ approximations to the hydrogen 1s and 2s radial orbitals.

This exercise introduces two new features of the EXCEL software the matrix multiplication and transpose macros **MMULT( )** and the **TRANSPOSE( )**. These are best activated using the '$= f_x$' icon on the TOOLBAR and the option to drag out the array data to be multiplied, visible in the function window. You select an equal area of blank cells on the spreadsheet and then activate the **MMULT( )** or **TRANSPOSE( )** macros. **Note too that** after selecting the particular area for each array, the sequence to enter these cell references requires CTL/SHIFT/ENTER. Otherwise, only one entry will appear in the top-left hand cell of the chosen active cells for the operation.

1. Open a new spreadsheet and name it fig3-15.xls.
2. Layout the usual input data and scaling adjustments for the $|$sto-3g$\rangle$ sets of Table 1.6 in cells $D$2 to $I$7.
3. Layout the Simpson's rule integration procedure in cells $A18 to $C3018.
4. Project the $|$sto-3g:2s$\rangle$ and $|$sto-3g:1s$\rangle$ approximations down columns D and E.
5. Calculate the normalization and overlap integrals for these projections in cells $E$8 to $E$10 and feed these results into the appropriate cells of the S matrix of cells $A$10 to $B$11.
6. Calculate the rotation angle for the Jacobi transformation of equation 3.25, with the conditional formula

$$\$B\$14 = \textbf{IF(ABS}(\$A\$10\text{-}\$B\$11)\langle 1E\text{-}04,\textbf{PI}(\,)/4,$$
$$0.5^*\textbf{ATAN}(2^*\$B\$10/(\$A\$10\text{-}\$B\$11)))$$

to avoid the singularity in the $\tan^{-1}$ function when the denominator in equation 3.25 is zero, while allowing for some imprecision in the integrals.

7. Calculate the eigenvalues of the overlap matrix form equation 3.27 and 3.28, with

$$\$B\$15 = \$A\$10^*\textbf{POWER(COS}(\$B\$14),2)+$$
$$\$B\$11^*\textbf{POWER(SIN}(\$B\$14),2)+\$B\$10^*\textbf{SIN}(2^*\$B\$14)$$

and

$$\$B\$16 = \$A\$10^*\mathbf{POWER}(\mathbf{SIN}(\$B\$14),2)+$$
$$\$B\$11^*\mathbf{POWER}(\mathbf{COS}(\$B\$14),2)-\$B\$10^*\mathbf{SIN}(2^*\$B\$14)$$

8. Construct the orthogonalizing matrix and its equal transpose from equations 3.29 and 3.30, with

$$\$D\$10 = \mathbf{COS}(\$B\$14)$$

$$\$D\$11 = \mathbf{SIN}(\$B\$14)$$

$$\$E\$10 = \mathbf{SIN}(\$B\$14)$$

$$\$E\$12 = \mathbf{-COS}(\$B\$14)$$

and, then the transpose O^T in cells $\$F\$10$ to $\$G\$11$, by selecting these cells and opting for the array multiplication procedure outlined above for the command

$$\$F\$10 = \mathbf{TRANSPOSE}(\$D\$10:\$E\$11)$$

9. Construct the canonical orthogonalization matrix for equations 3.29, with the matrix multiplication of the matrices O and the S, as described above, with the entry

$$\$D\$13 = \mathbf{MMULT}(\$D\$10:\$E\$11,\$H\$10:\$I\$11)$$

10. Use this matrix to project the canonically orthogonal functions from the original $|$sto-ng$\rangle$ sets, with for example,

$$\$F19 = -(\$D\$13^*(\$D\$7^*\mathbf{EXP}(-\$D\$6^*\$A19^2)+$$
$$\$E\$7^*\mathbf{EXP}(-\$E\$6^*\$A19^2)+\$F\$7^*\mathbf{EXP}(-\$F\$6^*\$A19^2))+$$
$$\$D\$14^*(\$G\$7^*\mathbf{EXP}(-\$G\$6^*\$A19^2)+$$
$$\$H\$7^*\mathbf{EXP}(-\$H\$6^*\$A19^2)+\$I\$7^*\mathbf{EXP}(-\$I\$6^*\$A19^2)))$$

and

$$\$G19 = -(\$E\$13^*(\$D\$7^*\mathbf{EXP}(-\$D\$6^*\$A19^2)$$
$$+\$E\$7^*\mathbf{EXP}(-\$E\$6^*\$A19^2)+\$F\$7^*\mathbf{EXP}(-\$F\$6^*\$A19^2))$$
$$+\$E\$14^*(\$G\$7^*\mathbf{EXP}(-\$G\$6^*\$A19^2)$$
$$+\$H\$7^*\mathbf{EXP}(-\$H\$6^*\$A19^2)+\$I\$7^*\mathbf{EXP}(-\$I\$6^*\$A19^2)))$$

filled down the active cells of columns F and G.

11. Calculate normalization and overlap integrals for the orthogonal functions as in cells $\$G\$15$ to $\$G\$17$.

12. Form normalized projections of the orthogonal functions in column H and I and check that orthonormal functions have been constructed using the **SUMPRODUCT** data in cells $\$I\$15$ to $\$I\$17$.

13. Use the CHART wizard to construct comparison graphs of the exact hydrogen 1s and 2s functions and the canonically orthonormal set of columns H and I, as is reproduced in Figure 3.15. Use the coarser radial mesh set out in cells $\$N\$18$ to $\$N\$49$.

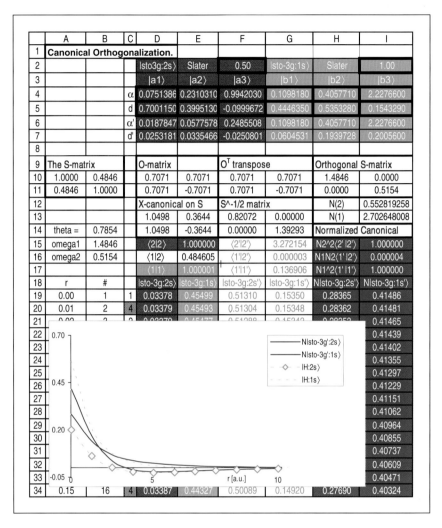

	A	B	C	D	E	F	G	H	I
1	Canonical Orthogonalization.								
2				|sto3g:2s⟩	Slater	0.50	|sto-3g:1s⟩	Slater	1.00
3				|a1⟩	|a2⟩	|a3⟩	|b1⟩	|b2⟩	|b3⟩
4			α	0.0751386	0.2310310	0.9942030	0.1098180	0.4057710	2.2276600
5			d	0.7001150	0.3995130	-0.0999672	0.4446350	0.5353280	0.1543290
6			α'	0.0187847	0.0577578	0.2485508	0.1098180	0.4057710	2.2276600
7			d'	0.0253181	0.0335466	-0.0250801	0.0604531	0.1939728	0.2005600
8									
9	The S-matrix			O-matrix		O^T transpose		Orthogonal S-matrix	
10	1.0000	0.4846		0.7071	0.7071	0.7071	0.7071	1.4846	0.0000
11	0.4846	1.0000		0.7071	-0.7071	0.7071	-0.7071	0.0000	0.5154
12				X-canonical on S		$S^{-1/2}$ matrix		N(2)	0.552819258
13				1.0498	0.3644	0.82072	0.00000	N(1)	2.702648008
14	theta =	0.7854		1.0498	-0.3644	0.00000	1.39293	Normalized Canonical	
15	omega1	1.4846		⟨2|2⟩	1.000000	⟨2'|2'⟩	3.272154	N2^2⟨2' |2'⟩	1.000000
16	omega2	0.5154		⟨1|2⟩	0.484605	⟨1'|2'⟩	0.000003	N1N2⟨1' |2'⟩	0.000004
17				⟨1|1⟩	1.000001	⟨1'|1'⟩	0.136906	N1^2⟨1' |1'⟩	1.000000
18	r	#		|sto-3g:2s⟩	|sto-3g:1s⟩	|sto-3g:2s'⟩	|sto-3g:1s'⟩	N|sto-3g:2s'⟩	N|sto-3g:1s'⟩
19	0.00	1	1	0.03378	0.45499	0.51310	0.15350	0.28365	0.41486
20	0.01	2	4	0.03379	0.45493	0.51304	0.15348	0.28362	0.41481
21	0.02	3		0.03379	0.45477	0.51288	0.15342	0.28353	0.41465
22									0.41439
23									0.41402
24									0.41355
25									0.41297
26									0.41229
27									0.41151
28									0.41062
29									0.40964
30									0.40855
31									0.40737
32									0.40609
33									0.40471
34	0.15	16	4	0.03387	0.44327	0.50089	0.14920	0.27690	0.40324

Figure 3.15   Application of the canonical procedure to render the 1s and 2s |sto-ng⟩ basis sets of Table 1.6 mutually orthogonal as is seen in the value of the overlap integral in cell $G$16 compared to the value 0.4846, cell $D$16 of the unmixed originals. Note that the new functions are not normalized, cells $G$15 and $G$17 and that multiplication by the appropriate normalization constants cells $I$12 and $I$13 is required.

The results in Figure 3.15 are to be compared with the similar results for the Schmidt orthogonalization based construction of the |sto-3g⟩ and, indeed, |sto-6g⟩ approximations to the hydrogenic 1s and 2s radial functions. The canonical orthogonalization returns much poorer agreement with the exact functions. We should not be surprised at this result. The Schmidt procedure is based on rendering orthogonal the second function to the first one leaving the first unchanged. Since this first function is a fair approximation to the exact 1s radial function, the rotation effected by the Schmidt transformation has to be a reasonable approximation, too, to the 2s exact function.

This conclusion is evident when we write the Schmidt transformation as the matrix equation

$$\begin{pmatrix} 1 & -S_{12}/(1 - S_{12}^2)^{1/2} \\ 0 & 1/(1 - S_{12}^2)^{1/2} \end{pmatrix} \begin{pmatrix} |1\rangle \\ |2\rangle \end{pmatrix} \qquad 3.30$$

One vector is kept the same, while the other is constructed as a linear combination of the two originals, with the second at right angles to the unchanged first.

In the remaining chapters, when we solve the Schrödinger equation in a canonically orthogonal basis, we will find that a further rotation of the canonical vectors is required to construct the eigenfunction approximations in the chosen basis set.

3.7   THE $S^{-1/2}$ 'TRICK'

There is a unified procedure by which the canonical orthogonalization procedure can be extended to return normalized functions. Normalization results, if the canonical transformation involves the matrix of inverse square roots of the eigenvalues of the overlap matrix. This is the $S^{-1/2}$ procedure applied in modern molecular orbital programmes.

For the $|$sto-ng$\rangle$ 2s and 1s basis sets of Table 1.6, equation 3.15 under the Jacobi transformation becomes the diagonal matrix of cells $H\$10$ to $I\$11$ in Figure 3.15 and 3.16,

$$S^{-1/2} = \begin{pmatrix} 1.4846 & 0 \\ & 0.5154 \end{pmatrix} \qquad 3.31$$

for which the inverse square root matrix is

$$S^{-1/2} = \begin{pmatrix} 0.82072 & 0 \\ & 1.39293 \end{pmatrix} \qquad 3.32$$

as in cells $J\$9$ to $K\$10$ of Figure 3.16.

As you see in Figure 3.16, canonical transformation now leads directly to the orthonormal canonical eigenfunctions of the overlap matrix as the entries in the active cells of columns J and K of fig3-16.xls.

3.8   SYMMETRIC ORTHONORMALIZATION

Symmetric orthogonalization, as vector rotation (47), involves the taking of linear combinations of two vectors so that the right angle disposition required is achieved by enlarging the angle between the originals to 90° by their equal and opposite rotation.

As a similarity transform, symmetric orthonormalization is the sequence

$$O S^{-1/2} O^T \qquad 3.33$$

and, for the present case, equations 3.29 are replaced by the transformations

$$(\cos(\theta_{12}) \quad \sin(\theta_{12})) \begin{pmatrix} 0.82072 & 0 \\ 0 & 1.39293 \end{pmatrix} (\cos(\theta_{12}) \quad \sin(\theta_{12})) \begin{pmatrix} |1\rangle \\ |2\rangle \end{pmatrix} \qquad 3.34$$

C	D	E	F	G	H	I	J	K
1	Canonical Orthonormalization							
2	|sto3g:2s⟩	Slater	0.50	|sto-3g:1s⟩	Slater	1.00		
3	|a1⟩	|a2⟩	|a3⟩	|b1⟩	|b2⟩	|b3⟩		
4 α	0.0751386	0.2310310	0.9942030	0.1098180	0.4057710	2.2276600		
5 d	0.7001150	0.3995130	-0.0999672	0.4446350	0.5353280	0.1543290		
6 α'	0.0187847	0.0577578	0.2485508	0.1098180	0.4057710	2.2276600		
7 d'	0.0253181	0.0335466	-0.0250801	0.0604531	0.1939728	0.2005600		
8							$S^{-1/2}$	
9	O-matrix		O^T transpose		Diagonal S-matrix		0.82072	0.00000
10	0.7071	0.7071	0.7071	0.7071	1.4846	0.0000	0.00000	1.39293
11	0.7071	-0.7071	0.7071	-0.7071	0.0000	0.5154	$S^{-1/2}$ canonical	
12	Canonical on S				N(2)	0.552819258	0.58033584	0.98495147
13	1.0498	0.3644			N(1)	2.702648008	0.58033584	-0.9849515
14	1.0498	-0.3644			Normalized Canonical		$S^{-1/2}$ direct	
15	⟨2|2⟩	1.000000	⟨2'|2'⟩	3.272154	N2^2⟨2'|2'⟩	1.000000	$S^{(-1/2)}$⟨2|2⟩	1.000000
16	⟨1|2⟩	0.484605	⟨1'|2'⟩	0.000003	N1N2⟨1'|2'⟩	0.000004	$S^{(-1/2)}$⟨1|2⟩	0.000001
17	⟨1|1⟩	1.000001	⟨1'|1'⟩	0.136906	N1^2⟨1'|1'⟩	1.000000	$S^{(-1/2)}$⟨1|1⟩	1.000000
18	|sto-3g:2s⟩	|sto-3g:1s⟩	|sto-3g:2s'⟩	|sto-3g:1s'⟩	N|sto-3g:2s'⟩	N|sto-3g:1s'⟩	|S^1/2:2s'⟩	|S^1/2:1s'⟩
19 1	0.03378	0.45499	0.51310	0.15350	0.28365	0.41486	0.28365	0.41486
20 4	0.03379							481
21 2	0.03379							465
22 4	0.03379							439
23 2	0.03379							402
24 4	0.03379							355
25 2	0.03380							297
26 4	0.03380							229
27 2	0.03381							151
28 4	0.03382							062
29 2	0.03382							963
30 4	0.03383							855
31 2	0.03384							737
32 4	0.03385							608
33 2	0.03386							471
34 4	0.03387							324
35 2	0.03388							168
36 4	0.03389							002

(Embedded chart — y-axis from −0.05 to 0.70; x-axis r [a.u.] from 0 to 10. Legend: N\|sto-3g':2s⟩, N\|sto-3g':1s⟩, \|H:2s⟩, \|H:1s⟩, \|S^1/2:2s'⟩, \|S^1/2:1s'⟩)

Figure 3.16 Direct orthonormalization of the 2s basis set using the $S^{-1/2}$ diagonal matrix to form the canonical transformation. Note the exact agreement with the normalized results based only on the diagonal overlap matrix.

and

$$(\sin(\theta_{12}) \quad -\cos(\theta_{12}))\begin{pmatrix} 0.82072 & 0 \\ 0 & 1.39293 \end{pmatrix}(\sin(\theta_{12}) \quad -\cos(\theta_{12}))\begin{pmatrix} |1\rangle \\ |2\rangle \end{pmatrix} \qquad 3.35$$

The results for symmetric orthonormalization of the two |sto-3g⟩ basis sets for the hydrogen 1s and 2s radial functions are shown in Figure 3.17.

Again, relatively poor agreement with the exact radial functions is found. However, although it will not be apparent within the limited scope of the examples discussed in

C	D	E	F	G	H	I	J	K
1	**Symmetric Orthonormalization**							
2	$\lvert sto3g:2s\rangle$	Slater	0.50	$\lvert sto\text{-}3g:1s\rangle$	Slater	1.00		
3	$\lvert a1\rangle$	$\lvert a2\rangle$	$\lvert a3\rangle$	$\lvert b1\rangle$	$\lvert b2\rangle$	$\lvert b3\rangle$		
4	α 0.0751386	0.2310310	0.9942030	0.1098180	0.4057710	2.2276600		
5	d 0.7001150	0.3995130	-0.0999672	0.4446350	0.5353280	0.1543290		
6	α 0.0187847	0.0577578	0.2485508	0.1098180	0.4057710	2.2276600		
7	d' 0.0253181	0.0335466	-0.0250801	0.0604531	0.1939728	0.2005600		
8							SO^T	
9	O-matrix			O^T transpose		$S^{-1/2}$		0.58033584	0.58033584
10	0.7071	0.7071	0.7071	0.7071	0.82072	0.00000	0.98495147	-0.9849515
11	0.7071	-0.7071	0.7071	-0.7071	0.00000	1.39293	OSO^T	
12			Diagonal S-matrix		N(2)	0.552819258	1.10682527	-0.2861065
13			1.4846	0.0000	N(1)	2.702648008	-0.2861065	1.10682527
14			0.0000	0.5154	Normalized Canonical		$S^{-1/2}$ direct	
15	$\langle 2\lvert 2\rangle$	1.000000	$\langle 2'\lvert 2'\rangle$	3.272154	N2^2$\langle 2'\lvert 2'\rangle$	1.000000	S(-1/2)$\langle 2\lvert 2\rangle$	0.999999
16	$\langle 1\lvert 2\rangle$	0.484605	$\langle 1'\lvert 2'\rangle$	0.000003	N1N2$\langle 1'\lvert 2'\rangle$	0.000004	S(-1/2)$\langle 1\lvert 2\rangle$	0.000000
17	$\langle 1\lvert 1\rangle$	1.000001	$\langle 1'\lvert 1'\rangle$	0.136906	N1^2$\langle 1'\lvert 1'\rangle$	1.000000	S(-1/2)$\langle 1\lvert 1\rangle$	1.000001
18	$\lvert sto\text{-}3g:2s\rangle$	$\lvert sto\text{-}3g:1s\rangle$	$\lvert sto\text{-}3g:2s'\rangle$	$\lvert sto\text{-}3g:1s'\rangle$	N$\lvert sto\text{-}3g:2s'\rangle$	N$\lvert sto\text{-}3g:1s'\rangle$	$\lvert S^{1/2}:2s'\rangle$	$\lvert S^{1/2}:1s'\rangle$
19 (1)	0.03378	0.45499	0.51310	0.15350	0.28365	0.41486	0.09278	0.49392
20 (4)	0.0....	0.45...	0.51204	0.15248	0.28...	0.41481	0.0....	0.49386
21 (2)								0.49369
22 (4)								0.49339
23 (2)								0.49298
24 (4)								0.49245
25 (2)								0.49181
26 (4)								0.49105
27 (2)								0.49017
28 (4)								0.48918
29 (2)								0.48808
30 (4)								0.48687
31 (2)								0.48554
32 (4)								0.48411
33 (2)								0.48257
34 (4)								0.48093
35 (2)								0.47918
36 (4)								0.47734
37 (2)	0.03391	0.43827	0.49568	0.14737	0.27402	0.39828	0.08786	0.47539

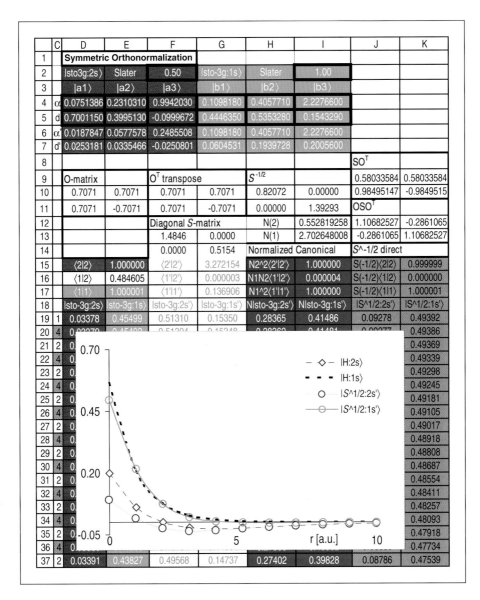

Figure 3.17 Symmetric orthonormalization of the $\lvert sto\text{-}3g\rangle$ basis sets for hydrogen 1s and 2s radial functions and the comparisons with the exact radial functions from solution of the Schrödinger equation.

the remaining chapters of this book, the Jacobi based transformations, both canonical and symmetric, to construct an orthonormal set of basis functions are much better for the typical calculations that one might carry out for large molecules, with many molecular orbitals. They do have technical problems (47,50) for calculations in which the eigenvalues of the overlap matrix approach zero. However, these can be overcome by appropriate programming involving the ordering of the matrix of eigenvalues in descending order along the diagonal and then the discarding of the near zero values.

4

The hydrogen atom — numerical solutions

This chapter discusses the application of Slater and Gaussian basis set theory to solve Schrödinger's equation for the hydrogen atom. In the beginning, again, the Schmidt procedure is used to generate orthogonal approximations to the hydrogen 2s atomic orbital. However, the direct diagonalization of the Fock matrix by canonical transformation to an orthonormal basis is developed later for both Slater and Gaussian basis sets.

Thus, in this chapter you will learn how to calculate the eigenvalues of the hydrogenic atomic orbitals on a spreadsheet, using

1. the actual atomic orbitals;
2. slater orbitals;
3. contracted and *split-basis* Gaussian basis sets;
4. and both Slater and Gaussian approximations to the hydrogenic 2s orbital using the canonical procedure.

In the early literature on basis set theory, there was considerable interest in the atomic problem. In this early work (17,45) it was recognized that Slater functions are exact eigenfunctions of a central-field problem in which an additional potential is added to the one-electron Hamiltonian. Thus, linear combinations of Gaussian functions could be tailored to provide good approximations to the Slater functions (15). This chapter reviews spreadsheet calculations of the hydrogen orbital energies using Slater functions and Gaussian basis sets.

In spherical polar coordinates, Figure 1.1 and Figure 1.6, the Schrödinger equation for an atom like hydrogen or any other atomic ion, of atomic number Z, ionized to leave only one extra nuclear electron, is

$$-\frac{h^2}{8\pi^2\mu}\left[\frac{1}{r^2}\frac{\partial}{\partial r}\left(r^2\frac{\partial}{\partial r}\right)+\frac{1}{r^2\sin\theta}\frac{\partial}{\partial\theta}\left(\sin\theta\frac{\partial}{\partial\theta}\right)+\frac{1}{r^2\sin^2\theta}\frac{\partial^2}{\partial\phi^2}\right]\phi(r,\theta,\phi)$$

$$-\frac{Ze^2}{4\pi\varepsilon_0 r}\phi(r,\theta,\phi)=\varepsilon\phi(r,\theta,\phi) \qquad\qquad 4.1$$

with $\mu = m_e m_p/(m_e + m_p)$, the *reduced* mass of the electron mass, m_e, and the nuclear mass, m_p; ε_0, the dielectric polarizability, 'e' the charge on the electron and 'h' Planck's constant of action. The eigenfunction solutions and the associated eigenvalues are identified using the symbols, $\phi(r, \theta, \phi)$ and ε.

In quantum chemistry, it is standard practice to define *atomic units*. Assuming that the nucleus is stationary, so that the reduced mass is simply the mass of the electron, equation 4.1 written in atomic units is the dimensionless form

$$-\frac{1}{2}\left[\frac{1}{r^2}\frac{\partial}{\partial r}\left(r^2\frac{\partial}{\partial r}\right) + \frac{1}{r^2\sin\theta}\frac{\partial}{\partial\theta}\left(\sin\theta\frac{\partial}{\partial\theta}\right) + \frac{1}{r^2\sin^2\theta}\frac{\partial^2}{\partial\phi^2}\right]\phi(r, \theta, \phi)$$

$$-\frac{Z}{r}\phi(r, \theta, \phi) = \varepsilon\phi(r, \theta, \phi) \tag{4.2}$$

For this transformation to be true it is necessary to define these atomic units in the identity

$$e = h/2\pi = m_e \tag{4.3}$$

to measure radial distance in Bohr units, a_0, the magnitude of the first Bohr orbit radius in the old atomic theory, with

$$a_0 = \left(\frac{4\pi\varepsilon_0 h^2}{m_e e^2}\right) = 1 \tag{4.4}$$

and the electronic energy, ε, in Hartrees, with

$$1\text{ Hartree} = \left(\frac{4\pi\varepsilon_0 h^2}{m_e e^2}\right) = 1 \tag{4.5}$$

It is important to be conversant with the relations, which convert these atomic units of quantum theory into the modern SI [*Système International*] or *mks* set of units (37) with length measured in metres, m, mass in kilograms, kg, while time is measured in seconds and energy is measured in joules, J, with

$$1\text{ J} = 1\text{ kg m}^{-1}\text{s}^{-2} \tag{4.6}$$

These relations are listed in Table 4.1 together with the conversion factors for some non-SI units still commonly in use.

Table 4.1 Some conversion factors from atomic units to their SI values (42,65).

Atomic unit		SI value	Other
a_0	Length	$5.2917706 \times 10^{-11}$ m	0.52917706 Å[Angströms]
m_e	e/mass	9.109534×10^{-31} kg	
e	e/charge	$-1.602189 \times 10^{-19}$ C	
Hartree	Energy	4.359814×10^{-18} J	27.21161 electron volts (eV)
			2.000000 Rydbergs (Ryd)
			2625.501003 kJ/mole
			2.194747×10^5 cm^{-1}
h	Planck's constant	6.626176×10^{-34} J s	4.135701×10^{-15} eV s

The separation of the equation into the distinct radial and angular parts, leads to, the radial equation,

$$-\frac{1}{2}\left[\frac{d^2}{dr^2} - \frac{2}{r}\frac{d}{dr}\right] R(r) + \frac{l(l+1)}{2r^2} R(r) - \frac{Ze^2}{r} R(r) = E R(r) \qquad 4.7$$

and the angular equation,

$$\left[\frac{1}{\sin\theta}\frac{\partial}{\partial\theta}\left(\sin\theta\frac{\partial}{\partial\theta}\right) + \frac{1}{\sin^2\theta}\frac{\partial^2}{\partial\phi^2}\right] Y_{lm}(\theta,\phi) + I(I+1) Y_{lm}(\theta,\phi) = E Y_{lm}(\theta,\phi) \qquad 4.8$$

The 1s atomic orbital for the hydrogen atom results as an exact solution, for the choice of the first Laguerre polynomial ($n = 1$) for the radial wave function and the lowest spherical harmonic ($l = 0$) Y_{00}, for the angular wave function. Thus, from Table 1.1, the normalized 1s atomic orbital for the hydrogen atom is,

$$\phi(r,\theta,\phi) = 2\left(\frac{Z}{a_0}\right) \exp\left(-\frac{Zr}{a_0}\right) \times \left(\frac{1}{4\pi}\right)^{1/2} = \left(\frac{1}{\pi}\right)^{1/2} \exp(-r) \qquad 4.9$$

and the orbital energy is found to be $-1/2$ Hartree.

The numerical solution procedure, as an exercise in basis set theory, is to choose the representation of the atomic orbital using a particular recipe for the basis set and then solve the equation for the energy, by numerical integration, subject to some condition to render the results as close to the correct results as possible. Within molecular orbital theory, the standard condition is to impose a variation principle minimum energy requirement, using any available disposable parameters. In basis set theory, such parameters are the scaling factors multiplying the Gaussian exponents for particular applications and, of course, any LCAO coefficients, undetermined by group theory.

For a trial function, φ, the standard procedure is to substitute the approximate function into the Schrödinger equation, and then obtain the energy expression, by left multiplying by the trial function and integrating over the whole of the atomic or molecular space,

$$E = \frac{\left\langle \varphi \left| -\frac{1}{2}\left(\frac{d^2}{dr^2} + \frac{2}{r}\frac{d}{dr}\right) - \frac{1}{r} + \frac{l(l+1)}{2r^2} \right| \varphi \right\rangle}{\langle \varphi | \varphi \rangle} \qquad 4.10$$

Note, for completeness, that the formality of complex conjugation is implicit in the Dirac notation, although the matter of complex functions does not arise in this book.

In the present context, two types of real trial function, $|\varphi\rangle$, are of interest. The simplest Slater orbital representation of the radial wave function of an atomic orbital is the general form given in equation 1.12

$$R(r:n\zeta) = (2\zeta)^{n^*+(1/2)}[(2n^*)!]^{-1/2} r^{n^*-1} \exp(-\zeta r) \qquad 1.12$$

while the sto-ng representation involves linear combinations of Gaussian functions in the form of equation 1.13

$$g(r:n,\alpha) = N_g r^{n-1} e^{-\alpha r^2} \qquad 1.13$$

which for simplicity in the evaluation of integrals, generally are restricted to 1s, 2p and 3d Gaussians.

Look at the expression for the energy in equation 4.10. The numerator requires various differentiations and terms involving $1/r^n$ to be integrated over the atomic space. The denominator requires the calculation of the norm. Both of these terms can be evaluated numerically for the choice of Slater orbitals or the $|$sto-ng$\rangle$ basis sets, although for real calculations the normal procedure is to evaluate the various integrals analytically. For the present purposes, the advantage of the numerical procedure is that we begin to understand clearly what the implications are of the use of the different approximate representations.

On multiplying out the terms in the energy expression, it becomes

$$
E = \frac{-\dfrac{1}{2}\left\langle \varphi \left| \dfrac{d^2}{dr^2} + \dfrac{2}{r}\dfrac{d}{dr} \right| \varphi \right\rangle - \left\langle \varphi \left| \dfrac{1}{r} \right| \varphi \right\rangle + \left\langle \varphi \left| \dfrac{l(l+1)}{2r^2} \right| \varphi \right\rangle}{\langle \varphi | \varphi \rangle}
\qquad 4.11
$$

Thus, the calculation of the energy, for the particular representation of the atomic orbital chosen, depends on the evaluation of all the terms in this equation, i.e. the kinetic energy term, the potential energy term reflecting the attraction to the nucleus and the centrifugal potential energy term because of the non-zero angular momentum of the electron for orbitals with the angular quantum number greater than zero.

4.1   EIGENVALUE CALCULATIONS FOR HYDROGEN BASED ON ANALYTICAL FUNCTIONS

First, consider the case of the 1s eigenstate of the hydrogen atom. The 'trial' function, in this case, is the actual 1s orbital of equation 4.9. In Chapter 6, the hydrogen orbital is used in the calculation of the molecular orbital energies for dihydrogen. So it is useful here to generalize the function to its Slater form. Thus, the differentiations required in two of the integrals in equation 4.11 go as

$$
\frac{d}{dr}\exp(-\zeta r) = -\zeta \exp(-\zeta r)
\qquad 4.12
$$

and

$$
\frac{d^2}{dr^2}\exp(-\zeta r) = \zeta^2 \exp(-\zeta r)
\qquad 4.13
$$

Using the analytical hydrogen 1s atomic orbital, slightly generalized to include the Slater exponent as a variable parameter, the energy expression simplifies to the form

$$
E = \frac{-\dfrac{1}{2}\left\langle \exp(-2\zeta r)\left(\zeta^2 - \dfrac{\zeta}{r}\right)\right\rangle - \left\langle \exp(-2\zeta r)\dfrac{1}{r}\right\rangle}{\langle \exp(-2\zeta r)}
\qquad 4.14
$$

Note that the centrifugal force term, in the angular momentum quantum number l, is zero for s orbitals. While each of these simple integrals has an analytical solution, we can now apply the numerical procedures of Chapter 2 directly to equation 4.14 and determine the energy of this 1s 'trial' function. Since this is the exact 1s orbital for the hydrogen atom, this exercise functions also as a good test of the integration procedure.

	A	B	C	D	E	F
1	Simpson's rule integration					
2	H1s eigenvalue.					
3						
4				ζ	1.0000	Z
5				N	2.0000	1
6				$\langle 1s\|1s\rangle$	1.00000	
7				T	0.500000	
8				V	-1.000000	
9				E_{1s}	-0.50000	
10						
11	r			$\|1s\rangle$	$r^2 grad^2\|1s\rangle$	
12	0.00	1	1	2.00000	0.00000	
13	0.01	2	4	1.98010	-0.03940	
14	0.02	3	2	1.96040	-0.07763	
15	0.03	4	4	1.94089	-0.11471	
16	0.04	5	2	1.92158	-0.15065	

Figure 4.1 Detail from spreadsheet fig4-1.xls on the CDROM for the calculation of the energy of the 1s orbital in hydrogen using the numerical approach. In this particular case, the Slater orbital function is the correct analytical wave function of equation 4.9, but the integration is over the radial coordinate only, so that the normalization constant is the value given in Table 1.1.

Fig4-1.xls provides for the various numerical integrations for the calculation of the 1s orbital energy given in equation 4.9. As usual, a part of the spreadsheet is shown in Figure 4.1 and you can see that the design follows directly from Chapter 2 and the Simpson's rule procedure for integration of exponential functions.

Exercise 4.1.   Calculation of the 1s orbital energy in hydrogen on a spreadsheet.

1. Most of the design details should be familiar. Provide for the Simpson's rule integration procedure in the usual manner in columns A to C from cell $A$12 to cell $C$3013.
2. Lay out the projections of the 1s function on the appropriate cells of column D, with, for example, in $D$12

$$\$D\$12 = E\$5*EXP(-E\$4*\$A12)$$

3. For the calculation of the kinetic energy term in equation 4.14 we need to project the $grad^2$ function, too, on the radial array. But, note, that there is a

singularity at the origin, because of the $1/r$ factor. So to avoid this enter the grad^2 function times the square of the radius needed for the integration, with, for example, in cell \$E\$12[1]

$$\$E\$12 = \$E\$5*EXP(-\$E\$4*\$A12)*((\$A12*\$E\$4)^2-2*\$E\$4*\$A12)$$

and note that this is the transformed function after the action of the grad^2 operator.

4. Check the normalization of the 1s function with the usual SUMPRODUCT entry in cell \$E\$6 with

$$\$E\$6 = \textbf{SUMPRODUCT}(\$A\$12:\$A\$3012, \$A\$12:\$A\$3012,$$
$$\$C\$12:\$C\$3012,\$D\$12:\$D\$3012,\$D\$12:\$D\$3012)*$$
$$(\$A\$13-\$A\$12)/3$$

5. Calculate the kinetic energy integral as the entry in cell \$E\$7 with

$$\$E\$7 = -0.5*\textbf{SUMPRODUCT}(\$C\$12:\$C\$3012,\$D\$12:\$D\$3012,$$
$$\$E\$12:\$E\$3012)*(\$A\$13-\$A\$12)/(3*\$E\$6)$$

and note the inclusion of the $\langle 1s|1s \rangle$ term in the denominator as required in equation 4.10. Note, too, the absence of the r^2 term in the product since this is already present in the cell entries in column E.

6. Calculate the potential energy term as the entry in cell \$E\$8 with

$$\$E\$8 = -\textbf{SUMPRODUCT}(\$A\$12:\$A\$3012,\$D\$12:\$D\$3012,$$
$$\$D\$12:\$D\$3012,\$C\$12:\$C\$3012)*$$
$$(\$A\$13-\$A\$12)/(3*\$E\$6)$$

and again note the inclusion of only one r term, to allow for the $1/r$ in the potential energy component of equation 4.10 and avoid the possible singularity at the origin.

7. Add the two energy terms to return the energy of the 1s atomic orbital in Hartrees in cell \$E\$9.

The next interesting calculations are to apply the numerical procedure to the higher energy orbitals of the hydrogen atom. In hydrogenic atoms or ions the orbital energies, ε_n, are determined by the ratio of the squares of the atomic number and the principal quantum number n, with

$$\varepsilon_n = -Z^2/n^2 \qquad\qquad 4.15$$

and so the $|2s\rangle$ and $|2p\rangle$ orbitals exhibit eigenvalues equal -0.125 H.

For the $|2s\rangle$ calculation, the grad^2 kinetic energy operator requires the first and second derivatives of the 2s radial function of Table 1.1, i.e. of the product $(2 - r)\exp(-r/2)$, with

$$\frac{d}{dr}(2-r)e^{(-0.5r)} = \left(\frac{r}{2}-2\right)e^{(-0.5r)} \qquad\qquad 4.16$$

and

$$\frac{d^2}{dr^2}(2-r)e^{(-0.5r)} = \left(\frac{3}{2}-\frac{r}{4}\right)e^{(-0.5r)} \qquad\qquad 4.17$$

[1] You may prefer to multiply by r here, rather than r^2. I have chosen r^2 since all cases discussed are included.

Exercise 4.2.   Calculation of the 2s orbital energy in hydrogen.

1. Make a copy of fig4-1.xls and rename it fig4-2.xls. change the formulae for the calculations with the 2s radial function of Table 1.1. So

    (i)   Enter the principal quantum number value in cell $F$7.

    (ii) Allow for the change in exponent in the exponential term, with

$$\$E\$4 = \$F\$5/\$F\$7$$

    (iii) Provide for normalization in cell $E$5, from Table 1.1

$$\$E\$5 = \mathbf{SQRT}(1/8)*\mathbf{POWER}(\$F\$5,3.5)$$

    (iv) Project the 2s radial function down the active cells of column D, for example

$$\$D\$12 = \$E\$5*\mathbf{EXP}(-\$E\$4*\$A12)*(2-\$A12)$$

2. Replace the entries in the column of data for the projections of the effect of grad^2 so that the entries in column E are of the form required by equations 4.16 and 4.17, with for example,

$$\$E12 == \$E\$5*\mathbf{EXP}(-\$E\$4*\$A12)*(2.5*A12^2-A12^3/4-4*A12)$$

3. The 2s orbital energy is the calculated result in cell $E$9 of Figure 4.2.

	A	B	C	D	E	F
1	Simpson's rule integration					
2	H2s eigenvalue.					
3						
4				ζ	0.5000	Z
5				N	0.3536	1
6				$\langle 2s\|2s\rangle$	1.00000	n
7				T	0.125000	2
8				V	-0.250000	
9				$E_{2s} =\rangle$	-0.125000	
10						
11	r			$\|2s\rangle$	$r^2\mathrm{grad}^2\|1s\rangle$	
12	0.00	1	1	0.70711	0.00000	
13	0.01	2	4	0.70006	-0.01398	
14	0.02	3	2	0.69307	-0.02765	
15	0.03	4	4	0.68613	-0.04101	
16	0.04	5	2	0.67924	-0.05407	

Figure 4.2   Calculation of the 2s orbital energy in hydrogen using the radial function of Table 1.1.

For calculations of the energy of p orbitals in hydrogen, the centrifugal term in the Hamiltonian (61), which prevents p and d electrons approaching the nucleus, is greater than zero. Figure 4.3 displays the total atomic potential term (64) for

$$V(r) = \frac{l(l+1)}{2r^2} - \frac{Z}{r} \qquad \qquad 4.18$$

as in equation 4.7.

Thus, in a calculation of the energy of the 2p orbital in hydrogen, we must allow for this extra term. Again, in the extra term, there is a denominator, which goes to zero with the radius 'r'. In addition, the spreadsheet design must include the different functional form for the 2p function and the effect of the grad^2 operator.

The complete 2p atomic orbitals are, from Table 1.1,

$$|2p_x\rangle = \left(\frac{1}{24}\right)^{1/2} \left(\frac{3}{4\pi}\right)^{1/2} re^{r/2}x \qquad \qquad 4.19$$

$$|2p_y\rangle = \left(\frac{1}{24}\right)^{1/2} \left(\frac{3}{4\pi}\right)^{1/2} re^{r/2}y \qquad \qquad 4.20$$

$$|2p_z\rangle = \left(\frac{1}{24}\right)^{1/2} \left(\frac{3}{4\pi}\right)^{1/2} re^{r/2}z \qquad \qquad 4.21$$

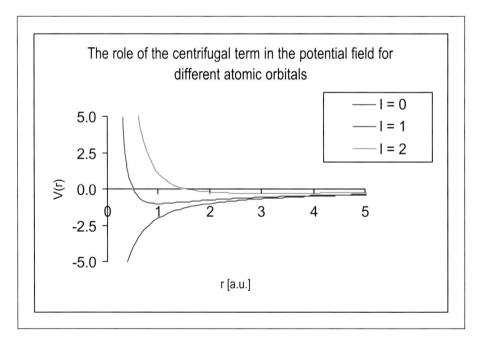

Figure 4.3   The effect of the centrifugal term in the potential field about an atomic nucleus. When the angular quantum number is greater than zero, this term dominates for small 'r' and accounts for the zero amplitudes of p, d and f radial functions at the nucleus. This diagram is the chart in the spreadsheet fig4-3.xls.

The integrations over the angular coordinates of these functions are of unit value since the angular functions' normalization constants are included in these equations.

The effect of grad^2 follows from the first and second derivatives of the radial parts of equations 4.19 to 4.21, with

$$\frac{d}{dr}re^{(-0.5r)} = \left(1.0 - \frac{r}{2}\right)e^{(-0.5r)} \qquad 4.22$$

and

$$\frac{d^2}{dr^2}re^{(-0.5r)} = \left(\frac{r}{4} - 1.0\right)e^{(-0.5r)} \qquad 4.23$$

Exercise 4.3.   Calculation of the 2p orbital energy in hydrogen.

1. As before, make a copy of fig4-1.xls and rename it to be fig4-4.xls.

    (i) Enter the appropriate corrections for the 2p function in the input cells, with

    $$\$E\$4 = \$F\$5/\$F\$7$$

    $$\$E\$5 = \textbf{SQRT}\ (1/24)$$

	A	B	C	D	E	F
1	Simpson's rule integration					
2	H2p eigenvalue.					
3						
4				z	0.5000	Z
5				N	0.2041	1
6				$\langle 2p|2p \rangle$	1.00000	n
7				T	0.04167	2
8				V	-0.25000	ℓ
9				$\ell(\ell+1)/r^{\wedge}2$	0.08333	1
10				E_{2p}	-0.12500	
11	r			$|2p\rangle$	$r^2grad^2|2p\rangle$	
12	0.00	1	1	0.00000	0.00000	
13	0.01	2	4	0.00203	0.00402	
14	0.02	3	2	0.00404	0.00792	
15	0.03	4	4	0.00603	0.01170	
16	0.04	5	2	0.00800	0.01537	

Figure 4.4   Calculation of the 2p orbital energy in hydrogen using the exact function of Table 1.1.

(ii) Project the 2s radial function down the active cells of column D, for example

$$\$D\$12 = E\$5^*\mathbf{EXP}(-E\$4^*\$A12)^*\$A12$$

2. Replace the entries in the column of data for the projections of the effect of grad^2 so that the entries in column E are of the form required by equations 4.22 and 4.23, with for example,

$$\$E12 = \$E\$5^*\$A\$12^*\mathbf{EXP}(-\$E\$4^*\$A12)^*$$
$$(2+0.25^*\$A12^2-2^*\$A12)$$

3. Calculate the non-zero centrifugal term in cell $E$9 with the entry

$$\$E\$9 = \$F\$9^*(F\$9+1)/2^*\mathbf{SUMPRODUCT}(\$C\$12:\$C\$3012,$$
$$\$D\$12:\$D\$3012,\$D\$12:\$D\$3012)^*$$
$$(\$A\$13-\$A\$12)/(3^*\$E\$6)$$

4. Finally, calculate the 2p orbital energy in cell $E$10, as the sum over the components in $E$7, $E$8 and $E$9, with the formula entry

$$\$E\$10 = \$E\$8+\$E\$9+\$E\$10$$

4.2 CALCULATIONS USING SLATER ORBITALS

The Slater function for the 1s orbital in hydrogen is the exact eigenfunction of Table1.1. So the first significant calculation involves the Slater 2s radial function. From equation 1.12, we have

$$|S:2s\rangle = (2\zeta)^{5/2}[(4)!]^{-1/2}r\exp(-\zeta r)(1/4\pi)^{1/2} \qquad 4.24$$

with

$$\frac{d}{dr}r\exp(-\zeta r) = -\zeta\exp(-\zeta r)r + \exp(-\zeta r) \qquad 4.25$$

and

$$\frac{d^2}{dr^2}r\exp(-\zeta r) = (\zeta^2 r - 2\zeta)\exp(-\zeta r) \qquad 4.26$$

allowing in 4.25 and 4.26 for the variation of the Slater exponent in any calculation.

Exercise 4.4. Calculation of the 2s orbital energy in hydrogen using the Slater orbital of equation 4.24.

1. Make a copy of one of the previous spreadsheets and rename the copy as fig4-5.xls.
(i) Modify the input to provide for the Slater function in the calculations, with

$$\$E\$4 = \zeta = 0.5000$$

$$\$E\$5 = \textbf{POWER}(2^*\$E\$4,2.5)^*\textbf{SQRT}(1/\textbf{FACT}(4))$$

$$\$D\$12 = \$E\$5^*\textbf{EXP}(-E\$4^*\$A12)^*\$A12$$

2. Provide for the action of the grad^2 operator and avoid the singularity at $r = 0.0$, with

$$\$E\$12 = \$E\$5^*\textbf{EXP}(-\$E\$4^*\$A12)^* \\ (2^*A12+\$A12^{\wedge}3^*\$E\$4^{\wedge}2-4^*\$E\$4^*A12^{\wedge}2)$$

and fill this formula down the active cells of column F. Note that this formula reduces to the equivalent one in fig4-4.xls for the 2p orbital when the Slater exponent, $\$E\$4$ is entered as the basic rule value 0.5. Again, note that the 'r' factor in the radial function is not present in the transformed function after the action of the grad^2 operator.

3. Add the kinetic and potential energy terms to return the electronic energy in cell $\$E\$10$ of Figure 4.5.

	A	B	C	D	E	F
1	Simpson's rule integration					
2	H2s eigenvalue with Slater 2s					
3						
4				ζ	0.5000	
5				$d(\zeta)$	0.2041	
6				$\langle S{:}2s|S{:}2s\rangle$	1.00000	
7				T	0.04167	
8				V	-0.25000	
9						
10				$E_{2s} =\rangle$	-0.20833	
11	r			$|S{:}2s\rangle$	$r^2\text{grad}^2|S{:}2s\rangle$	
12	0.00	1	1	0.00000	0.00000	
13	0.01	2	4	0.00203	0.00402	
14	0.02	3	2	0.00404	0.00792	
15	0.03	4	4	0.00603	0.01170	
16	0.04	5	2	0.00800	0.01537	

Figure 4.5   Calculation of the 2s orbital energy in hydrogen using the Slater function. Note the poor result, which arises because the approximate calculation is not based on a function orthonormal with respect to the 1s function.

As you see, the calculated 2s orbital energy is very unsatisfactory. But, note, that the kinetic and kinetic energy terms are the same as are returned in the calculation for the 2p orbital, when this is based on the exact function, which has the same form as the Slater approximation.

In the early literature (17,45), you will find this lack of agreement between the approximate and exact results explained on the basis that Slater orbitals are eigenfunctions of a modified Schrödinger equation in which there is an extra 'centrifugal' term of the form

$$V'(r) = \frac{n(n-1)}{2r^2}$$

4.27

with n the principal quantum number of the state to be calculated using the approximate function. It is straightforward to write this extra term into the Slater-H2s.xls spreadsheet and the result is shown in Figure 4.6, which is based on fig4-4.xls, with the label for the extra term changed to the form above. As you can see, the modified Schrödinger equation returns the E_{2s} value very close to the exact result given by equation 4.15.

For our purposes, this rescue of the Slater approach is not satisfactory. It would mean having to concoct a different Schrödinger equation, for each application of the approximate functions, in order to get agreement with exact data or experimental results. However, from the investigation, in Chapter 3, of the orthonormality deficit in Slater functions, we know why the Slater orbitals do not return the correct orbital energies in calculations.

It is appropriate therefore to attempt calculations based on the use of mutually orthonormal linear combinations of the simple Slater functions. In particular, let us make a 2s Slater function orthonormal to the 1s function and calculate the 2s orbital energy again. The calculation on a spreadsheet involves two stages, the construction of the 2s function orthogonal to the 1s function and then the calculation of the energy terms using our numerical procedure.

	A	B	C	D	E	F
1	Simpson's rule integration					
2	H2s eigenvalue with Slater 2s					
3						
4				ζ	0.5000	Z
5				$d(\zeta)$	0.2041	1.00
6				$\langle S{:}2s\|S{:}2s\rangle$	1.00000	n
7				t11 =$\rangle$	0.04167	2.00
8				v11 =$\rangle$	-0.25000	/
9				n(n-1)/2*r^2	0.08333	1.00
10				E_{2s} =$\rangle$	-0.12500	
11	r			$\|S{:}2s\rangle$	$r^2grad^2\|S{:}2s\rangle$	
12	0.00	1	1	0.00000	0.00000	
13	0.01	2	4	0.00203	0.00402	
14	0.02	3	2	0.00404	0.00792	
15	0.03	4	4	0.00603	0.01170	
16	0.04	5	2	0.00800	0.01537	

Figure 4.6   The recovery of the correct 2s orbital energy, by including the extra term, 4.27, in \$E\$9.

Exercise 4.5.   Calculation of the 2s hydrogen orbital energy using Slater functions and the Schmidt orthonormality requirement.

1. Lay out the integration procedure in columns A, B and C of a new spreadsheet as in Figure 4.7.

	A	B	C	D	E	F	G	H	I	J
1	H2s eigenvalue with |Slater 2s⟩ rendered orthogonal to |1s⟩.									
2	Basic calculation using Slater exponents.								V.C.	1.269775946334640
3				|S:1s⟩	|S:2s⟩	⟨S:1s|S:1s⟩	1.00000	t11 =⟩	0.50000	T
4		ζ		1.0000	0.5000			t22 =⟩	0.04167	0.207238805501981
5		d(ζ)		2.0000	0.2041	⟨S:2s|S:2s⟩	1.00000	t12 =⟩	0.00000	V
6		Z		1.0000						-0.326417910341112
7		n		1.0000	2.0000	⟨S:1s|S:2s⟩	0.48385	v11 =⟩	-1.00000	E₂ₛ
8		ℓ		0.0000	0.0000			v22 =⟩	-0.25000	-0.119179104839131
9						⟨2s -λ1s⟩ ^2	0.76589	v12 =⟩	-0.24192	
10										
11	r			|S:1s⟩	|S:2s⟩	|2s-λ1s⟩	N|2s-1s⟩	r²grad²|S:1s⟩	r²grad²|S:2s⟩	r²grd²N|2s-1s⟩
12	0.000	1	1	2.00000	0.00000	0.96770	1.26350	0.00000	0.00000	0.0000
13	0.010	2	4	1.98010	0.00203	0.95604	1.24827	-0.03940	0.00402	-0.0264
14	0.020	3	2	1.96040	0.00404	0.94450	1.23320	-0.07763	0.00792	-0.0520
15	0.030	4	4	1.94089	0.00603	0.93307	1.21828	-0.11471	0.01170	-0.0768
16	0.040	5	2	1.92158	0.00800	0.92175	1.20351	0.15065	0.01537	-0.1009
17	0.050	6	4	1.90246						-0.1242
18	0.060	7	2	1.88353						-0.1468
19	0.070	8	4	1.86479						-0.1686
20	0.080	9	2	1.84623						-0.1898
21	0.090	10	4	1.82786						-0.2103
22	0.100	11	2	1.80967						-0.2301
23	0.110	12	4	1.79167						-0.2492
24	0.120	13	2	1.77384						-0.2677
25	0.130	14	4	1.75619						-0.2856
26	0.140	15	2	1.73872						-0.3028
27	0.150	16	4	1.72142						-0.3195
28	0.160	17	2	1.70429						-0.3355
29	0.170	18	4	1.68733						-0.3509
30	0.180	19	2	1.67054						-0.3658
31	0.190	20	4	1.65392	0.03527	0.76498	0.99881	-0.56878	0.05745	-0.3801
32	0.200	21	2	1.63746	0.03694	0.75535	0.98623	-0.58949	0.05947	-0.3939

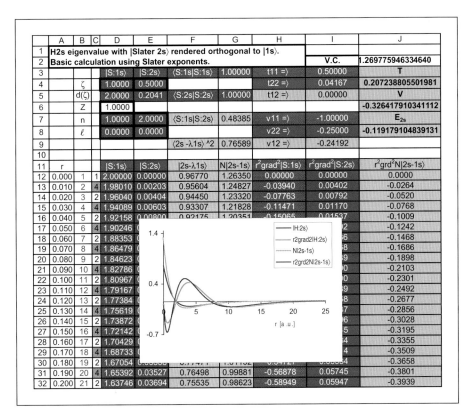

Figure 4.7   Calculation of the hydrogen E_{2s} energy using the Slater 2s orbital rendered orthogonal to the hydrogenic 1s orbital and comparisons of the matches between the functions and their transforms under the action of the kinetic energy operator, ∇^2. Poor agreement is returned in the calculation of the orbital energy, reflecting the significant mismatches in the functions and the results after the action of the Laplacian [actually as $r^2 \nabla^2 \phi(r)$].

2. Enter the input data for the 1s and 2s Slater functions in cells $D$4 to $E$8. For the 1s function the input follows from Table 1.1. The Slater exponent in $D$5 follows from Table 1.1 and the normalization constant in $E$5 follows from equation 1.12, with

$$\$E\$5 = \textbf{POWER}(2*\$E\$4,2.5)*\textbf{SQRT}(1/\textbf{FACT}(4))$$

3. Project the functions in columns D and E, with for example,

$$\$D\$12 = \$D\$5^*\textbf{EXP}(-\$D\$4^*\$A12)$$

$$\$E\$12 = \$E\$5^*\textbf{EXP}(-E\$4^*\$A12)^*\$A12$$

4. As in Exercise 3.2 construct the Schmidt orthogonalized 2s function in column F and the normalized projection on the radial array in column G), based on the various **SUMPRODUCT** results in $\$G\$3$, $\$G\$5$, $\$G\$7$ and $\$G\$9$ for the overlap and normalization integrals required with

$$\langle S{:}1s|S{:}1s\rangle = \$G\$3 = \textbf{SUMPRODUCT}(\$A\$12{:}\$A\$3012,\$A\$12{:}\$A\$3012,$$
$$\$C\$12{:}\$C\$3012,\$D\$12{:}\$D\$3012,$$
$$\$D\$12{:}\$D\$3012)^*(\$A\$13{-}\$A\$12)/3$$

$$\langle S{:}2s|S{:}2s\rangle = \$G\$5 = \textbf{SUMPRODUCT}(\$A\$12{:}\$A\$3012,\$A\$12{:}\$A\$3012,$$
$$\$C\$12{:}\$C\$3012,\$E\$12{:}\$E\$3012,\$E12{:}\$E3012)^*$$
$$(\$A\$13{-}\$A\$12)/3$$

$$\langle S{:}1s|S{:}2s\rangle = \$G\$7 = \textbf{SUMPRODUCT}(\$A\$12{:}\$A\$3012,\$A\$12{:}\$A\$3012,$$
$$\$C\$12{:}\$C\$3012,\$D\$12{:}\$D\$3012,\$E\$12{:}\$E3012)^*$$
$$(\$A\$13{-}\$A\$12)/3$$

$$\langle 2s{-}\lambda 1s\rangle\char`^2 = \$G\$9 = \textbf{SUMPRODUCT}(\$A\$12{:}\$A\$3012,\$A\$12{:}\$A\$3012,$$
$$\$C\$12{:}\$C\$3012,\$F\$12{:}\$F\$3012,\$F\$12{:}\$F\$3012)^*$$
$$(\$A\$13{-}\$A\$12)/(3)$$

5. Generate the projections of $r^2\nabla^2\phi(r)$ on the radial mesh for the $|S{:}1s\rangle$, $|S{:}2s\rangle$ and $N|2s - 1s\rangle$ functions in the active cells for column H to J with, for example,

$$r^2\nabla^2|S{:}1s\rangle = \$H\$12 = \$D\$5^*\textbf{EXP}(-\$D\$4^*\$A12)^*(\textbf{POWER}$$
$$(\$A12^*\$D\$4,2){-}2^*\$A12^*\$D\$4)$$

$$r^2\nabla^2|S{:}2s\rangle = \$I\$12 = \$E\$5^*\textbf{EXP}(-\$E\$4^*\$A12)^*(2^*\$A12$$
$$+\$A12\char`^3^*\$E\$4\char`^2{-}4^*\$E\$4^*A12\char`^2)$$

$$r^2\nabla^2N|2s{-}\lambda 1s\rangle = \$J\$12 = {-}\textbf{SQRT}(1/\$g\rangle\$9)^*(\$I12{-}\$g\rangle\$7^*\$H12)$$

based on the earlier procedures for the calculations using only the individual functions.

6. Calculate the components of the kinetic energy and potential energy integrals in cells $\$I\$3$ to $\$I\$9$, with

$$-0.5^*\langle 1s|\nabla^2|1s\rangle = \$I\$3 = {-}0.5^*\textbf{SUMPRODUCT}(\$C\$12{:}\$C\$3012,$$
$$\$D\$12{:}\$D\$3012,\$H\$12{:}\$H\$3012)^*$$
$$(\$A\$13{-}\$A\$12)/(3)$$

$$-0.5^*\langle 2s|\nabla^2|2s\rangle = \$I\$4 = {-}0.5^*\textbf{SUMPRODUCT}(\$C\$12{:}\$C\$3012,$$
$$\$E\$12{:}\$E\$3012,\$I\$12{:}\$I\$3012)^*$$
$$(\$A\$13{-}\$A\$12)/(3)$$

$-0.5^*\langle 1s|\nabla^2|2s\rangle = \$I\$5 = -0.5^*\textbf{SUMPRODUCT}(\$C\$12:\$C\$3012,$
$\$E\$12:\$E\$3012,\$H\$12:\$H\$3012)^*$
$(\$A\$13-\$A\$12)/(3)$

$-\langle 1s|1/r|1s\rangle = \$I\$7 = -\textbf{SUMPRODUCT}(\$C\$12:\$C\$3012,$
$\$D\$12:\$D\$3012,\$D\$12:\$D\$3012,\$A\$12:\$A\$3012)^*$
$(\$A\$13-\$A\$12)/(3)$

$-\langle 2s|1/r|2s\rangle = \$I\$8 = -\textbf{SUMPRODUCT}(\$C\$12:\$C\$3012,$
$\$E\$12:\$E\$3012,\$E\$12:\$E\$3012,\$A\$12:\$A\$3012)^*$
$(\$A\$13-\$A\$12)/(3)$

$-\langle 1s|1/r|2s\rangle = \$I\$9 = -\textbf{SUMPRODUCT}(\$C\$12:\$C\$3012,$
$\$D\$12:\$D\$3012,\$E\$12:\$E\$3012,\$A\$12:\$A\$3012)^*$
$(\$A\$13-\$A\$12)/(3)$

7. Form the total contributions and the calculated 2s orbital energy in cells $\$J\$4$, $\$J\$6$ and $\$J\$8$, with

$$-0.5^*\langle 2s\text{-}\lambda 1s|\nabla^2|2s\text{-}\lambda 1s\rangle = T = \$J\$4 = (\$I\$4+\$g\rangle\$7\text{\textasciicircum}2^*\$I\$3$$
$$-2^*\$g\rangle\$7^*\$I\$5)/\$g\rangle\$9$$

$$-\langle 2s\text{-}\lambda 1s|1/r|2s\text{-}\lambda 1s\rangle = V = \$J\$6 = (\$I\$8+\$g\rangle\$7\text{\textasciicircum}2^*\$I\$7$$
$$-2^*\$g\rangle\$7^*\$I\$9)/\$g\rangle\$9$$

$$E_{2s} = \$J\$8 = \$J\$4 + \$J\$6$$

8. Following the procedure in fig4-2.xls project the analytical H-2s function and its concavity on the active cells of columns K and L.
9. Finally, use the CHART wizard to construct the comparisons graphs in Figure 4.7.

As you see in Figure 4.7, the results obtained using the orthogonalized function are not very satisfactory. The calculated energy terms are very different to the correct ones and the comparisons between the graphs of the orthogonal function and its concavity [*times* r^2] with the exact behaviour of the hydrogen 2s radial orbital, while exhibiting similar behaviours differ considerably, especially in the comparison of the concavities. We can improve the calculation in several ways. We can investigate whether varying the Slater exponent for the 2s orbital improves the calculated results. Alternatively, we can attempt to minimize the disagreement between the graphs of the orthogonal function and its concavity with the graphs obtained using the analytical hydrogen 2s radial orbital. Both of these procedures can be attempted manually by using the inbuilt variation macros in EXCEL or, indeed, by using the least-squares integral approach to the design of Gaussian basis sets of Hehre, Stewart and Pople, equations 2.9 and 2.10, discussed in Chapter 2.

Figure 4.8 displays the 'best' manual result obtained by simply varying the Slater exponent, cell $\$E\$4$ on the spreadsheet. Note, that the agreement between the Slater-based function and the analytical hydrogen 2s radial function now is good and an accurate value is returned in cell $\$J\$8$ for the potential energy. The improved, but still relatively unsatisfactory, match of the $r^2\text{grad}^2$ transformations appears to be the main reason for the error in the calculation of the 2s energy. The main problem with attempting an improvement by changing the input is that we are limited by having only one variable parameter. This is not sufficiently flexible to facilitate the matching of the variations in the functions over different radial ranges.

Figure 4.9 displays the results found for the application of the canonical orthonormalization procedure of Section 3.6.

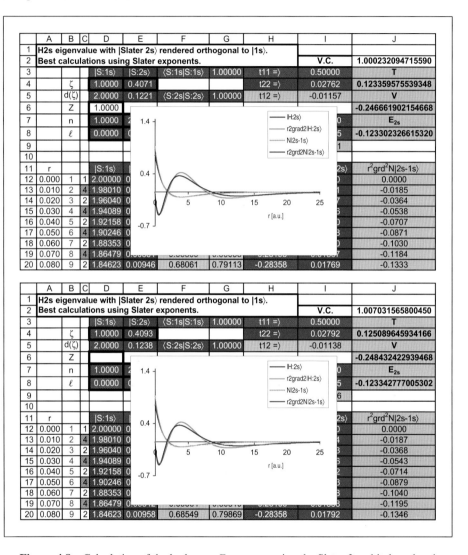

Table 1 (top)

	A	B	C	D	E	F	G	H	I	J
1	H2s eigenvalue with |Slater 2s⟩ rendered orthogonal to |1s⟩.									
2	Best calculations using Slater exponents.								V.C.	1.000232094715590
3				|S:1s⟩	|S:2s⟩	⟨S:1s|S:1s⟩	1.00000	t11 =⟩	0.50000	T
4		ζ		1.0000	0.4071			t22 =⟩	0.02762	0.123359575539348
5		d(ζ)		2.0000	0.1221	⟨S:2s|S:2s⟩	1.00000	t12 =⟩	-0.01157	V
6		Z		1.0000						-0.246661902154668
7		n		1.0000		1.4		IH:2s)	0	E₂ₛ
8		ℓ		0.0000				r2grad2IH:2s)	5	-0.123302326615320
9								NI2s-1s)	1	
10								r2grd2NI2s-1s)		
11	r			|S:1s⟩		0.4			2s)	r²grd²N|2s-1s⟩
12	0.000	1	1	2.00000	0				0	0.0000
13	0.010	2	4	1.98010	0	5	10	15	20	-0.0185
14	0.020	3	2	1.96040	0			25	1	-0.0364
15	0.030	4	4	1.94089	0				7	-0.0538
16	0.040	5	2	1.92158	0	-0.7		r [a.u.]	6	-0.0707
17	0.050	6	4	1.90246	0					-0.0871
18	0.060	7	2	1.88353	0				7	-0.1030
19	0.070	8	4	1.86479	0					-0.1184
20	0.080	9	2	1.84623	0.00946	0.68061	0.79113	-0.28358	0.01769	-0.1333

Table 2 (bottom)

	A	B	C	D	E	F	G	H	I	J
1	H2s eigenvalue with |Slater 2s⟩ rendered orthogonal to |1s⟩.									
2	Best calculations using Slater exponents.								V.C.	1.007031565800450
3				|S:1s⟩	|S:2s⟩	⟨S:1s|S:1s⟩	1.00000	t11 =⟩	0.50000	T
4		ζ		1.0000	0.4093			t22 =⟩	0.02792	0.125089645934166
5		d(ζ)		2.0000	0.1238	⟨S:2s|S:2s⟩	1.00000	t12 =⟩	-0.01138	V
6		Z								-0.248432422939468
7		n		1.0000		1.4		IH:2s)	0	E₂ₛ
8		ℓ		0.0000				r2grad2IH:2s)	5	-0.123342777005302
9								NI2s-1s)	6	
10								r2grd2NI2s-1s)		
11	r			|S:1s⟩		0.4			2s)	r²grd²N|2s-1s⟩
12	0.000	1	1	2.00000	0				0	0.0000
13	0.010	2	4	1.98010	0	5	10	15	4	-0.0187
14	0.020	3	2	1.96040	0			20	3	-0.0368
15	0.030	4	4	1.94089	0			25		-0.0543
16	0.040	5	2	1.92158	0	-0.7		r [a.u.]	2	-0.0714
17	0.050	6	4	1.90246	0				3	-0.0879
18	0.060	7	2	1.88353	0				3	-0.1040
19	0.070	8	4	1.86479	0					-0.1195
20	0.080	9	2	1.84623	0.00958	0.68549	0.79869	-0.28358	0.01792	-0.1346

Figure 4.8   Calculation of the hydrogen E_{2s} energy using the Slater 2s orbital rendered orthogonal to the hydrogenic 1s orbital with the 'best' choice for the Slater exponent subject to two conditions. In the first diagram, the SOLVER based solution is determined by the requirement that the *Virial theorem coefficient* be 1.00. In the second diagram, the best values for the kinetic and potential energy terms have been determined for the choice of the Slater 2s exponent. Note, the substantial cancellation of mismatches of the $r^2\text{grad}^2$ transformed Slater function compared to the variation of the transformed exact function.

	A	B	C	D	E	F	G	H
1	Canonical orthonormalization to calculate the 2s orbital energy in hydrogen							
2	using Slater functions.							
3	S			theta =	0.78540		Slater ζ 1s	1.0000
4	1.00000	0.37623		omega1	1.37623		Slater ζ 2s	0.4091
5	0.37623	1.00000		omega2	0.62377			
6								
7	U			U - adjoint			S-diagonal	
8	0.70711	0.70711		0.70711	0.70711		1.37623	0.00000
9	0.70711	-0.70711		0.70711	-0.70711		0.00000	0.62377
10							The S^1/2 matrix	
11							0.85242	0.00000
12							0.00000	1.26616
13	Fock matrix			X-canonical			X-canon transpose	
14	-0.50000	-0.18812		0.60275	0.89531		0.60275	0.60275
15	-0.18812	-0.17666		0.60275	-0.89531		0.89531	-0.89531
16								
17	theta	0.59253		Transformed Fock matrix				
18	omega1	-0.50000		-0.41476	-0.27923	==>	-0.38253	-0.17449
19	omega2	-0.12334		-0.21987	-0.01026		-0.17449	-0.24081
20								
21	Coefficients			Fock - diagonal			C - originals	
22	0.82953	0.55847		-0.50000	0.00000		1.00000	-0.40607
23	0.55847	-0.82953		0.00000	-0.12334		0.00000	1.07930

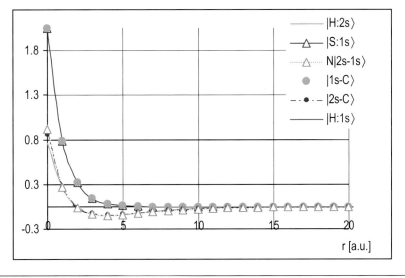

Figure 4.9 Application of the canonical orthonormalization procedure of Section 3.6 to the calculation of the 1s and 2s eigenfunctions and eigenvalues approximations for the 1s and 2s orbitals in hydrogen over Slater functions. Note the exact fit of the 1s Slater, which is an eigenfunction of the Fock matrix for the hydrogen atom and the relatively close agreement of the 1s/2s linear combinations based on simple canonical orthogonalization and also direct orthonormalization using the $S^{-1/2}$ matrix procedure of Section 3.7.

Exercise 4.6.   Application of the canonical orthonormalization transformation to the calculation of the 2s orbital energy in hydrogen using Slater functions.

1. Make a copy of fig4-8.xls, rename it fig4-9.xls.
2. Open fig3-16.xls and on a new worksheet in fig4-9.xls, enter the transformations required to effect the canonical orthonormalization as displayed in Figure 4.9, based on the appropriate entries in the cells of fig3-16.xls. Make sure to link the overlap input data to the normalization and overlap integral data in the basis! worksheet by activating the formula bar, with the entry '=' in each of cells $A$4 to $B$5, and then simply selecting 'basis!'g3, 'basis!'g5 and 'basis!'$G$7. It is good practice after each of these operations to change the relative referencing result to an absolute one using the '$' prefix.
3. Cells $A$14 to $B$15 are the linked sums of the kinetic and potential energy terms over the Slater functions defined in columns 'basis!'D and E. For later consistency, these sums of the energy terms over the bases, in the evaluation of the Schrödinger equation, for the hydrogen atom orbitals are identified as the Fock matrix elements on this worksheet.
4. Now provide for the diagonalization of the Fock matrix in the new orthogonal basis formed using the canonical coefficient over the original Slater functions in cells $D$14 to $E$15 of the 'canonical!' worksheet.
5. Retransform the coefficients for the orthonormal linear combinations of the canonically orthogonalized Slater linear combinations, which diagonalize the Fock matrix, into linear coefficients over the original Slater functions, with

$$= \mathbf{MMULT}(D14{:}E15, A22{:}B23)$$

as the array multiplication formula for the cells 'canonical!'g22 to $H$23.
6. Provide for the chart in Figure 4.9 on the restricted radial array in 'basis!' column P and include the graphs of the 'eigenfunctions' of the Fock matrix in the display.

There are two interesting features apparent in Figure 4.9.

The Slater 1s function is the actual hydrogen radial orbital. Thus the coefficients over the original Slater basis, to form the 'eigenfunctions' of the Fock matrix, simply return the 1s Slater functions as one eigenfunction [the coefficient values 1.0 and 0.0 in 'canonical!' g22 and g23], while the second eigenfunctions is formed as a linear combination of the 1s and 2s Slater function [the coefficients −0.40756 and 1.08327 in 'canonical!' $H$22 and $H$23].

Secondly, the canonical orthonormalization procedure to diagonalize the overlap matrix and then the application of the Jacobi transformation to diagonalize the Fock matrix in the 'eigenfunctions' of the overlap matrix, returns two eigenvalues, the values −0.50000 and −0.12352 Hartrees, in 'canonical!' $B$18 and $B$19. This is the important elementary point that we can make two linear combinations of two functions and so there are two possible eigenvalues to be calculated. These eigenvalues, of course, are present in the calculation set out in the other worksheet, based on the Schmidt procedure. The 1s eigenvalue can be extracted, by adding only the kinetic and potential energy contributions in 'basis!' $I$3 and $I$7.

Finally, note the extra detail in the chart of Figure 4.9. The linear combinations to form the approximation to the 2s orbital in hydrogen are presented for the choice of canonical orthogonalization based only on the Jacobi diagonalized overlap matrix and for the choice leading to direct orthonormalization, involving the $S^{-1/2}$ matrix discussed in Section 3.7

As you can see, in both Figures 3.16 and 4.9, both procedures, in this case, lead to the same results. We shall see that this is not always the case. The agreement here occurs because the 1s Slater function is the exact 1s hydrogen orbital, so the coefficients of the eigenfunctions approximations formed, cell $G$22 and $G$23, are 1.0 and 0.0 for the eigenvalues. In general, however, the Jacobi rotation and the final canonical rotation to diagonalize the Fock matrix need not leave the first function unchanged.

Exercise 4.7.   Use the least-squares matching technique to establish the 'best' result for this calculation. Attempt to match both the functions and their transformation under the action of the kinetic energy operator, over the radius range of the calculation. The results are to be found in fig4.9a.xls.

Because of the possibility of cancellation of differences in different ranges of the radius no improvement in the calculation is found. Only marginal improvements in the matches of the functions and their transformations after the application of the kinetic energy operator are found in the least-squares minimization. There is not sufficient flexibility in the approximate Slater function, even when orthogonalized to the 1s function, to improve on the 'best' value for the 2s orbital energy, returned by variation of the basic Slater exponent.

4.3   CALCULATIONS WITH GAUSSIAN FUNCTIONS

There is an important difference with Gaussian basis sets. While, Slater designed his functions to match the numerical radial data for the valence regions of electronic structure in different atoms, many optimized Gaussian basis sets were designed originally on the basis of best fits to the appropriate Slater functions. That good estimates of atomic eigenvalues should follow was not a necessary condition of this optimization. The dominant objective in this early Gaussian basis set theory was the construction of the smallest representations of the Slater functions, which would render the integral evaluations required as simple as possible and minimizes the storage requirements for all the integral components. Remember the rule that the number of integrals to be calculated increases as $n^4/8$, with 'n' the number of components in the basis set. Modern Gaussian basis sets invariably, however, are based on energy optimization criteria and no attempt is made to match to Slater orbitals, since, in particular, these are very poor approximations to atomic d-orbitals.

We have seen that the TABLE macro in EXCEL provides a very efficient format for calculations involving Gaussian basis sets, since the design for the integrations required to solve the hydrogen atom problem is based only on two projections, the quadratic exponential and its transform under the action of the Laplacian over the radial array. For

	A	B	C	D	E	F	G	H	I
1	sto-ng H-1s energy.			alpha(i)	alpha(j)			$\langle i\|j\rangle$	1.000000
2				1.00000	1.00000			V	-0.999646
3		|sto-ng⟩	Slater ζ	1.00000				T	0.499819
4		|1⟩	|2⟩	|3⟩	|4⟩	|5⟩	|6⟩	E_{1s}	-0.499827
5	α	0.06511	0.15809	0.40710	1.18506	4.23592	23.10300		
6	d	0.13033	0.41649	0.37056	0.16854	0.04936	0.00916	v.c.	0.999992
7	αζ²	0.06511	0.15809	0.40710	1.18506	4.23592	23.10300		
8	d'	0.01197	0.07442	0.13460	0.13643	0.10387	0.06882		
9		[1|1]	[2|2]	[3|3]	[1|2]	[1|3]	[2|3]	[4|4]	[5|5]
10	d'*d	0.00014	0.00554	0.01812	0.00089	0.00161	0.01002	0.01861	0.01079
11	α'+α'	0.13022	0.31618	0.81420	0.22320	0.47221	0.56519	2.37012	8.47184
12	5.5683	118.499	31.321	7.57931	52.80692	17.16027	13.10499	1.52605	0.22582
13	sij	0.01699	0.17347	0.13732	0.04705	0.02765	0.13127	0.02841	0.00244
14		[1|1]	[2|2]	[3|3]	[1|2]	[1|3]	[2|3]	[4|4]	[5|5]
15		0.13022	0.31618	0.81420	0.22320	0.47221	0.56519	2.37012	8.47184
16	6.2832	48.2509	19.8724	7.7170	28.1508	13.3060	11.1170	2.65100	0.74166
17	vij	-0.0069	-0.1101	-0.13981	-0.02508	-0.02144	-0.11136	-0.04934	-0.00800
18		[1|1]	[2|2]	[3|3]	[1|2]	[1|3]	[2|3]	[4|4]	[5|5]
19		-23.146	-14.854	-9.25659	-14.61154	-5.77944	-8.95354	-5.42539	-2.86964
20	tij	0.00166	0.04113	0.08385	0.00651	0.004657	0.044844	0.05049	0.015482
21	0.0000	0.13022	0.31618	0.81420	0.22320	0.47221	0.56519	2.37012	8.47184
22	0.0651	-23.146							
23	0.1581		-14.854						

	A	B	C	D	E
44	r		#	[i|j]	r²grad²
45	0.0000	1	1	1.00000	-6.0000
46	0.0100	2	4	0.99990	-5.9996
47	0.0200	3	2	0.99960	-5.9984
48	0.0300	4	4	0.99910	-5.9964
49	0.0400	5	2	0.99840	-5.9936
50	0.0500	6	4	0.99750	-5.9900
51	0.0600	7	2	0.99641	-5.9856
52	0.0700	8	4	0.99511	-5.9804
53	0.0800	9	2	0.99362	-5.9744
54	0.0900	10	4	0.99193	-5.9676

Figure 4.10   Implementation of the TABLE macro facility in EXCEL to calculate the energy components of an \|sto-ng> approximation to the hydrogen 1s orbital.

the normalization and potential energy integrals a one-variable table is sufficient. For the kinetic energy integral the first and second-derivatives of the 1s primitive Gaussians, used to approximate 1s and 2s orbitals, go as

$$\frac{d}{dr}e^{(-\alpha r^2)} = -2\alpha r e^{(-\alpha r^2)} \qquad\qquad 4.28$$

and

$$\frac{d^2}{dr^2}e^{(-\alpha r^2)} = (4\alpha^2 r^2 - 2\alpha)e^{(-\alpha r^2)} \qquad\qquad 4.29$$

For the slightly more complicated differentiations on the p Gaussians, the chain rule results, for the derivatives, are

$$\frac{d}{dr}re^{(-\alpha r^2)} = e^{(-\alpha r^2)} - 2\alpha r^2 e^{(-\alpha r^2)} \qquad 4.30$$

and

$$\frac{d^2}{dr^2}re^{(-\alpha r^2)} = [4\alpha^2 r^3 - 6\alpha r]e^{(-\alpha r^2)} \qquad 4.31$$

Thus, in all cases

$$t_{ij} = t_{ji} = -\frac{1}{2}\langle j|\frac{d^2}{dr^2} + \frac{2}{r}\frac{d}{dr}|i\rangle \propto -\frac{1}{2}\int_0^\infty e^{(-(\alpha_j+\alpha_i)r^2)} f(\alpha_i, r^m)dr \qquad 4.32$$

with $f(\alpha_j, r^m)$ the pre-exponential factor of the result of applying the Laplacian, as in equation 4.32, to the quadratic exponential of each primitive Gaussian. These integral components require a two-variable table design as is shown in Figure 4.10 for the $|$sto-6g$\rangle$ calculation of the 1s orbital energy in hydrogen[2].

Exercise 4.8.   Calculation of the hydrogen 1s orbital energy using an $|$sto-6g:1s$\rangle$ basis set.

1. Enter the Pople, Hehre and Stewart parameters from Table 1.6 for the sto-6 g basis in rows 5 and 6 of fig4.10.xls.
2. Provide for scaling with the Slater exponent in cell $D$3 of these parameters in the active cells of rows 7 and 8. Thus,

$$\$B\$7 = B\$5^*\textbf{POWER}(\$D\$3,2)$$

$$\$B\$8 = B\$6^*\textbf{POWER}(2^*B7/\textbf{PI}(\,), 0.75)$$

3. All the integrals involve the products of the primitive Gaussians. So we prepare for the products of coefficients and sums of exponents in rows 10 and 11, with

$$\$B\$10 = B\$8^*B\$8$$

$$\$B\$11 = (\$B\$7+B\$7)$$

   and similarly across the active cells of all these rows.
4. The rest of the spreadsheet design involves the implementation of the three TABLE macros for the calculation of the components of the normalization, potential and kinetic energy integrals. We need master variable parameters to drive the TABLE macro and these are entered in cells $D$4 and $E$4 and feed the standard integration procedure laid out from cell $A$45 to $C$3045, with a suitable choice for the radial mesh and its extent in column A.
5. The normalization integral, which is used also in the potential energy calculation, over the Gaussian primitives requires the projections of quadratic

---

[2]In the design of the *two-variable* table in this application, the self-adjoint property of the Hamiltonian operator and its components of potential and kinetic energy are assumed. But it is useful to construct the table without using the identity $\langle i|H|j\rangle = \langle i|H^\dagger|j\rangle = \langle i|H|j\rangle^*$, i.e. calculate all $\langle i|H|j\rangle$ and $\langle j|H|i\rangle$ and demonstrate this seemingly strange relationship (47).

exponentials on the radial array using each possible sum of pairs of the exponents of the primitives. So use the dummy parameter $H$4 to generate these projections in column D, with the typical entry,

$$\$D45 = \textbf{EXP}(-\$D\$4*\textbf{POWER}(\$A45,2))$$

6. Form the entries for the normalization TABLE calculation as the master formula

$$\$A\$12 = \textbf{SUMPRODUCT}(\$A\$45:\$A\$3045,\$A\$45:\$A\$3045,$$
$$\$C\$45:\$C\$3045,\$D\$45:\$D\$3045)*(\$A\$46-\$A\$45)*4*\textbf{PI}(\,)/3$$

Use the wizard to display the individual results in cells $B$12 to $V$12 and multiply by the pre-exponential products in row 10 to return the components of the integral in row 13.

7. Similarly, form the potential energy integral in row 17 based on the **SUMPRODUCT** including only one multiplication of the radial array, with the master formula

$$\$A\$16 = \textbf{SUMPRODUCT}(\$A\$45:\$A\$3045,\$C\$45:\$C\$3045,$$
$$\$D\$45:\$D\$3045)*(\$A\$46-\$A\$45)*4*\textbf{PI}(\,)/3$$

in cell $A$16 and the variables, the sums of the input exponents, $H$4, in each calculation returned by the TABLE elements in the remaining active cells of the row.

8. The kinetic energy integral requires the evaluation of the effects of the Laplacian operator, from equations 4.28 and 4.29, on the Gaussian primitives and then the compiling of the integral components as the integral of the products of these functions with the starting functions over the radial mesh. So,

(i)   Form the projections of the pre-exponential factor of the transformed functions in terms of the master input of one in $E$4, with, for example,

$$\$E\$45 = 4*\$E\$4^2*\$A45^2-6*\$I\$4$$

(ii) Enter the master formula for the integration in cell $A$21 for the two-dimensional TABLE with variables the individual values for $D$4 and $E$4 of the formulae cells $D$46 to $E$3046, with

$$\$A\$21 = \textbf{SUMPRODUCT}(\$A\$45:\$A\$3045,\$A\$45:\$A\$3045,$$
$$\$C\$45:\$C\$3045,\$D\$45:\$D\$3045,\$E\$45:\$E\$3045)*$$
$$(\$A\$46-\$A\$45)*4*\textbf{PI}(\,)/3$$

(iii) Copy the variables into the adjacent cells of row 21 and column A. Then activate the two-variable TABLE wizard and render invisible all but the diagonal element entries, by setting the off-diagonal cell font colour as 'white'.

(iv) Finally, sum the integral components and determine the electronic energy of the 1s orbital in the basis set in the sequence,

$$V = \$I\$2 = (\textbf{SUM}(\$B\$17:\$D\$17)+\textbf{SUM}(\$H\$17:\$J\$17)$$
$$+2*(\textbf{SUM}(\$E\$17:\$G\$17)+\textbf{SUM}(\$K\$17:\$M\$17)$$
$$+\textbf{SUM}(\$N\$17:\$V\$17)))/\$I\$1$$

$$T = \$I\$3 = (\mathbf{SUM}(\$B\$20:\$D\$20)+\mathbf{SUM}(\$H\$20:\$J\$20)$$
$$+2^*(\mathbf{SUM}(\$E\$20:\$G\$20)+\mathbf{SUM}(\$K\$20:\$M\$20)$$
$$+\mathbf{SUM}(\$N\$20:\$V\$20)))/\$I\$1$$

$$E_{1s} = \$I\$4 = \$I\$2+\$I\$3$$

Little modification is required to apply fig4-10.xls to other calculations in basis set theory. It is straightforward to calculate energies for basis sets of different sizes and to compare the results obtained using linear combinations recommended by different groups. These applications are left as exercises for the interested reader or instructor to use in classes.

Exercise 4.9.   Calculation of the 1s orbital energy in hydrogen using the $|$sto-ng$\rangle$ basis sets, Table 1.6, proposed by Hehre, Pople and Stewart.

1. Make a copy of fig4-10.xls. In TOOLS, choose OPTIONS/CALCU-LATIONS/MANUAL.
2. Enter the appropriate values for the exponents of the primitive Gaussians and the coefficients of the linear combinations, for each basis set, using 0.0 as the entry wherever appropriate.
3. Press F9 to cancel the manual calculation option and activate the calculation.
4. Compare the results with the values given in Table 4.2.

Exercise 4.10.   Calculation of the 1s orbital energy in hydrogen using the $|$sto-ng$\rangle$ basis sets, Table 1.5, proposed by Reeves.

1. Make a copy of fig4-10.xls.
2. Remember that Reeves included the normalization factor in the coefficients of his linear combinations of primitive Gaussians and so enter his data in rows 7 and 8 of the copied spreadsheet.
3. Compare the results with the values given in Table 4.3.

Table 4.2   Values returned by the application of fig4-10.xls to the calculation of the 1s orbital energy terms in hydrogen, using the Hehre, Pople and Stewart basis sets of Table 1.6.

$|\phi\rangle$	T	V	E	v.c.
$|$sto-2g:1s$\rangle$	0.477942	−0.959098	−0.481156	0.996649
$|$sto-3g:1s$\rangle$	0.494298	−0.989205	−0.494907	0.999384
$|$sto-4g:1s$\rangle$	0.498353	−0.996834	−0.498481	0.999872
$|$sto-5g:1s$\rangle$	0.499475	−0.998981	−0.499506	0.999969
$|$sto-6g:1s$\rangle$	0.499819	−0.999646	−0.499827	0.999992

Exercise 4.11.   Calculation of the 1s orbital energy in hydrogen using the |sto-ng⟩ basis sets, Table 1.7, proposed by Huzinaga.

1.   Make a copy of fig4-10.xls.
2.   Enter the Huzinaga data of Table 1.7 in cells $B$5 to $G$6 or zeros as required and control the calculation as 'manual' using the option 'MANUAL' under the TOOLS menu until all the data are in!
3.   Compare the results with the values given in Table 4.4.

Exercise 4.12.   Extend the design of the spreadsheet fig4-10.xls and calculate the 1s orbital energy in hydrogen using the remaining |sto-ng⟩ basis sets in Table 1.7, proposed by Huzinaga.

Check your answers against the remaining data in Table 4.4.

More significantly, perhaps, little modification is needed to investigate the modelling of 2s and 2p orbitals in hydrogen and the utility of split-basis sets.

Table 4.3   Values returned by the application of fig4-10.xls to the calculation of the 1s orbital energy terms in hydrogen, using the Reeves basis sets of Table 1.5.

$|\phi\rangle$	T	V	E	v.c.
$|$sto-1g:1s$\rangle$	0.424410	−0.848823	−0.424413	0.999996
$|$sto-2g:1s$\rangle$	0.485813	−0.971574	−0.485813	0.999947
$|$sto-3g:1s$\rangle$	0.495872	−0.992840	−0.496967	0.998897
$|$sto-4g:1s$\rangle$	0.499448	−0.998724	−0.499276	1.000172

Table 4.4   Values returned by the application of fig4-10.xls to the calculation of the 1s orbital energy terms in hydrogen, using the Huzinaga $|$sto-ng:1s$\rangle$, n = 2 − 10, basis sets of Table 1.7.

$|\phi\rangle$	T	V	E	v.c.
$|$sto-2g:1s$\rangle$	0.485812	−0.971625	−0.485813	1.000004
$|$sto-3g:1s$\rangle$	0.496979	−0.993962	−0.496982	1.000003
$|$sto-4g:1s$\rangle$	0.499437	−0.998715	−0.499277	1.000160
$|$sto-5g:1s$\rangle$	0.499703	−0.999513	−0.499809	0.999894
$|$sto-6g:1s$\rangle$	0.499840	−0.999780	−0.499940	0.999899
$|$sto-7g:1s$\rangle$	0.499903	−0.999879	−0.499976	0.999927
$|$sto-8g:1s$\rangle$	0.499961	−0.999953	−0.499991	0.999970
$|$sto-9g:1s$\rangle$	0.499990	−0.999986	−0.499997	0.999993
$|$sto-10g:1s$\rangle$	0.499974	−0.999973	−0.499999	0.999976

Exercise 4.13.   Calculation of the 2s orbital energy in hydrogen using the |sto-ng⟩ basis sets of Table 1.6.

1.   Make a copy of fig4-10.xls. Rename the copy fig4-11.xls
2.   There are six active cells for input of the basic data, so divide these into two sets of three and colour code as in Figure 4.11.
3.   Modify the scaling by the Slater exponents so that the active cells in row 7 are scaled by cells $D$3 and g3.
4.   Divide the overlap integral cell $I$5 of fig4-10.xls into components to check the normalization of the two |sto-3g⟩ sets and their overlap, with

$$\langle\text{sto-3g:2s}|\text{sto-3g:2s}\rangle = \$I\$3 = \textbf{SUM}(\$B\$13{:}\$D\$13){+}2^*\textbf{SUM}(\$E\$13{:}\$\$G13)$$

$$\langle\text{sto-3g:1s}|\text{sto-3g:1s}\rangle = \$I\$4 = \textbf{SUM}(\$H\$13{:}\$J\$13){+}2^*\textbf{SUM}(\$K\$13{:}\$M\$13)$$

$$\langle\text{sto-2g:2s}|\text{sto-3g:1s}\rangle = \$I\$3 = \textbf{SUM}(\$N\$13{:}\$V\$13)$$

5.   Similarly, divide the components of the potential energy into contributions from the primitive Gaussians from each basis sets, with

$$\langle\text{sto-3g:2s}|1/r|\text{sto-3g:2s}\rangle = \$J\$3 = \textbf{SUM}(\$B\$17{:}\$D\$17)$$
$$+2^*\textbf{SUM}(\$E\$17{:}\$g\rangle\$17)$$

$$\langle\text{sto-3g:1s}|1/r|\text{sto-3g:1s}\rangle = \$J\$4 = \textbf{SUM}(\$H\$17{:}\$J\$17)$$
$$+2^*\textbf{SUM}(\$K\$17{:}\$M\$17)$$

$$\langle\text{sto-3g:2s}|1/r|\text{sto-3g:1s}\rangle = \$J\$5 = \textbf{SUM}(\$N\$17{:}\$V\$17)$$

and the components of the kinetic energy, with

$$\left\langle\text{sto-3g:2s}\left|-\tfrac{1}{2}\nabla^2\right|\text{sto-3g:2s}\right\rangle = \$K\$3 = \textbf{SUM}(\$B\$20{:}\$D\$20)$$
$$+2^*\textbf{SUM}(\$E\$20{:}\$g\rangle\$20)$$

$$\left\langle\text{sto-3g:1s}\left|-\tfrac{1}{2}\nabla^2\right|\text{sto-3g:1s}\right\rangle = \$K\$4 = \textbf{SUM}(\$H\$20{:}\$J\$20)$$
$$+2^*\textbf{SUM}(\$K\$20{:}\$M\$20)$$

$$\left\langle\text{sto-3g:1s}\left|-\tfrac{1}{2}\nabla^2\right|\text{sto-3g:1s}\right\rangle = \$K\$4 = \textbf{SUM}(\$N\$20{:}\$V\$20)$$

6.   Combine these individual contributions from each basis set and their overlap to form the total potential and kinetic energies for the linear combination of the 2s and 2s functions, which has been rendered orthogonal to the 1s functions, remembering to divide by the overall normalization integral, with

$$\$J\$6 = (J\$3{+}(\$I\$5/\$I\$4)\hat{\;}2^*J\$4{-}2^*(\$I\$5/\$I\$4)^*J\$5)/\$F\$44$$

$$\$K\$6 = (K\$3{+}(\$I\$5/\$I\$4)\hat{\;}2^*K\$4{-}2^*(\$I\$5/\$I\$4)^*K\$5)/\$F\$44$$

7. Finally, obtain the 2s orbital energy in hydrogen in this representation as the sum of these two contributions to the energy, with

$$\$K\$7 = \$J\$6 + \$K\$6$$

8. Open fig4-9.xls and move a copy of the worksheet 'canonical!' to the open spreadsheet fig4-11.xls.
9. Modify the links in this worksheet to collect the overlap and Fock matrix data for the direct diagonalization of the Fock matrix over the |sto-ng⟩ basis sets.

Figures 4.11 and 4.12 display important features of the results. The basic calculation, the first diagram in Figure 4.11, for the 2s orbital energy has not returned a value close to the exact value of −0.125 Hartrees. There are considerable disagreements between the variations of the functions and the transforms over the radial mesh, as you see in the chart of these graphs. Perhaps, we should have anticipated this lack of agreement, since the calculation with the 'parent' Slater orbitals [Figures 4.7 and 4.8] had to be 'improved' before a reasonable result for the 2s orbital was obtained. The next step, to improve the calculation using the Gaussian basis sets, is to optimize the calculation and try for as good fits as possible of the orthogonal linear combination of the Gaussians and the Laplacian transform of this linear combination with the exact hydrogen 2s function and its transform under the action of the Laplacian.

So, for example, set the TARGET cell in SOLVER to be $L$5 and require the minimum to be found by variation of the Slater 2s exponent, the cell entry $D$3. Have a cup of coffee and, hopefully, your fig4.11.xls should have changed into the results displayed in the second diagram of fig4-11.xls!

Note that again the cancellation of errors in the mismatches of the transformed functions is the obvious reason for the better agreement obtained. But, the results are quite encouraging, with a kinetic energy of 0.12478 H, a potential energy of −0.24848 H and the 2s orbital energy returned as −0.1237 H.

Figure 4.12 displays the results obtained for application of the canonical orthonormalization procedure to the calculation of the 2s orbital energy in hydrogen. There is an important but subtle new feature evident in cells $D$22 to $E$23. The direct orthonormalization procedure has returned somewhat different values for the 1s and 2s energies to those found for the calculation based on Schmidt orthonormalization. The difference arises because the canonical procedure leads to the diagonalization of the Fock matrix and the 'eigenfunctions' both involve linear combinations of the original |sto-3g⟩ basis sets, as you see in the coefficient matrix, cells $G$22 to $H$23. The Schmidt procedure based calculation does not involve the modification of the 1s basis set. It appears that we achieve a better approximation to the 1s orbital energy because, effectively, a large basis set is used, even though only a small degree of mixing of the two basis sets is involved in the construction of the 1s 'eigenfunction' of the Fock matrix.

This distinction in the two procedures is illustrated, too, in the comparisons of the variations of the linear combinations approximating to the 1s and 2s eigenfunctions set

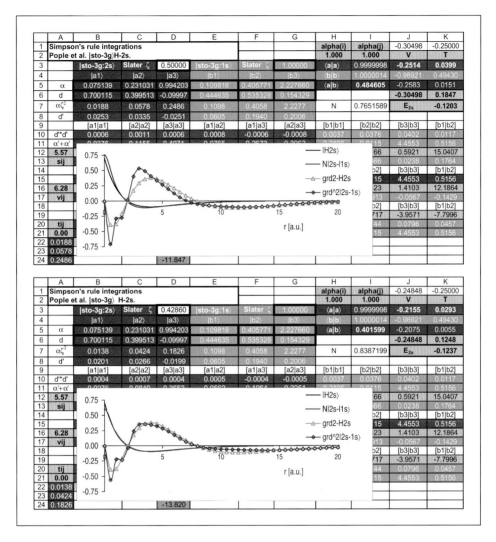

Figure 4.11   Calculation of the 2s orbital energy in hydrogen with the |sto-3g:1s⟩ basis set, Table 1.6, for the 2s Slater function rendered orthogonal to the |sto-3g:1s⟩ function. The initial calculation returns a poor estimate of the energy terms and E_{2s} for the minimization condition on the least-squares integral of Chapter 3. Optimization based on the minimization of the energy, using SOLVER on the Slater exponent, returns closer agreement with the exact results.

out in Figure 4.13 and when the 1s energy is calculated using the |sto-3g:1s⟩ basis set. Table 4.2 records the value -0.494907 Hartree, while the present calculation result is -0.49494 Hartree.

Table 4.5 lists the results obtained for the calculation of the 2s and 2p orbital energies in hydrogen using Reeves' basis sets listed in Table 1.5. The 2s calculations are direct

	A	B	C	D	E	F	G	H
1	Canonical orthonormalization to calculate the energies of the 1s and 2s orbitals in hydrogen							
2	using |sto-3g⟩ basis sets.							
3	S			theta =	0.785398		Slater ζ(1s)	0.91454
4	1.000000	0.464220		omega1	1.464221		Slater ζ(2s)	0.44082
5	0.464220	1.000001		omega2	0.535780			
6								
7	U			U - adjoint			S-diagonal	
8	0.70711	0.70711		0.70711	0.70711		1.46422	0.00000
9	0.70711	-0.70711		0.70711	-0.70711		0.00000	0.53578
10							The S^1/2 matrix	
11							0.82641	0.00000
12							0.00000	1.36618
13	Fock matrix			X-canonical			X-canon transpose	
14	-0.190661	-0.214140		0.58436	0.96603		0.58436	0.58436
15	-0.214140	-0.491246		0.58436	-0.96603		0.96603	-0.96603
16								
17	theta	-0.58673		Transformed Fock matrix				
18	omega1	-0.49192		-0.23655	0.02268	==>	-0.37910	0.16968
19	omega2	-0.12388		-0.41220	0.26769		0.16968	-0.23669
20								
21	Coefficients			Fock - diagonal			C - originals	
22	0.83275	-0.55364		-0.491917	0.000000		-0.04821	-1.12799
23	-0.55364	-0.83275		0.000000	-0.123878		1.02147	0.48094
24								
25								
26	r	|H:1s⟩	|H:2s⟩	|2s:canonical⟩	|1s:canonical⟩	|sto-3g:2s⟩	|sto-3g:1s⟩
27	0.0	0.564190	0.199471	0.405122	0.159832	0.027967	0.39793

Chart (rows 28–40), legend:
— |H:1s⟩
—□— |H:2s⟩
—△— |2s:canonical⟩
——— |1s:canonical⟩
—✳— N|2s-1s⟩
—○— N|1s⟩

Vertical axis: 0.7, 0.5, 0.2, -0.1
Horizontal axis: 0, 5, 10, 15 with label r [a.u.]

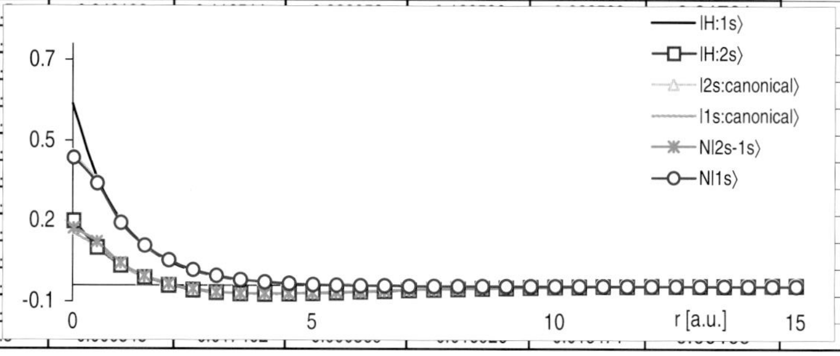

Figure 4.12 Calculation of the 1s and 2s orbital energies in hydrogen using the |sto-3g⟩ basis sets of Table 1.6 and canonical orthonormalization. A 'better' 1s energy compared with the result found for Schmidt orthonormalization, since the |sto-3g⟩ 1s basis is not 'improved' in that calculation. On the scale of the chart, there appears to be no difference in the approximate functions obtained. However, when the scale is enlarged, a small difference is evident and accounts for the different 1s orbital energy calculated. Note, in this calculation both Slater exponents, cells $H$4 and $H$5, have been allowed to vary in the SOLVER routine.

applications of fig4-10.xls, with, as before, the difference that Reeves defined his basis sets to include the normalization factor for the primitive Gaussians.

The 2p results involve the extra calculation of the centrifugal term and a different form, from equation 1.13 and 1.14, for example,

$$g(r{:}\alpha) = (128\alpha^5/\pi^3)^{0.25} r \exp(-\alpha r^2/a_0^3) \cos\theta \qquad 4.33$$

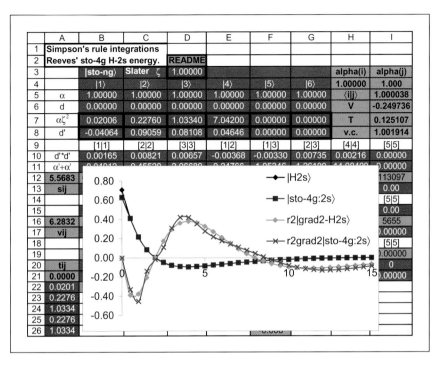

Figure 4.13 The modified design of fig4-10.xls leading to the calculation of the H$_{2s}$ orbital energy using Reeves' |sto-4g⟩ linear combination and the comparisons with the analytical data. Remember the Reeves' data, Table 1.5, are normalized and do not require scaling.

Table 4.5 The terms in the calculations of the 2s and 2p energies for the H-atom, based on Reeves' Gaussian sets listed in Table 1.5.

|φ⟩	T	V	$l(l+1)/r^2$	E	v.c.
|sto-2g:2s⟩	0.116801	−0.233634	—	−0.11683	0.999863
|sto-3g:2s⟩	0.123733	−0.247400	—	−0.123667	1.000263
|sto-4g:2s⟩	0.125107	−0.249736	—	−0.124629	1.001914
|sto-1g:2p⟩	0.052815	−0.226352	0.060360	−0.11318	0.999992[3]
|sto-2g:2p⟩	0.045258	−0.246623	0.078076	−0.123289	1.000184
|sto-3g:2p⟩	0.046252	−0.249580	0.082201	−0.124728	1.000502
|sto-4g:2p⟩	0.041801	−0.249696	0.082943	−0.124951	0.999171

for the primitive Gaussians in the 2p$_x$-like linear combination. In addition, equation 4.30 and 4.31, lead to a different result for the action of the Laplacian, with

$$r^2\nabla^2|\text{sto-ng:2p}\rangle = (4\alpha r^4 - 10\alpha r^2 + 2)|\text{sto-ng:2p}\rangle \qquad 4.34$$

Equation 4.34 is written in this manner, including multiplication by the square of the radial distance mainly, to avoid the singularity in the Laplacian transform of the 'p'

[3]Estimated as $-2(T + l(l+1)/r^2)/V$

Gaussian, but also to preserve the full Gaussian function in the algebra. Thus, the full details are as follows.

Exercise 4.14.   Calculation of the 2p orbital energy in hydrogen using an |sto-ng⟩ basis set.

1. Make a copy of fig4-10.xls and rename to fig4-14.xls. As usual set CALCU-LATION to MANUAL in TOOLS/OPTIONS.
2. Provide for the appropriate normalization factor in the 'p' primitive Gaussians in row 8, with for example,

$$\$B\$8 = B\$6^*\textbf{POWER}(128,0.25)^*\textbf{POWER}(B\$7,5/4)/\textbf{POWER}(\textbf{PI}(\ ),0.75)$$

3. An extra *one-variable* table is required for the calculation of the centrifugal term, which is non-zero for p orbitals. Provide for this part of the calculation by inserting four rows after row 17 of the copied spreadsheet.
4. The 2p Gaussians, equation 4.33, include a pre-exponential r factor, so modify the projections of the product of the quadratic exponent in column D to include this change, with, for example,

$$\$D\$49 = \$A49\char`^2^*\textbf{EXP}(-\$H\$3^*\textbf{POWER}(\$A49,2))$$

and note that the insertion causes a displacement of the radial array down the spreadsheet, so that the origin, $r = 0.0$, occurs at cell \$A\$49.
5. From equation 4.34, enter the form for the product of the square of the radial distance times the term in brackets in column E, with

$$\$E\$49 = 4^*\$I\$3\char`^2^*\$A49\char`^4-10^*\$I\$3^*\$A49\char`^2+2$$

6. Now allow for the integration over the angular wave function, for example, $\cos\theta$, for the $2p_z$ orbital, for which the normalization constant is $(3/4\pi)^{1/2}$, which requires changes in the master formula for the TABLE macros, with

$$\langle i|j\rangle = \$A\$12 = (4^*\textbf{PI}(\ )/9)^*\textbf{SUMPRODUCT}(\$A\$49{:}\$A\$3049,$$
$$\$A\$49{:}\$A\$3049,\$C\$49{:}\$C\$3049,\$D\$49{:}\$D\$3049)^*$$
$$(\$A\$50{-}\$A\$49)$$

$$\langle i|1/r|j\rangle = \$A\$16 = (4^*\textbf{PI}(\ )/9)^*\textbf{SUMPRODUCT}(\$A\$49{:}\$A\$3049,$$
$$\$C\$49{:}\$C\$3049,\$D\$49{:}\$D\$3049)^*$$
$$(\$A\$50{-}\$A\$49)$$

$$\langle i|-\tfrac{1}{2}\nabla^2|j\rangle = \$A\$25 = (4^*\textbf{PI}(\ )/9)^*\textbf{SUMPRODUCT}(\$C\$49{:}\$C\$3049,$$
$$\$D\$49{:}\$D\$3049,\$E\$49{:}\$E\$3049)^*$$
$$(\$A\$50{-}\$A\$49)$$

Note, in this last equation, how, as usual, to avoid the singularity by including the radial factor multiplication in column E, this term is absent in the **SUMPRODUCT** function.

7. Construct the *one-variable* table for the centrifugal potential in the active cells of the inserted rows, with the master formula defined in $A$20

$$\langle I|l(l+1)/(2r^2)|j\rangle = \$A\$20 = (4^*PI(\,)/9)^*\textbf{SUMPRODUCT}(\$C\$49:\$C\$3049,$$
$$\$D\$49:\$D\$3049)^*(\$A\$50-\$A\$49)$$

8. Finally, use the *drag and drop* facility to rearrange the locations of the total components of the energy over the basis set and include the centrifugal term

$$l(l+1)/r+2 = \$I\$7 = \textbf{SUM}(\$B\$21:\$D\$21)+2^*\textbf{SUM}(\$E\$21:\$g\$21)$$

in the sum returning the orbital energy, with

$$E_{2p} = \$K\$7 = \$I\$5+\$I\$6+\$I\$7$$

The $|sto\text{-}3g\rangle$ basis set, Figure 4.14, does not return very good values for the energy terms in the 2p orbital energy of hydrogen, when compared with the exact results in Figure 4.4. As you see from Table 4.6, much closer agreement with the exact data follows when larger basis sets are used. The defect, as you might now expect, rests with the poorer representation of the effect of the Laplacian [*times* r^2] required to calculate the kinetic energy term.

Figure 4.15 demonstrates that for the $|sto\text{-}6g\rangle$ basis set, there is much better agreement between the model function, its transform and the exact behaviour and so the energy terms calculated approach their exact values.

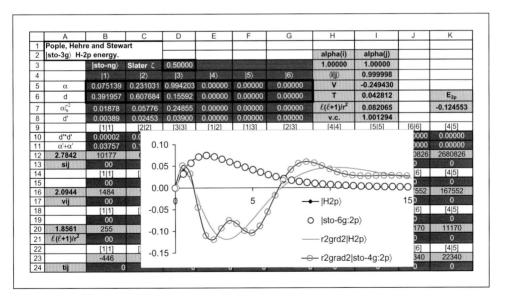

Figure 4.14 Further modification of fig4-10.xls for the calculation of the 2p orbital energy in hydrogen using an $|sto\text{-}ng\rangle$ and here, an $|sto\text{-}3g\rangle$ basis set. For the Pople *et al.* sets of Table 1.6, enter the input parameters in rows 5 and 6. For the Reeves' data, in which the pre-exponential factors are complete, enter these in rows 7 and 8.

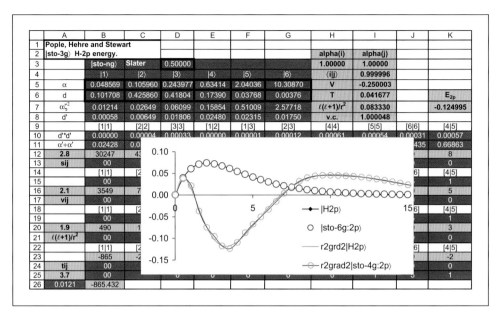

	A	B	C	D	E	F	G	H	I	J	K
1	Pople, Hehre and Stewart										
2	|sto-3g) H-2p energy.							alpha(i)	alpha(j)		
3		|sto-ng)	Slater	0.50000				1.00000	1.00000		
4		|1)	|2)	|3)	|4)	|5)	|6)	(i|j)	0.999996		
5	α	0.048569	0.105960	0.243977	0.63414	2.04036	10.30870	V	-0.250003		
6	d	0.101708	0.425860	0.41804	0.17390	0.03768	0.00376	T	0.041677		E_{2p}
7	$\alpha\zeta^2$	0.01214	0.02649	0.06099	0.15854	0.51009	2.57718	$\ell(\ell+1)/r^2$	0.083330		-0.124995
8	d'	0.00058	0.00649	0.01806	0.02480	0.02315	0.01750	v.c.	1.000048		
9		[1|1]	[2|2]	[3|3]	[1|2]	[1|3]	[2|3]	[4|4]	[5|5]	[6|6]	[4|5]
10	d'*d'	0.00000	0.00004	0.00033	0.00000	0.00001	0.00012	0.00061	0.00054	0.00031	0.00057
11	α'+α'	0.02428	0.0							435	0.66863
12	2.8	30247	4								8
13	sij	00									0
14		[1|1]	[2							6]	[4|5]
15		00									1
16	2.1	3549	7								5
17	vij	00									0
18		[1|1]	[2							156]	[4|5]
19		00									1
20	1.9	490	1								3
21	$\ell(\ell+1)/r^2$	00									0
22		[1|1]	[2							6]	[4|5]
23		-865	-2								-2
24	tij	00									0
25	3.7	00									1
26	0.0121	-865.432									

Figure 4.15   The close agreement achieved between the exact and basis-set modelled radial function and its Laplacian transform [*times* r^2] for the Pople, Hehre and Stewart |sto-6g:2p⟩ linear combination of Table 1.6.

Table 4.6   The terms in the calculation of the 2p orbital energy for the H-atom, based on Pople's Gaussian sets, based on the application of fig4.14xls.

|φ⟩	T	V	$l(l+1)/r^2$	E	v.c.
|sto-2g:2p⟩	0.046128	−0.243345	0.074840	−0.122376	0.994212
|sto-3g:2p⟩	0.042811	−0.249430	0.082064	−0.124554	1.001290
|sto-4g:2p⟩	0.041876	−0.249942	0.083155	−0.124910	1.000484
|sto-5g:2p⟩	0.041706	−0.249994	0.083308	−0.124980	1.000137
|sto-6g:2p⟩	0.041707	−0.250074	0.083371	−0.124995	1.00032

It is interesting to note, from these calculations that the main defect in the modelling of the hydrogenic functions using the Gaussian basis sets appears to rest with the representation of the effect of the Laplacian, which determines the kinetic energy contribution to the total orbital energy. This observation is emphasized in Figures 4.16 and 4.17, wherein are displayed the charts for the comparisons of the matches of the functions and their Laplacian transforms for the Huzinaga |sto-ng) basis sets for hydrogen 1s, with $1 < n < 11$ and for the case of the hydrogen 2s orbital with $1 < n < 7$ using the data in Tables 1.7 and 1.8.

In the figures, the components of the orbital energies are listed, together with the orbital energy and the least-squares differences following equations 2.9 and 2.10 for each of the calculations over the |sto-ng) basis sets. As you see, dramatic improvements occur in the matching of the approximate $r^2\text{grad}^2$ to that found for the exact hydrogen orbital and this is reflected in the close agreement of all the energy components for the calculations using the large basis sets.

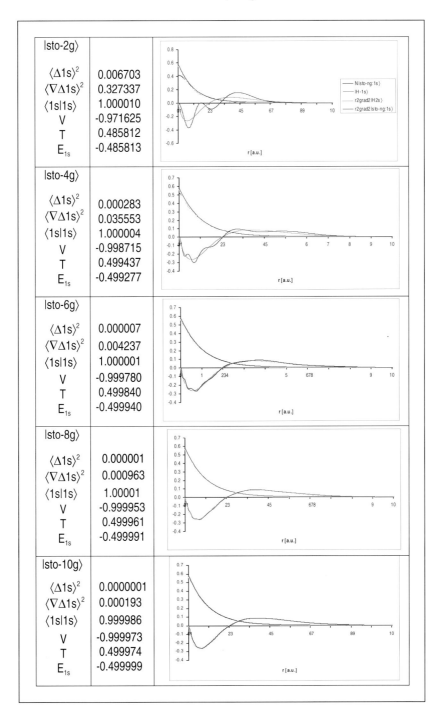

$|\text{sto-2g}\rangle$		
$\langle\Delta 1s\rangle^2$	0.006703	
$\langle\nabla\Delta 1s\rangle^2$	0.327337	
$\langle 1s|1s\rangle$	1.000010	
V	-0.971625	
T	0.485812	
E_{1s}	-0.485813	

$|\text{sto-4g}\rangle$		
$\langle\Delta 1s\rangle^2$	0.000283	
$\langle\nabla\Delta 1s\rangle^2$	0.035553	
$\langle 1s|1s\rangle$	1.000004	
V	-0.998715	
T	0.499437	
E_{1s}	-0.499277	

$|\text{sto-6g}\rangle$		
$\langle\Delta 1s\rangle^2$	0.000007	
$\langle\nabla\Delta 1s\rangle^2$	0.004237	
$\langle 1s|1s\rangle$	1.000001	
V	-0.999780	
T	0.499840	
E_{1s}	-0.499940	

$|\text{sto-8g}\rangle$		
$\langle\Delta 1s\rangle^2$	0.000001	
$\langle\nabla\Delta 1s\rangle^2$	0.000963	
$\langle 1s|1s\rangle$	1.00001	
V	-0.999953	
T	0.499961	
E_{1s}	-0.499991	

$|\text{sto-10g}\rangle$		
$\langle\Delta 1s\rangle^2$	0.0000001	
$\langle\nabla\Delta 1s\rangle^2$	0.000193	
$\langle 1s|1s\rangle$	0.999986	
V	-0.999973	
T	0.499974	
E_{1s}	-0.499999	

Figure 4.16   The results found for the calculations on the energy of the 1s orbital in hydrogen, using Huzinaga's 1s basis sets listed in Table 1.7.

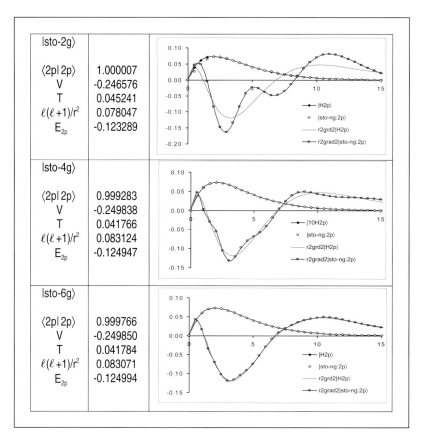

|sto-2g⟩		
⟨2p| 2p⟩	1.000007
V	-0.246576
T	0.045241
$\ell(\ell+1)/r^2$	0.078047
E_{2p}	-0.123289

|sto-4g⟩	
⟨2p| 2p⟩	0.999283
V	-0.249838
T	0.041766
$\ell(\ell+1)/r^2$	0.083124
E_{2p}	-0.124947

|sto-6g⟩	
⟨2p| 2p⟩	0.999766
V	-0.249850
T	0.041784
$\ell(\ell+1)/r^2$	0.083071
E_{2p}	-0.124994

Figure 4.17   The results found for the calculations on the energy of the 2p orbital in hydrogen using Huzinaga's 2p basis sets listed in Table 1.8. The calculations of the 'least-squares' equivalents of the entries in Figure 4.16 are left as an exercise for the interested reader.

While care is needed in the making of comparisons over regions of the radial array used in the diagrams, because of the factor r^2 multiplying the Laplacian transforms, it is particularly remarkable to note that the substantial disagreements for small basis sets, do not occur very close to the nucleus, where there is the problem about the mismatch of the Gaussian and the cusp of the atomic orbital. The major differences occur at radial distances important in chemical bonding, which provides a reason that the calculation of properties based on small basis sets can be deficient.

4.4   CALCULATIONS WITH *SPLIT-BASIS* [*SPLIT-VALENCE*] SETS

Given the third diagram in Figure 4.16, we cannot expect the (4s/2s) Dunning *split-basis* set for the calculation of the 1s orbital energy to return a better estimate. However, as an exercise in the application of the concerted orthonormalization procedure, standard in molecular orbital theory calculations, we gain valuable insight into this methodology.

Firstly let us consider the Dunning modification simply as a renormalization of the contracted basis set. From Huzinaga's original linear combination of four primitives Gaussians, (45), as in Table 1.7,

$$|\text{sto-4g:1s}\rangle = 0.50907g_{1s}(0.123317, \mathbf{r}) + 0.47449g_{1s}(0.453757, \mathbf{r})$$
$$+0.13424g_{1s}(2.01330, \mathbf{r})+0.01906g_{1s}(13.3615, \mathbf{r}) \qquad 1.23$$

Dunning (46) proposed the splitting of the basis into the two distinct components

$$\phi_1 = |\text{sto-1g:1s}\rangle = 1.0000^*g_{1s}(0.123317, \mathbf{r}) \qquad 1.24$$

and

$$\phi_2 = |\text{sto-3g:1s}\rangle = 0.817238g_{1s}(0.453757, \mathbf{r})$$
$$+0.231208g_{1s}(2.01330, \mathbf{r})+0.032828g_{1s}(13.3615, \mathbf{r}) \qquad 1.25$$

In Chapter 3, we used the Dunning basis to match the 2s orbital in hydrogen and showed how the match remained good for the scaling required to match the 2s orbital in lithium, Figures 3.10 and 3.11. That analysis supported the view that the *double-zeta or split-basis* approach is flexible and facilitates procedures dependent on particular variational conditions. So, let us, now, compare the results for the 1s orbital energy in hydrogen using the original Huzinaga basis and the Dunning contraction.

It is straightforward to work out the 1s orbital energy in hydrogen in the Huzinaga $|\text{sto-4g}\rangle$ basis and the spreadsheet calculation, used to generate the third diagram in Figure 4.16, is set out in Figure 4.18. From Figure 4.16, we see that the major deficiency in the representation of the exact function behaviour occurs in the modelling of the variation in the Laplacian transform [*times* r^2] with radial distance.

To perform the calculation using Dunning splitting of the basis, we can follow the analysis in Chapter 3, apply equations 3.17 and 3.18 in the identity

$$|\text{sto-4--31g:1s}\rangle = c_1|\text{sto-1g:1s}\rangle + c_2|\text{sto-3g:1s}\rangle = c_1\phi_1 + c_2\phi_2 \qquad 4.35$$

subject to the obvious constraint, equation 3.13, which is now

$$\langle\text{sto-4--31g:1s}|\text{sto-4--31g:1s}\rangle = 1 \qquad 4.36$$

i.e.

$$c_1^2\langle\text{sto-1g}|\text{sto-1g}\rangle + c_2^2\langle\text{sto-3g}|\text{sto-3g}\rangle + 2c_1c_2\langle\text{sto-1g}|\text{sto-3g}\rangle = 1 \qquad 4.37$$

Moreover, given c_1, we can solve for c_2 as in equation 3.14, with

$$c_2 = \frac{-2c_1S + \left[2c_1^2S^2 - 4N_2\left(N_1c_1 - 1\right)\right]}{2N_1} \qquad 4.38$$

	A	B	C	D	E	F	G	H	I	J	K
1	Simpson's rule integrations										
2	Huzinaga |sto-4g⟩ H-1s energy.										
3		|sto-ng⟩	Slater ζ	1.00000				alpha(i)	alpha(j)		
4		|1⟩	|2⟩	|3⟩	|4⟩	|5⟩	|6⟩	1.00000	1.000		
5	α	0.12332	0.45376	2.01330	13.36150	0.00000	0.00000	⟨i|j⟩	1.000004		
6	d	0.50907	0.47449	0.13424	0.01906	0.00000	0.00000	V	-0.998715		E$_{1s}$
7	αζ²	0.12332	0.45376	2.01330	13.36150	0.00000	0.00000	T	0.499437		-0.499277
8	d'	0.07550	0.18696	0.16170	0.09493	0.00000	0.00000	v.c.	1.000160		
9		[1|1]	[2|2]	[3|3]	[1|2]	[1|3]	[2|3]	[4|4]	[5|5]	[6|6]	[4|5]
10	d'*d'	0.00570	0.03496	0.02615	0.01412	0.01221	0.03023	0.00901	0.0000	0.0000	0.00000
11	α+α	0.24663	0.90751	4.02660	0.57707	2.13662	2.46706	26.72300	0.0000	0.0000	13.36150
12	5.5683	45.46167	6.44087	0.68916	12.70216	1.78293	1.43700	0.04031	113097	113097	0.11401
13	sij	0.25915	0.22514	0.01802	0.17930	0.02177	0.04344	0.00036	0	0	0.00000
14		[1|1]	[2|2]	[3|3]	[1|2]	[1|3]	[2|3]	[4|4]	[5|5]	[6|6]	[4|5]
15		0.24663	0.90751	4.02660	0.57707	2.13662	2.46706	26.72300	0	0	13.36150
16	6.2832	25.4757	6.9235	1.5604	10.8880	2.9407	2.5468	0.23512	5655	5655	0.47025
17	vij	-0.14522	-0.24201	-0.04080	-0.15369	-0.03590	-0.07700	-0.00212	0	0	0.00000
18		[1|1]	[2|2]	[3|3]	[1|2]	[1|3]	[2|3]	[4|4]	[5|5]	[6|6]	[4|5]
19		-16.81859	-8.76778	-4.16243	-7.38999	-1.24306	-3.19271	-1.61575	0.00000	0.0000	0.00000
20	tii	0.0479368	0.153239	0.05442	0.052158	0.007588	0.048262	0.00728	0	0.0000	0
21	0.0000	0.24663	0.90751	4.02660	0.57707	2.13662	2.46706	26.72300	0.00000	0.0000	13.36150
22	0.1233	-16.819									
23	0.4538		-8.768								

Figure 4.18   The Huzinaga |sto-4g:1s⟩ calculation to return the 1s orbital energy in hydrogen.

wherein N_1 and N_2 are the two normalization integrals, while S is the overlap integral between the two linear combinations.

The components of the energy in equation 4.7 now take the forms

$$\langle c_1\phi_1 + c_2\phi_2 | -\tfrac{1}{2}\nabla^2 | c_1\phi_1 + c_2\phi_2\rangle = c_1^2 N_1 \langle\phi_1| -\tfrac{1}{2}\nabla^2|\phi_1\rangle + c_2^2 N_2 \langle\phi_2|$$

$$-\tfrac{1}{2}\nabla^2|\phi_2\rangle + 2c_1 c_2 S\langle\phi_2| -\tfrac{1}{2}\nabla^2|\phi_2\rangle \qquad 4.39$$

$$\langle c_1\phi_1 + c_2\phi_2 | 1/r | c_1\phi_1 + c_2\phi_2\rangle = c_1^2 N_1 \langle\phi_1|1/r|\phi_1\rangle + c_2^2 N_2 \langle\phi_2|1/r|\phi_2\rangle$$

$$+ 2c_1 c_2 S\langle\phi_2|1/r|\phi_2\rangle \qquad 4.40$$

and the normalization integral, the denominator of equation 4.7, reduces to

$$\langle c_1\phi_1 + c_2\phi_2 | c_1\phi_1 + c_2\phi_2\rangle = c_1^2 N_1 + c_2^2 N_2 + 2c_1 c_2 S \qquad 4.41$$

To carry out the *split-basis* calculation in this manner we need to design a spreadsheet to handle the integrations for the two separate approximations to the radial function and combine these to return the overall results. It is straightforward to modify a copy of fig4.11.xls for this purpose, given that the essential part of this design was needed in Chapter 3 in fig-3-10.xls and its later copies. The detail of the spreadsheet design is shown in Figure 4.19 with, as the second diagram, the 'SOLVER' optimized result, using c_1, the entry in cell $L$4, as the variational parameter.

Now, consider the Dunning split as an exercise in the application of the Jacobi procedure and canonical orthonormalization. The Dunning recipe introduces the possibility to represent the 1s orbital by either of the two components identified in equations 1.24 and

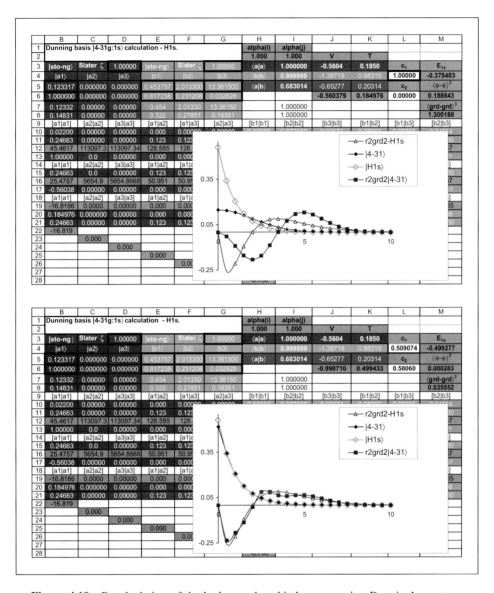

Figure 4.19 Recalculation of the hydrogen 1s orbital energy using Dunning's contraction of Huzinaga's |sto-4g:1s⟩ basis set of Table 1.7. The first diagram presents the initial situation, the assumption that only the single Gaussian contributes. The second diagram shows the results returned after application of the SOLVER macro to minimize the 1s energy calculated by varying c_1 using the SOLVER macro.

1.25, since these are the approximate functions for the choices $c_1 = 1.0$ and so $c_2 = 0.0$ or $c_1 = 0.0$ and so $c_2 = 1.0$. However, neither of these extreme choices diagonalize the Fock matrix, but as we expect the concerted procedure of Sections 3.6 to 3.8 returns the same 'best' values for the coefficients of the linear combination as displayed in Figure 4.20a, which is constructed as follows.

	A	B	C	D	E	F	G	H
1	Canonical orthonormalization applied to Dunning (42|2s) basis set							
2								
3	S			theta =	0.785398			
4	1.000000	0.683014		omega1	1.683014			
5	0.683014	0.999999		omega2	0.316986			
6								
7	U			U - adjoint			S-diagonal	
8	0.70711	0.70711		0.70711	0.70711		1.68301	0.00000
9	0.70711	-0.70711		0.70711	-0.70711		0.00000	0.31699
10							The S^1/2 matrix	
11							0.77083	0.00000
12							0.00000	1.77615
13	Fock matrix			X-canonical			X-canon transpose	
14	-0.375403	-0.449628		0.54506	1.25593		0.54506	0.54506
15	-0.449628	-0.404043		0.54506	-1.25593		1.25593	-1.25593
16								
17	theta	-0.02848		Transformed Fock matrix				
18	omega1	-0.49928		-0.44969	0.09322	==)	-0.49872	0.01961
19	omega2	0.18954		-0.46530	-0.05725		0.01961	0.18898
20								
21	Coefficients			Fock - diagonal			C - originals	
22	0.99959	-0.02847		-0.499277	0.0000000		0.50907	-1.27094
23	-0.02847	-0.99959		0.0000000	0.1895419		0.58060	1.23990

	B	C	D	E	F	G	H	I	J	K	L	M	N
1	Dunning basis |sto-431g:1s) calculation - H1s.						alpha(i)	alpha(j)					
2							1.000	1.000	V	T			
3	|sto-ng) Slater ζ		1.00000	|sto-ng)	Slater ζ	1.000000	(a|a)	1.0000000000	-0.56038	0.1850	c₁	E₁ₛ	E₂ₛ
4	|a1)	|a2)	|a3)	|b1)	|b2)	|b3)	(b|b)	0.9999992920	-1.38719	0.98315	0.509074	-0.499277	0.18954
5	0.123317	0.000000	0.000000	0.453757	2.013300	13.361500	(a|b)	0.6830141413	-0.65277	0.20314	c₂	(φ -φ)³	(grd-grd)²
6	1.000000	0.000000	0.000000	0.817238	1.231208	0.032828			-0.99871	0.49943	0.58060	0.00028	0.03555
7	0.12332	0.00000	0.00000	0.45376	2.013300	13.361500			-0.98046	1.1700004			
8	0.14831	0.00000	0.00000	0.32202	0.278512	0.163511							
9	[a1|a1]	[a2|a2]	[a3|a3]	[a1|a2]	[a1|a3]	[a2|a3]							
10	0.02200	0.00000	0.00000	0.00000	0.000000	0.000000							
11	0.24663	0.00000	0.00000	0.12332	0.123317	0.000000							
12	45	113097	113097	129	129	113097							
13	1.00000	0.00000	0.00000	0.00000	0.00000	0.00000							
14	[a1|a1]	[a2|a2]	[a3|a3]	[a1|a2]	[a1|a3]	[a2|a3]							
15	0.24663	0.00000	0.00000	0.12332	0.12332	0.000000							
16	25.476	5654.867	5654.867	50.951	50.951	5654.867							
17	-0.56038	0.00000	0.00000	0.00000	0.000000	0.000000							
18	[a1|a1]	[a2|a2]	[a3|a3]	[a1|a2]	[a1|a3]	[a2|a3]							
19	-16.8186	0.0000	0.00000	0.00000	0.000000	0.000000							
20	0.184976	0.000000	0.00000	0.00000	0.000000	0.000000							
21	0.24663	0.00000	0.00000	0.12332	0.123317	0.000000							
22	-16.819												
23		0.000											
24			0.000										
25				0.000									
26					0.000000								
27						0.000000							

Chart legend: r2grd2-H1s, |4-31), |H1s), r2grd2|4-31), |4_3-1), |H:2s)

Figure 4.20a Application of the Jacobi transformation and the canonical orthonormalization procedure to the calculation of the 1s orbital energy in hydrogen using the Dunning (4s|2s) Gaussian basis (46). Note that the calculation returns a second set of coefficients in cells $H$22 and $H$23, which are simply a product of the calculation procedure and give rise to a primitive approximation to the 2s orbital (22).

Exercise 4.15.   Application of the Jacobi transformation and canonical ortho-normalization to the calculation of the 1s orbital energy in hydrogen using the Dunning (4s|2s) basis.

1.   Make a copy of fig4-19.xls and rename it fig-4-20.xls.
2.   Open fig4-9.xls and copy the worksheet 'canonical'! into fig4-20.xls.
3.   Modify the input to the overlap and Fock matrix cells of this worksheet, with

$$\$A\$4 = \text{norm!}\$I\$3$$

$$\$B\$4 = \text{norm!}\$I\$5$$

$$\$A\$5 = \text{norm!}\$I\$5$$

$$\$B\$5 = \text{norm!}\$I\$4$$

$$\$A\$14 = \text{norm!J3+norm!K3}$$

$$\$B\$14 = \text{norm!J5+norm!K5}$$

$$\$A\$15 = \text{norm!J4+norm!K4}$$

$$\$B\$15 = \text{norm!J5+norm!K5}$$

4.   Link the output coefficients for the approximations to the eigenfunctions of the Fock matrix into the cells for the coefficients on the 'norm'! worksheet, with

$$\text{'norm'}!\$L\$4 = \text{canonical!}\$G\$22$$

$$\text{'norm'}!\$L\$6 = \text{canonical!}\$G\$23$$

5.   Create the similar link to display the lower approximate eigenvalue of the Fock matrix on this worksheet, with

$$\text{'norm'}!\$M\$4 = \text{canonical!}\$D\$22$$

6.   In turn, create the control cells for the Slater exponents on the 'canonical'! worksheet with the links,

$$\zeta_1 = \text{canonical!}\$H\$3 = \text{'norm'}!\$D\$3$$

$$\zeta_2 = \text{canonical!}\$H\$4 = \text{'norm'}!\$G\$3$$

7.   As you see, the remainder of this worksheet and, in particular, the chart on the 'norm'! worksheet readjusts to reflect this change in control of the calculation.

The results of the calculation are displayed in Figure 4.20a. Because the Dunning (4s|2s) basis corresponds to a two-term linear combination over the Gaussians in equations 1.24 and 1.25, the Jacobi transformation procedure returns two 'eigenvalues' and corresponding 'eigenfunctions'. This second solution is a very primitive approximation to the 2s orbital in hydrogen, to the extent that it is orthonormal to the calculated 1s function. This observation is demonstrated in Figure 4.20b, which displays the projections of the two linear combinations over the radial array chosen.

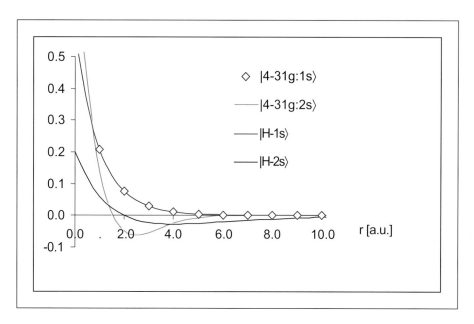

Figure 4.20b Comparisons of the exact radial projections of the hydrogen 1s and 2s radial functions with those of the two linear combinations resulting from the calculations in fig4-20ab.xls. The second linear combination is a primitive approximation to the hydrogen 2s radial function to the extent that it is orthonormal to the 1s function [cells 'canonical' !$K$24 to $M$24].

Figure 4.21a shows the results obtained if we assume that the Dunning (4s|2s) basis is scaleable to return a good approximation to the 2s orbital in hydrogen. As you see, a considerable improvement in the fit to the hydrogen 2s orbital has been achieved without complete loss of register in the match to the 1s and this is emphasized in Figure 4.21b, which is the equivalent of Figure 4.20b for the calculation using the Slater exponent 0.5, cells 'canonical' !$H$3 and $H$4.

4.5   REVIEW OF RESULTS FOR THE 1S AND 2S ORBITAL ENERGIES IN HYDROGEN

A large number of calculations, of the energies of the hydrogen 1s and 2s orbitals, have been set out in this chapter and considerable attention has been paid to the matchings of the approximate functions and their second derivatives with the behaviours of the exact functions on the radial array. A flavour for some of the complexity of modern molecular orbital theory is to be had from some review of this material.

The first calculations, set out in Section 4.1, concern the substitution, simply, of the exact hydrogen 1s, 2s and 2p eigenfunctions into the Schrödinger equation for the hydrogen atom and then numerical integration to return the eigenvalues. In all cases, exact results for the eigenvalues are returned in the numerical integrations. These calculations are important, in the present context, since they establish the validity and accuracy of the solution procedure [*the program*]. Therefore, we can conclude that any loss of accuracy in the calculations with Slater orbitals and Gaussian basis sets relate to the deficiencies of these approximations and do not relate to the numerical integrations.

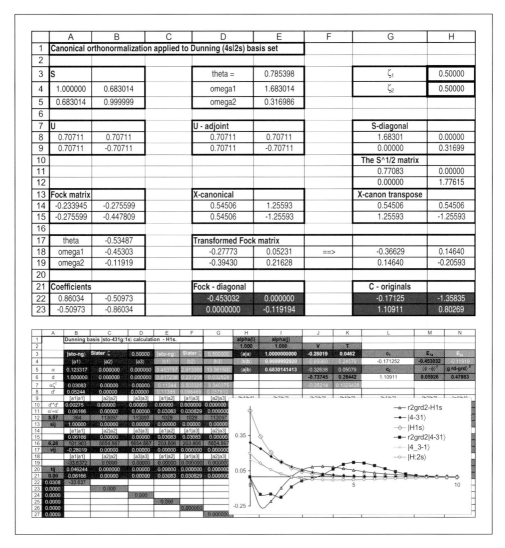

Figure 4.21a   The results for the assumption that the Dunning split can be scaled to return a good representation of the hydrogen 2s orbital.

One Hartree is 27.21 eV, which is [Table 4.1] some 2625 kJ/mole. Thus an error in calculation of 1×10^{-4} Hartree is 0.2625 kJ/mole. For the lighter elements of the Periodic Table Herman and Skillman (4) published calculated atomic orbital energies to four decimal places in Rydbergs. This corresponds to just under 0.15 kJ/mole, which is still a typical acceptable accuracy in modern calculations (50).

The Slater orbital calculations of Section 4.2 return the results summarized in Table 4.7. Since the Slater 1s function is the exact eigenfunctions for the 1s orbital in hydrogen, only the results for the 2s calculations are listed in Table 4.7.

Only the results returned by the calculations, Figure 4.6, involving the addition of the extra term to the Schrödinger equation (17,45) are acceptable. In the other two sets of

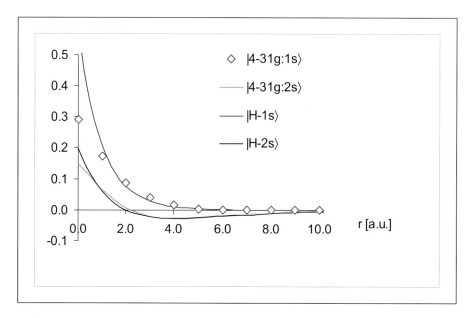

Figure 4.21b   Comparisons of the exact radial projections of the hydrogen 1s and 2s radial functions with those of the two linear combinations resulting from the calculations in fig4-21ab.xls. In this calculation, the Slater exponents were chosen to scale the Dunning basis for the calculation of the 2s orbital energy in the hydrogen atom. As you see more clearly in this figure a good match to the 2s radial function for hydrogen has been achieved without total loss of registry with the 1s radial function.

Table 4.7   Comparisons of the results from various calculations using Slater functions to approximate the 1s and 2s orbitals in hydrogen.

	Exact Figure 4.2	Slater Figure 4.6
T	0.125000000000000	0.041666667130167
V	−0.250000000000000	−0.250000000775663
$l(l+1)/r^2$	—	0.083333333639981
E_{2s}	−0.125000000000000	−0.125000000005515
V.C.	1.000000000000000	1.000000003102650
	Slater [Schmidt-'best'] Figure 4.7	Slater [Canonical-'best'] Figure 4.9
T	0.124931852379022	0.123359575079563
V	−0.248271233438144	−0.246661901694882
E_{2s}	−0.123339381059121	−0.123523505795558
V.C.	1.006414240175340	1.000232092852000

results, shown in the table, the minimum error for the energy terms is about 5 kJ/mole, which is not satisfactory by the standard specified above.

For the various 1s and 2p Gaussian basis sets proposed by Huzinaga (45) comparisons of this kind have been made in figures 4.16 and 4.17. As you can see Huzinaga's large basis set returns a good approximation to the 1s eigenvalue, with an error of only 0.0026 kJ/mole for the 1s energy. While the error rises only to some 0.016 kJ/mole for the 2p eigenvalues, for which Huzinaga published only an |sto-6g⟩ basis.

Figure 4.22 displays the 'canonical'! sheet of the spreadsheet fig4-22.xls, which provides for the calculation of the energy terms and the virial coefficients for the 1s and 2s orbitals in hydrogen using up to six primitives to represent each orbital. The spreadsheet design is left as a good exercise for the reader and you should be able to amalgamate details from the earlier spreadsheets described in the chapter, should you decide to make your own spreadsheet. Note, that the individual worksheets devoted to calculations with 1s and 2s Gaussian need not be included, if file size is a problem.

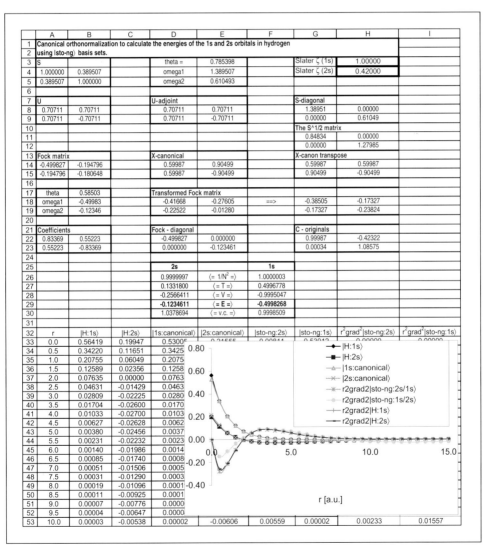

Figure 4.22 The 'best' calculation of the 2s orbital energy in hydrogen using the Pople, Hehre and Stewart (33) |sto-6g⟩ basis sets for the 1s and 2s orbitals. The error in the calculation amounts to just over 4 kJ/mole.

Table 4.8   The results obtained using various optimized choices for the coefficients in the Dunning *split-basis* calculations to determine the hydrogen 1s orbital energy.

	Figure 1.30 results	Figure 4.19 results	Figure 4.20 results
c_1	0.5093050	0.509073734008122	0.509074201942506
c_2	0.580370615043834	0.580596379275388	0.580595907908989
E_{1s}	−0.499277292300924	−0.499277315104380	−0.499277301329702
V	−0.998504114253260	−0.998710250340187	−0.99870982700991
T	0.499226821952336	0.499432935235807	0.49943251190562
v.c.	0.999949454040431	1.00015582110164	1.00015539729071

The data in Figure 4.22 represent the 'best' results for the choice that the Slater 2s exponent be varied to minimize the energy of the 2s orbital. This choice was made because, as you will find, in all the calculations possible on this spreadsheet, the main defect is the modelling of the 2s orbital, even with the 'best' choice of Slater exponent. For the choice of 'best' Slater 2s exponent the calculated 2s orbital energy in the basis of six contracted primitives is returned as −0.1234611 Hartree, which is some 4 kJ/mole in error from the exact value of −0.1250 Hartree.

It is appropriate, too, in this assessment of the quality of the results set out in this chapter, to evaluate the Dunning *split-basis* calculation as set out in Figure 4.19. We know that the two components of the splitting of Huzinaga's original data do not provide for good modelling of the 1s and 2s orbitals and this is investigated in Figures 4.20 and 4.21. However, several criteria have been applied in different applications to generate 'best' results using the *split-basis*. Thus, for the simple matching of the 1s basis-set modelled function, using the least-squares approach of Chapter 3, Section 3.5, Figure 3.11 and compare Figure 1.30, the values $c_1 = 0.509305$ and $c_2 = 0.58037$ were found. For the requirement, in this chapter, that the difference in energies in the approximate and exact calculation be minimized we find the values $c_1 = 0.509374$ and $c_2 = 0.580596$ to be best. It is important to appreciate that this is not a contradiction. First, we have not examined the precision of our calculations. Secondly, and, in any event, we are asking different questions and so we should not be surprised to obtain different answers. Anyway, as you will find by carrying out the calculations using the 'best' coefficients determined in these applications, the results for which are presented in Table 4.8, only small differences occur.

All the comments made carry over into real calculations. On the one hand, it is pleasing to obtain results for the orbital energies and, in molecules for total energies as well. In isolation it is not sufficient, generally, to obtain what appears to be a good answer, with only a few kJ/mole of error. The aim of theory is to represent the total physical situation and this becomes harder the more complex the molecule and the more elusive the property, being modelled. Even within the limitations of the spreadsheet approach, you will see in the remaining chapters how the complexity of the problem increases as we begin to consider other atoms and even the simplest molecules.

5

The helium atom and the self-consistent field

This chapter extends the numerical methods of the previous chapters to the case of the simplest many-electron atom, helium. Then, at the end of the chapter, the calculation of the electronic energy of helium is carried out as a first example of the standard modern form of the self-consistent field method, in which the integrals over Gaussian primitives are evaluated exactly.

The Englishman, Hartree (1,60) the Russian, Fock (2,3) and the American, Slater (5–7), in the early development of modern quantum mechanics, pioneered the calculation of atomic electronic structure. Hartree based his method on the variation principle and this led naturally to the development of the *self-consistent field* method, which is at the heart of the design of modern molecular orbital programs.

In this chapter you will learn about:

1. Hartree's solution of the helium atom problem and how to repeat the calculation on a spreadsheet;
2. the Hall–Roothaan equations and the orbital approximation of Hartree–Fock–Slater *self-consistent* field theory;
3. how to carry out a HFS *double-zeta* calculation for helium and so, on a spreadsheet, repeat Clementi's famous calculation;
4. the direct calculation of two-electron integrals over Gaussian functions and again the calculation of the energy of helium as an HFS exercise;
5. the special case calculation of the energies of the electron configurations $1s^1 2s^1$ in the helium atom using the Pople–Nesbit extension of the Hall–Roothaan equations.

5.1 HARTREE'S ANALYSIS OF THE HELIUM ATOM PROBLEM

Hartree identified two ways, to apply the variation principle, to solve the helium atom problem. He suggested applying the variation principle to functions of a particular analytical form, and then assessing how close the calculated energy might be to the expected value. Alternatively, Hartree proposed to determine the best approximate energy for a choice of arbitrary functions, chosen to comply with particular physical interpretations of what the wave function should represent.

It is straightforward to write down and solve the many-electron Schrödinger equation if it is assumed that the electrons do not interact, or interact only to a very small extent. Indeed, it is on this premise that the fabric of modern qualitative molecular orbital theory is based. For the two electrons in a helium atom $[Z = 2]$ for example, this *independent particle* model Schrödinger equation is simply

$$H(r_1, r_2)\Phi(r_1, r_2) = \left[\left(-\frac{1}{2}\nabla^2(r_1) - \frac{2}{r_1}\right) + \left(-\frac{1}{2}\nabla^2(r_2) - \frac{2}{r_2}\right)\right]\Phi(r_1, r_2) = E\Phi(r_1, r_2)$$

5.1

wherein the non-interacting electrons and their coordinates are identified as (1) and (2). The wave functions are the product of two solutions to the distinct one-electron parts within the brackets, with

$$\Phi(r_1, r_2) = \phi(r_1)\phi(r_2)$$

5.2

and

$$E = \varepsilon_1 + \varepsilon_2$$

5.3

In equation 5.3, each ε is a solution of a one-electron equation, for example

$$h(r_1)\phi(r_1) = \left[-\frac{1}{2}\nabla^2(r_1) - \frac{2}{r_1}\right]\phi(r_1) = \varepsilon_1\phi(r_1)$$

5.4

So, now, let us follow Hartree's analysis within the framework of our spreadsheet formulation of the numerical method.

This was Hartree's starting point in his analysis of the helium problem. Hartree chose to write the radial part of equation 5.2 in the form

$$\Phi(r_1, r_2) = P(r_1)P(r_2)/r_1 r_2$$

5.5

We recognize, in this expression of the two-electron wave function, the numerical radial functions[1] of Section 1.2. As is usual, we can require that the individual radial functions be normalized, so that

$$\int_0^\infty P^2(r)\,dr = 1$$

5.6

and we can include, in the normalization, the contribution [here $(4\pi)^2$ for any electron configurations over the 1s and 2s orbitals] from the integrations over the angular coordinates involving the s-type spherical harmonic, since

$$\iiint \phi(r_1, \theta_1, \phi_1)\,d\tau_1 \iiint \Phi(r_2, \theta_2, \phi_2)\,d\tau_2$$

$$= \int_0^\infty P^2(r_1)r_1^2\,dr_1 \int_0^\infty P^2(r_2)r_2\,dr_2 \int_0^\pi \sin(\theta_1)\,d\theta_1 \int_0^{2\pi} d\phi_1 \int_0^\pi \sin(\theta_2)\,d\theta_2 \int_0^{2\pi} d\phi_2$$

$$= 1 \times 1 \times (4\pi) \times (4\pi)$$

5.7

---

[1]Do not be confused by the symbol, $P(r)$, here. It is not the radial distribution, rather $r R(r)$, the numerical radial function as in the output of the Herman–Skillman program, hs.exe. $P(r)$ is used, here, to stay with Hartree's original derivation.

Hartree's next step was to include electron–electron repulsion in equation 5.1

$$H(\mathbf{r}_1, \mathbf{r}_2) = -\frac{1}{2}\nabla^2(\mathbf{r}_1) - \frac{1}{2}\nabla^2(\mathbf{r}_2) - \frac{2}{r_1} - \frac{2}{r_2} + \frac{1}{r_{12}}$$ 5.8

but still to keep the independent particle form for the radial wave function.[2]

Each term in equation 5.8 leads to an integral of equation 4.10, but now involving the two-electron radial function of equation 5.5. Hartree chose to calculate terms involving the kinetic energy operator component, ∇^2, using the form,

$$\nabla^2 f = \frac{1}{r}\frac{d^2}{dr^2}(rf)$$ 5.9

since, with the angular integrations in equation 5.7 complete, again, there are left only simple functions of the radial coordinates[3]. Of course, equation 5.9 multiplies out to the form used throughout Chapter 4

$$\nabla^2 f = \frac{2}{r}\frac{df}{dr} + \frac{d^2 f}{dr^2}$$

Hence, the kinetic energy terms, in equation 5.8, are

$$\nabla^2(r_1)\Phi(r_1, r_2) = \frac{1}{r_1}\frac{\partial^2}{\partial r_1^2}(r_1\Phi(r_1, r_2)) = \frac{1}{r_1 r_2}\frac{\partial^2 P(r_1)}{\partial^2 r_1} = \frac{1}{r_1 r_2}P''(r_1)P(r_2)$$ 5.10

and

$$\nabla^2(r_2)\Phi(r_1, r_2) = \frac{1}{r_2}\frac{\partial^2}{\partial r_2}r_2(r_2\Phi(r_1, r_2)) = \frac{1}{r_2 r_1}P(r_1)\frac{\partial^2 P(r_2)}{\partial r_2} = \frac{1}{r_1 r_2}P(r_1)P''(r_2)$$ 5.11

Therefore,

$$r_1^2 r_2^2 \Phi(r_1, r_2)H(r_1, r_2)\Phi(r_1, r_2) = -\frac{1}{2}P(r_1)\left[P''(r_1) + \frac{4}{r_1}P(r_1)\right]P^2(r_2)$$

$$- \frac{1}{2}P(r_2)\left[P''(r_2) + \frac{4}{r_2}P(r_2)\right]P^2(r_1) + \frac{P(r_1)P(r_1)P(r_2)P(r_2)}{r_{12}}$$ 5.12

and the numerator of the energy expectation value, equation 4.10, involving the integral,

$$\iiiiii r_1^2 r_2^2 \Phi(r_1, \theta_1, \phi_1)H(r_1, \theta_1, \phi_1, r_2, \theta_2, \phi_2)\Phi(r_2, \theta_2, \phi_2)\,d\tau_1\,d\tau_2$$ 5.13

divides into distinct integrals over the separate radial and angular components with individual and equal terms of the form,

$$I = -\frac{1}{2}\int_0^\infty P(r)\left[P''(r) + \frac{4}{r}P(r)\right]dr \times (4\pi)^2$$ 5.14[4]

---

[2]Ignoring the general requirement to write the many-electron wave function as a Slater determinant.
[3]Note the distinction between the vector symbol, $\mathbf{r}$, and the radial distance, r, in these equations.
[4]Remember the $(4\pi)^2$ term here, which will cancel at the end with the denominator in the energy expectation value equation and note the cancellation, too, of the terms in r_1 and r_2!

leaving the remaining term in equation 5.12 as

$$F_0 = \frac{1}{(4\pi)^2} \iiiint \frac{1}{\mathbf{r}_{12}} \frac{P(\mathbf{r}_1)P(\mathbf{r}_1)}{(4\pi \mathbf{r}_1^2)} \left[\frac{P(\mathbf{r}_2)P(\mathbf{r}_2)}{(4\pi \mathbf{r}_2^2)} \right] d\mathbf{r}_1 \, d\mathbf{r}_2 \qquad 5.15$$

Hartree identified this to be the mutual potential energy of two spherically symmetrical charge distributions, each being of volume density

$$\rho = 4\frac{P^2(r)}{(4\pi r^2)} \qquad 5.16$$

everywhere on the surface of the sphere of radius, r, about the origin. At each position, r_1, in 5.15 for the first electron, there is the electron–electron repulsion potential due to all of the second[5] satisfying the equation for a hollow charged sphere

$$\nabla^2 V = -4\pi\rho = -\frac{P^2(r)}{r^2} \qquad 5.17$$

Again, Hartree applied equation 5.9 to this relation and writing

$$V(r) = \frac{Y(r)}{r} \qquad 5.18$$

obtained equation 5.17 as the second-order differential equation, which he could integrate for various choices of the radial wave function, with

$$\frac{d^2 Y(r)}{dr^2} = -\frac{P^2(r)}{r} \qquad 5.19$$

Thus, to obtain this contribution to the energy of the two-electron system, we have to evaluate this integral for each value of the radial coordinate of the other electron and then integrate again to obtain 5.15 in the form,

$$F_0 = \int_0^\infty P^2(r)\frac{1}{r}Y(r)\,dr \qquad 5.20$$

with the total energy of the helium atom given by

$$E' = 2I + F_0 \qquad 5.21$$

for particular choices of $P(r)$. This statement establishes the *self-consistent* field method. Equation 5.21 can be evaluated only when we have decided upon a functional form for $P(r)$. The calculation returns a form for the function, which then can be used to carry out another calculation and so on until the difference between the input choice and the output calculated function is as small as possible, that is to say, the potential field is *self-consistent*.

5.2   CALCULATIONS WITH MODIFIED HYDROGEN ATOM WAVE FUNCTIONS

The simplest model of the helium atom is to assume that it is like the hydrogen atom, and that, apart from the two neutrons, there are two nuclear protons and two electrons, the interactions of which lead to a partial screening of the nuclear charge and an extra electron–electron

---

[5]Not quite, in general, because of the Fock exchange term, the self-interaction of each electron distribution.

repulsion. Thus, Hartree sought solutions of the Schrödinger equation for the helium atom as modified hydrogenic wave functions, with

$$P(r) = \left[\frac{k^3}{2}\right]^{1/2} r e^{(-1/2kr)} \tag{5.22}$$

and the variable parameter k to be determined. Hartree included the factor $\frac{1}{2}$, in the exponent, in equation 5.22, since this choice provides for cancellations in the expressions for the various integrals involving P^2 and PP''. Like our keeping of the symbol $P(r)$, it is useful to keep this factor so that we can compare the results of our numerical integrations with the analytical expressions in Hartree's book (39). Direct integration to determine the norm for the radial function choice in equation 5.22, over the radial coordinates only, returns $[k^3/2]^{1/2}$ as the normalization constant, so this is the pre-exponential term in equation 5.22. For k equal to unity this normalization constant reduces to $[1/2]^{1/2}$.

To complete Hartree's calculation, we need to evaluate the integrals set out in equations 5.14 and 5.20. For the wave function choice, equation 5.22, the integral in equation 5.14 includes the kinetic energy and nuclear-electron potential energy terms calculated in Chapter 4 for the Slater approximation to the 2s eigenstate in hydrogen.

The electron−electron repulsion term, equation 5.20, requires us to know the functional form for $Y(r)$. We cannot avoid the calculus, but it is not too demanding for the choice of $P(r)$ in equation 5.22 and the integral is to be found in any listings of indefinite integrals (66).

$$\int x e^{ax} \, dx = \frac{e^{ax}}{a^2} (ax - 1) \tag{5.23}$$

so that on integration twice, as required by equation 5.20, we obtain[6]

$$Y(r) = - \left(1 + \tfrac{1}{2}kr\right) e^{-(kr)} + c_1 r + c_2 \tag{5.24}$$

in which expression, c_1 and c_2 are the constants of integration. These are determined by the boundary conditions, which Hartree chose, that the potential be finite at the origin and of the order of $1/r$, i.e. $O(1/r)$ at infinity, which conditions are sufficient to require that c_1 be zero and c_2 be equal 1.

Thus, $Y(r)$ reduces to

$$Y(r) = 1 - \left(1 + \tfrac{1}{2}kr\right) e^{-(kr)} \tag{5.25}$$

Exercise 5.1.   Calculation of the energy of the helium atom using modified hydrogen functions.

1. Enter the basic data for the calculation in cells $D$5 to $D$7 for the wave function defined in equation 5.22.
2. Lay out the Simpson's rule procedure, for the projections of equation 5.22 on the radial array of column A, in columns B and C.
3. Since the analysis leading to equation 5.25 is based on normalized expressions for the wave functions, calculate the normalization integral for this wave

---

[6]Remember to include the square of the normalization constant in the integration!

function in cell $D$10 using the **SUMPRODUCT** function. As you can see, the result, in cell $D$10, includes Hartree's factor $\frac{1}{2}$ as the entry for the exponent while the scaler is set equal to 2, for this normalization test at the start of the calculation[7].

	A	B	C	D	E	F
1	He 1s² energy					
2	Hartree's calculation.					
3						
4						
5			scaler	2.000E+00	V	-2.0000
6			exponent	1.000E+00	T	0.5000
7			coefficient	2.000E+00	ee	0.6250
8					E (He)	-2.3750
9			Normcheck			
10			⟨1|1⟩	0.250000	ε₁ₛ	-0.8750
11						
12	r		|1⟩	r*|1⟩	grad^2	Y
13	0.000	1	1.00000	0.00000	1.00000	0.00000
14	0.010	4	0.99005	0.01980	0.99500	0.01000
15	0.020	2	0.98020	0.03921	0.99000	0.01999
16	0.030	4	0.97045	0.05823	0.98500	0.02998
17	0.040	2	0.96079	0.07686	0.98000	0.03996
18	0.050	4	0.95123	0.09512	0.97500	0.04992
19	0.060	2	0.94176	0.11301	0.97000	0.05986
20	0.070	4	0.93239	0.13054	0.96500	0.06979

Figure 5.1   Part details of the spreadsheet, fig5-1to5-3.xls, for the calculation of the energy of the helium atom, based on Hartree's analysis in Section 5.1. The assumption of simple hydrogenic behaviour does not return good energies.

4. The form for the ∇^2 operator in the kinetic energy terms for the two electrons follows directly from equation 4.16 and 4.17, for example, in cell $F$13 of Figure 5.1 the entry is

$$\$F13 = \text{-}0.5*(\textbf{POWER}(\$D\$6,2)*\$A13\text{-}2*\$D\$6)$$

and the kinetic energy, contribution to the total energy, is calculated in cell $F$6, with the **SUMPRODUCT** function in the form,

$$T = \$F\$6 = \textbf{SUMPRODUCT}(\$B\$13:\$B\$3013,\$C\$13:\$C\$3013,$$
$$\$D\$13:\$D\$3013,\$E\$13:\$E\$3013)*(\$A\$14\text{-}\$A\$13)/$$
$$(3*\textbf{SQRT}(\$D\$10))$$

for each electron in the two-electron atom.

---

[7] Note, because no normalization factor has been included, the normalization integral equals 0.25, rather than 1.00 for the choice of exponent describing the hydrogenic orbital function, compare Table 1.1.

5. The potential energy contribution from the electrostatic attraction between the electrons and the helium nucleus is calculated in cell $F$5 for the two electrons

$$V = \$F\$5 = -2*\textbf{SUMPRODUCT}(\$B\$13:\$B\$3013,\$C\$13:\$C\$3013,$$
$$\$D\$13:\$D\$3013)*(\$A\$14-\$A\$13)/(3*\textbf{SQRT}(\$D\$10))$$

6. The next step is to include the electron–electron repulsion term, equation 5.25. Remember that the normalization constant is included. Thus the entries in column F take the form

$$Y(r) = \$F13 = 1-(1+\$D\$6*\$A13)*\$C13*\$C13$$

and the electron–electron repulsion term for this approximation to the radial function is the **SUMPRODUCT** function

$$ee = \textbf{SUMPRODUCT}(\$B\$13:\$B\$3013,\$C\$13:\$C\$3013,$$
$$\$D\$13:\$D\$3013,\$F\$13:\$F\$3013)*(\$A\$14-\$A\$13)/(3*\textbf{SQRT}(\$D\$10))$$

to avoid counting in the normalization factor twice and the 1/r singularity at the origin.

7. Finally, Hartree determined the 'best' k, in equation 5.22, by applying the variation principle. In EXCEL, this stage of the calculation is reproduced very straightforwardly using the SOLVER macro. So activate the SOLVER macro under the TOOLS menu and minimize the total energy in cell $F$8 using the scaler, cell $D$5, as the variable.

The results of the variation principle calculation are shown in Figure 5.2. The stationary value of the total energy of the helium atom within the variation principle calculation occurs for the parameter 'k' of equation 5.22 equal to 3.375 a.u. (54/16) of inverse of length squared. Hartree pointed out that this result

$$P(r) = re^{(-(27/16)r)} = re^{-(2-(5/16))r} \qquad 5.26$$

presents a satisfactory model for the ground state of the helium atom. The effect of the extra electron, compared with the one-electron situation in hydrogen, is to screen and so 'reduce' the nuclear charge of +2 protons to the extent that Z_{eff} is 1.6875.

The first ionization energy, the difference in total energy between a neutral atom or molecule and the corresponding ion, formed by removing one-electron, is a well-defined parameter, which can be determined very accurately. For the helium atom, this first ionization energy is 0.9037 Hartree. The energy of the helium ion, the one-electron atomic species, with two protons in the nucleus, is exactly -2.0000 a.u. using equation 4.15. Hartree's 'best' energy of helium has been calculated to be -2.8477 Hartree[8], cell H13 in Figure 5.2. The difference, 0.8477 Hartree, is not a very satisfactory calculation of the first ionization energy, since the error with respect to the experimental value amount to almost 150 kJ/mole. While part of the error relates to the assumption, (equation 5.2) that

---

[8]Hartree presented the results of his calculation in Rydbergs (1 Hartree = 2 Rydbergs) and so you will find double the number in Figures 5.1 and 5.2 for the energy in his paper.

	A	B	C	D	E	F
1	He 1s^2 energy					
2	Hartree's calculation.					
3						
4						
5			scaler	3.375E+00	V	-3.3750
6			exponent	1.688E+00	T	1.4238
7			coefficient	4.384E+00	ee	1.0547
8					E (He)	-2.8477
9			Normcheck			
10			$\langle 1|1 \rangle$	0.052025	ε_{1s}	-0.8965
11						
12	r		$|1\rangle$	r*$|1\rangle$	grad^2	Y
13	0.000	1	1.00000	0.00000	1.68750	0.00000
14	0.010	4	0.98327	0.04311	1.67326	0.01687
15	0.020	2	0.96681	0.08478	1.65902	0.03373
16	0.030	4	0.95064	0.12503	1.64479	0.05054
17	0.040	2	0.93473	0.16392	1.63055	0.06731
18	0.050	4	0.91909	0.20148	1.61631	0.08401
19	0.060	2	0.90371	0.23772	1.60207	0.10062
20	0.070	4	0.88858	0.27270	1.58783	0.11715

Figure 5.2   The variation principle based 'best' calculation of the energy of the helium atom using the hydrogenic function of equation 5.3 and the SOLVER macro.

the electron motions are independent of each other, the major error arises from the primitive description of the electronic wave function achieved using equation 5.22 to represent the one-electron orbitals. In the rest of this chapter you will see that calculations using linear combinations of Slater functions or Gaussian functions to represent the atomic orbitals return better results. However, for wave functions formed in the manner of equation 5.2, the minimum error that can be achieved is still some 110 kJ/mole different to the experimental value and this limit in the capacity of such calculations is known as the *Hartree–Fock* limit.

Figure 5.3 displays the degree of agreement between the radial function from the Hartree calculation and that returned by running the Herman–Skillman program, which calculation is based on the Hartree product wave function, equation 5.2 and the averaged electron–electron repulsion potential. The comparison suggests reasonable agreement between the two calculations and it is comforting, to the extent that we can retrieve the radial function for hydrogen by setting the exponent at 2.0 [remember Hartree divided by 2] for the function in equation 5.22.

Finally, we can work out the orbital energy for the 1s electrons in helium based on this level of approximation. Each electron is subject to the same interaction with the helium nucleus and possesses the same kinetic energy, so the orbital energy is the sum of these

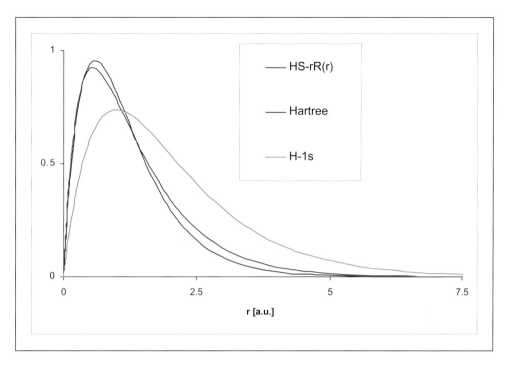

Figure 5.3   The 'best' result from Figure 5.2 as a comparison of the numerical radial function from the Herman–Skillman program and the Hartree calculation. All the graphs are of the function $rR(r)$ against the radial distance 'r' in atomic units. Use the figure to check the consistency of the scaling, by recovering the hydrogen result.

terms and, for each electron, the contribution from the two-electron repulsion term, all of which add to −0.895 Hartrees. The Herman–Skillman value for the orbital energy is −0.8605 Hartree.

5.3   THE HALL–ROOTHAAN EQUATIONS, THE ORBITAL APPROXIMATION AND THE MODERN HARTREE–FOCK *SELF-CONSISTENT* FIELD METHOD

The formal analysis of the mathematics required incorporating the linear combination of atomic orbitals molecular orbital approximation into the *self-consistent* field method was a major step in the development of modern Hartree–Fock–Slater theory. Independently, Hall (57) and Roothaan (58) worked out the appropriate equations in 1951. Then, Clementi (8,9,63) applied the procedure to calculate the electronic structures of many of the atoms in the Periodic Table using linear combinations of Slater orbitals. Nowadays linear combinations of Gaussian functions are the standard approximations in modern *LCAO-MO* theory, but the Clementi atomic calculations for helium are recognized to be very instructive examples to illustrate the fundamentals of this theory (67–69).

It is appropriate therefore to present the mathematical details, here too, using the simple case of a *double-zeta* Slater basis function calculation of the electronic energy of the helium atom. This mathematics is sufficient for the calculations in Chapter 6 on two-electron atoms like dihydrogen and the helium hydride ion, HeH^+ using Slater and Gaussian basis sets to approximate the 1s atomic orbitals of the basic LCAO-MO approximate function.

The Hartree product wave function, equation 5.2, for helium does not comply with the anti-symmetry requirement of the Pauli Principle (42,47) that electronic wave functions must change sign on exchange of the coordinates for a pair of electrons. Fock identified this defect in the overestimation of the electron–electron repulsion term, which occurs for Hartree product wave functions, while Slater showed how to overcome this problem by writing the product wave function in the form of a determinant (6,7,42,47,64).

The Hartree–Fock–Slater wave function for the singlet state of helium is the single determinant

$$\Phi(1,2) = \sqrt{\frac{1}{2}} \begin{vmatrix} \phi_{1s}^+(1) & \phi_{1s}^-(1) \\ \phi_{1s}^+(2) & \phi_{1s}^-(2) \end{vmatrix} \qquad 5.27$$

in which expression, each one-electron *spin orbital* is identified with a columns and each electron with a rows. Equation 5.27 is a solution to equation 5.4 because the Hamiltonian operator does not contain any term depending on electron spin, and the *spin orbitals* are simply the independent products of distinct space and spin functions, *viz*

$$\phi_{1s}^+(r_i, \sigma_i) = \phi_{1s}(r_i)\alpha(\sigma_i) \qquad 5.28$$

or

$$\phi_{1s}^-(r_i, \sigma_i) = \phi_{1s}(r_i)\beta(\sigma_i) \qquad 5.29$$

with the electron coordinate index 'i' either 1 or 2 for the case of helium.

The energy of the two-electron atom, now, for the wave function of equation 5.27, is the expectation value

$$E = \langle \Phi(1,2)|H(1,2)|\Phi(1,2) \rangle \qquad 5.30$$

since $\Phi(1,2)$, equation 5.27, is normalized.

The calculation is complete when we have found the 'best' choice of spin orbitals to minimize the energy expression.

To proceed, we substitute equation 5.27 into the equation 5.30 and use the Hamiltonian of equation 5.1 extended by the two-electron term $1/r_{12}$ to take account of the repulsive interaction of the electrons. For each one-electron component of equation 5.1 we have

$$\langle \Phi(1,2)|h(1)|\Phi(1,2) \rangle = \langle 2^{-1/2}[\phi_{1s}^+(1)\phi_{1s}^-(2) - \phi_{1s}^+(2)\phi_{1s}^-(1)]|$$
$$\times h(r_1)|2^{-1/2}[\phi_{1s}^+(1)\phi_{1s}^-(2) - \phi_{1s}^+(2)\phi_{1s}^-(1)]\rangle \quad 5.31$$

for example, for the electron identified by the coordinates (r_1, σ_1) and in which expression the dependence of the one-electron Hamiltonian on only the spatial coordinates is acknowledged. On carrying out the multiplications inside the Dirac brakets, four terms result for each one-electron Hamiltonian

$$\langle \Phi(1,2)|h(1)|\Phi(1,2) \rangle = \tfrac{1}{2}\langle \phi_{1s}^+(1)\phi_{1s}^+(1)|h(r_1)|\phi_{1s}^-(2)\phi_{1s}^-(2) \rangle$$
$$+ \tfrac{1}{2}\langle \phi_{1s}^-(1)\phi_{1s}^-(1)|h(r_1)|\phi_{1s}^+(2)\phi_{1s}^+(2) \rangle$$

$$-\tfrac{1}{2}\langle\phi_{1s}^{+}(1)\phi_{1s}^{-}(1)|h(\mathbf{r}_1)|\phi_{1s}^{-}(2)\phi_{1s}^{+}(2)\rangle$$

$$-\tfrac{1}{2}\langle\phi_{1s}^{+}(1)\phi_{1s}^{-}(1)|h(\mathbf{r}_1)|\phi_{1s}^{-}(2)\phi_{1s}^{+}(2)\rangle \qquad 5.32^{9}$$

Now, because of the opposed spins of the electrons in the helium ground state, the integrations over the spins reduce the first two integrals in equation 5.32 to integrals over the spatial coordinates of the electrons, while the same integrations over the spins reduces the third and fourth integrals in equation 5.32 to zero. Thus, equation 5.32 reduces to

$$\langle\Phi(1,2)|h(1)|\Phi(1,2)\rangle = \langle\phi_{1s}(\mathbf{r}_1)|h(\mathbf{r}_1)|\phi_{1s}(r_1)\rangle \qquad 5.33$$

We find the same result for the other one-electron Hamiltonian, so the contribution to the energy of the helium atom from the one-electron terms in equation 5.32 is

$$\langle\Phi(1,2)|h(\mathbf{r}_1)+h(\mathbf{r}_1)|\Phi(1,2)\rangle = 2\langle\phi_{1s}(\mathbf{r})|h(\mathbf{r})|\phi_{1s}(r)\rangle \qquad 5.34$$

There remains the two-electron part of equation 5.30,

$$\langle\Phi(1,2)|\frac{1}{r_{12}}|\Phi(1,2)\rangle = \langle 2^{-1/2}[\phi_{1s}^{+}(1)\phi_{1s}^{-}(2)-\phi_{1s}^{+}(2)\phi_{1s}^{-}(1)]|$$

$$\times \frac{1}{r_{12}}|2^{-1/2}[\phi_{1s}^{+}(1)\phi_{1s}^{-}(2)-\phi_{1s}^{+}(2)\phi_{1s}^{-}(1)]\rangle \qquad 5.35$$

Again, we can multiply out the terms in the integral to obtain

$$\left\langle\Phi(1,2)\left|\frac{1}{r_{12}}\right|\Phi(1,2)\right\rangle = \frac{1}{2}\left\langle\phi_{1s}^{+}(1)\phi_{1s}^{+}(1)\left|\frac{1}{r_{12}}\right|\phi_{1s}^{-}(2)\,\phi_{1s}^{-}(2)\right\rangle$$

$$+ \frac{1}{2}\left\langle\phi_{1s}^{-}(1)\,\phi_{1s}^{-}(1)\left|\frac{1}{r_{12}}\right|\phi_{1s}^{+}(2)\,\phi_{1s}^{+}(2)\right\rangle$$

[9] In Hartree–Fock–Slater theory it is common to refer to this manner of writing these integrals, with the coordinates ordered as in equation 5.32 as 'chemist's' notation and so to identify each term in equation 5.3 over the spatial coordinates only in the form

$$(ij|kl) = (\phi_i(1)\phi_j(1)|\frac{1}{r_{12}}|\phi_k(2)\phi_l(2))$$

This is an important statement for later reference, since the following identities are true

$$(kl|ij) = (ij|kl)$$

and also for real spin orbitals, which is generally true in Hartree–Fock–Slater theory

$$(ij|kl) = (ji|kl) = (ij|lk) = (ji|lk)$$

The physicist's notation, in two-electron theory, is to order the spin orbitals with their complex conjugates on the left of the electron repulsion operator and in the order starting with the coordinates of electron one, with equation 5.38 written as

$$\langle ij|kl\rangle = \langle ji|lk\rangle = \left\langle\phi_i(1)\phi_j(2)\left|\frac{1}{r_{12}}\right|\phi_k(r_1)\phi_l(r_2)\right\rangle$$

and with $\langle ij|kl = \langle kl|ij\rangle^{*}$. Both notations are the same for the one-electron integrals.

$$-\frac{1}{2}\left\langle \phi_{1s}{}^{+}\,(1)\,\phi_{1s}{}^{-}\,(1)\left|\frac{1}{r_{12}}\right|\phi_{1s}{}^{-}\,(2)\,\phi_{1s}{}^{+}\,(2)\right\rangle$$

$$-\frac{1}{2}\left\langle \phi_{1s}{}^{+}\,(1)\,\phi_{1s}{}^{-}\,(1)\left|\frac{1}{r_{12}}\right|\phi_{1s}{}^{-}\,(2)\,\phi_{1s}{}^{+}\,(2)\right\rangle \qquad 5.36$$

The integrals divide into equal pairs since the coordinates can be interchanged with the result

$$\left\langle \Phi\,(1,2)\left|\frac{1}{r_{12}}\right|\Phi\,(1,2)\right\rangle = \left\langle \phi_{1s}{}^{+}\,(1)\,\phi_{1s}{}^{+}\,(1)\left|\frac{1}{r_{12}}\right|\phi_{1s}{}^{-}\,(2)\,\phi_{1s}{}^{-}\,(2)\right\rangle - \left\langle \phi_{1s}{}^{+}\,(1)\,\phi_{1s}{}^{-}\,(1)\right.$$

$$\times \left|\frac{1}{r_{12}}\right|\phi_{1s}{}^{-}\,(2)\,\phi_{1s}{}^{+}\,(2)\right\rangle \qquad 5.37$$

The first term on the right-hand side of equation 5.37 identifies the Coulomb integral (J) of Hartree–Fock–Slater theory; the second term identifies the Exchange integral (K). For the closed shell, spin-paired electronic configuration of the ground state of helium, integration over the spin coordinates renders the Exchange integral equal to zero and so the total energy of the ground state of the helium atom is[10]

$$E = \langle \Phi\,(1,2)|H\,(1,2)\,|\Phi\,(1,2)\rangle = 2\,(\phi_{1s}\,(\mathbf{r})\,|h\,(\mathbf{r})\,|\phi_{1s}\,(\mathbf{r}))$$

$$+ \left(\phi_{1s}\,(\mathbf{r}_1)\,\phi\,(\mathbf{r}_1)\,|r_{12}^{-1}|\phi_{1s}\,(\mathbf{r}_2)\,\phi_{1s}\,(\mathbf{r}_2)\right) \qquad 5.38$$

Using the result in equation 5.38 we can recast the calculation of the energy of the helium atom as the problem to determine the eigenfunctions of the one-electron Hartree–Fock–Slater Hamiltonian. This identifies universally the Fock operator in the equation

$$F\,(\mathbf{r},\mathbf{r}')\,\phi_{1s}\,(\mathbf{r}) = \left[-\frac{1}{2}\nabla^2\,(\mathbf{r}) - \frac{2}{r} + \int |\phi_{1s}\,(\mathbf{r}')|^2\,\frac{1}{r_{12}}d\mathbf{r}'\right]\phi_{1s}\,(\mathbf{r}) \qquad 5.39$$

and we recover the orbital approximation in Hartree–Fock–Slater theory, since we have arrived at an expression for the energy of one electron moving in the potential field of the other, whose distribution is described, now, over the dummy coordinate system $\mathbf{r}'$.

The Hall–Roothaan equations for the case of the helium ground state appear, when equation 5.39 is written out in an appropriate linear combination of functions upon which the variation principle procedure can be applied to return the 'best' energy in a calculation. For example, for the *double-zeta* basis of Slater functions used in the previous section, we have

$$\phi_{1s}\,(r) = \sum_{i=1}^{2} c_i \zeta_i\,(r) \qquad 5.40$$

Substitution of equation 5.40 into equation 5.39, left-multiplication, in turn, by each term in the basis set and integration over the one-electron coordinate leads to the set of integro-differential equations upon which the variation principle is applied. Since the

---

[10]Note, to derive a general case relationship here, it is customary to count the number of possible J and K integrals without making a decision as to whether the K integrals are zero because of spin orthogonality. Thus, the general closed-shell expression, in place of this special case, equation 5.38, for two electrons only, takes the form

$$E = 2\sum_{n=1}^{N/2}\varepsilon_n - \sum_{m=1}^{N/2}\sum_{n=1}^{N/2}(2J_{mn} - K_{mn})$$

for N electrons and with J and K as in equation 5.37.

parameters in the equations are the coefficients of linear combination in equation 5.40, differentiation with respect to these coefficients and setting the results equal to zero leads to the matrix form of the Hall–Roothaan equations

$$\begin{vmatrix} f_{11} & f_{12} \\ f_{21} & f_{22} \end{vmatrix} \begin{vmatrix} c_{11} & c_{12} \\ c_{21} & c_{22} \end{vmatrix} = \begin{vmatrix} \Delta_{11} & \Delta_{12} \\ \Delta_{21} & \Delta_{22} \end{vmatrix} \begin{vmatrix} c_{11} & c_{12} \\ c_{21} & c_{22} \end{vmatrix} \begin{vmatrix} \varepsilon_1 & 0 \\ 0 & \varepsilon_2 \end{vmatrix} \qquad 5.41$$

Because there are two components in the linear combination in equation 5.40, we find two solutions to the Fock equation. Since these are eigenfunctions of the Fock operator, it is required that the solutions be orthonormal, yet this is not a condition in equation 5.40. However, this is exactly the problem discussed in Sections 3.6 and 3.7 and the canonical orthonormalization procedure can be applied in the diagonalization of equation 5.41 in the following manner to solve the *self-consistent* field problem.

The calculation of the two-electron potential in equation 5.39 involves the calculation of the charge density, $\rho(\mathbf{r})$, from the 1s orbital approximated to in equation 5.40. Since the two 1s spin orbitals are occupied in the helium atom this charge density is

$$\int \rho(\mathbf{r}) \, d\mathbf{r} = 2 \sum_{i=1}^{n/2} |\phi_{1s}(\mathbf{r})|^2 \qquad 5.42$$

Substitution of equation 5.41 into equation 5.42 leads to the defining of the density matrix formulation of equation 5.41, upon which the standard procedure for solving equation 5.41 is based. We have

$$\rho(\mathbf{r}) = 2 \sum_{i=1}^{N/2} |\phi_i(\mathbf{r})|^2 = 2 \sum_{i=1}^{N/2} \sum_j c_{ji} \zeta_j(\mathbf{r}) \sum_k c_{ki} \zeta_k(\mathbf{r})$$

$$= \sum_j \sum_k \left[2 \sum_{i=1}^{N/2} c_{ji} c_{ki} \right] \phi_j(\mathbf{r}) \phi_k(\mathbf{r}) \qquad 5.43$$

$$= \sum_j \sum_k P_{jk} \phi_j(\mathbf{r}) \phi_k(\mathbf{r})$$

and a definition of the density matrix P_{jk}

$$P_{jk} = 2 \sum_{i=1}^{N/2} c_{ji} c_{ki} \qquad 5.44$$

The Fock matrix elements, now, can be assembled from the one-electron kinetic and nuclear potential term and the two-electron field constructed using equation 5.43, with

$$F_{ij} = t_{ij} + v_{ij} + g_{ij} \qquad 5.45$$

with t_{ij} and v_{ij} the one-electron kinetic and potential energy terms of the matrix element over the basis set components 'i' and 'j' and g_{ij} the two-electron potential over all the terms arising from the application of equation 5.45 into the Coulomb integral of equation 5.38[11], *viz*

$$g_{11} = \tfrac{1}{2}[P_{11}(11|11) + P_{12}(11|12) + P_{21}(11|21) + P_{22}(11|22)] \qquad 5.46$$

$$g_{12} = \tfrac{1}{2}[P_{11}(12|11) + P_{12}(12|12) + P_{21}(12|21) + P_{22}(12|22)] \qquad 5.47$$

$$g_{22} = \tfrac{1}{2}[P_{11}(22|11) + P_{12}(22|12) + P_{21}(22|21) + P_{22}(22|22)] \qquad 5.48$$

5.4   CALCULATIONS USING SLATER DZ FUNCTIONS

To repeat Clementi's calculation for helium in a *double-zeta* Slater basis, we need only to modify fig5-1to3.xls to facilitate the extra component of the basis function and to add the standard worksheet for the Jacobi and canonical transformation of the Fock matrix.

Clementi's proposal to use a Slater *double-zeta* basis for the *self-consistent* field calculation on the helium atom means that the components of equation 5.40 are

$$\chi_i(r) = \left(\frac{\zeta_i^3}{\pi}\right)^{1/2} e^{-(\zeta_i r)} \qquad 5.49$$

which is equation 1.13 written out for 1s Slater radial orbital normalized over all the coordinates.

Exercise 5.2.   Calculation of the energy of the helium atom using the Clementi *double-zeta* basis.

This calculation is set out in fig5-4.xls. Create this spreadsheet by making a copy of fig5-1to3.xls and then adding as a copy any worksheet from Chapter 4 for the carrying out of the Fock matrix diagonalization, for example, fig4-21.xls.

1.   The 'worksheet' !Dzcalc contains the Simpson's rule procedure for carrying out the various integrations to return the terms in the Fock matrix, for the *double-zeta* basis defined in cells $D$6 and $E$6 [Table 1.3] and normalized as in equation 5.49, cells $D$7 and $E$7. Note now the Y(r) factors are handled in columns G to I, with the sums of the exponents for integration following equation 5.20 entered as appropriate sums in cells $G$17 to $I$17.

2.   The overlap, kinetic and potential energy terms over the components of the basis set are displayed as the usual **SUMPRODUCT** relations in cells $E10 to $E$12 for overlap/normalization, $H$2 to $H$4 for potential energy and $H$6 to $H$8 for kinetic energy. The forms for the grad^2 entries in columns E and F follow from equation 4.16 and 4.17 and, for example, the formula in cell $E$19 is

$$\$E\$19 = -0.5*\textbf{POWER}(\$D\$6*\$B19,2)-2*\$D\$6*\$B19$$

---

[11]The factor $\tfrac{1}{2}$, often, is subsumed into the *J* integral terms in these equations.

	A	B	C	D	E	F	G	H	I	J	K	L	M
1	He 1s² energy.									h₁₁	-1.846597		K.E.
2	Slater DZ calculation.						V₁₁	-2.89220		h₁₂	-1.885959		2.86134
3							V₁₂	-3.62932		h₂₂	-1.628307		N.E.
4			Z	2.00000			V₂₂	-5.72440					-6.74878
5			Slater DZ	|1⟩	|2⟩					g₁₁	0.959031		(g|g)
6			zeta	1.44610	2.86220		t₁₁	1.04560		g₁₂	0.979683		1.025766
7			d	0.981121	2.7319644		t₁₂	1.74336		g₂₂	1.292414		ε
8			c_i	0.83416	0.19059		t₂₂	4.09609					-0.917952
9										f₁₁	-0.887566		
10				Δ₁₁	1.00000		(11|11)	0.90381		f₁₂	-0.906277		E
11				Δ₁₂	0.84240		(22|22)	1.78888		f₂₂	-0.335893		-2.86167
12				Δ₂₂	1.00000		(11|22)	1.17494					
13							(12|12)	0.95542		Δ₁₁	1.000000		
14							(11|12)	0.90406		Δ₁₂	0.842403		
15							(12|22)	1.28909		Δ₂₂	1.000000		
16			Norms	1.0000			2*$E$11	2*$F$11	$E$11*$F$11				
17			1.0000	1.0000			-2.89220	-5.72440	-4.30830				
18	r		|1⟩	|2⟩	grad`|1⟩	grad`|2⟩	Y[11]	Y[12]	Y[22]	r|1⟩	r|2⟩	nr|1⟩	nr|2⟩
19	0.000	1	0.9811	2.7320	0.00000	0.00000	0.00000	0.00000	0.00000	0.00000	0.00000	0.00000	0.00000
20	0.010	4	0.9670	2.6549	-0.02871	-0.05642	0.00115	0.00228	0.00144	0.00967	0.02655	0.00986	0.00972
21	0.020	2	0.9532	2.5800	-0.05701	-0.11121	0.00230	0.00455	0.00288	0.01906	0.05160	0.01943	0.01889

	A	B	C	D	E	F	G	H	I	J	K
27	c₁	c₂		c₁	c₂		T	N.E.	ee	ε	E - He
28	0.000000	0.000000	#	0.642833	0.396567					-1.980713	
29	0.642833	0.396567	0	0.871321	0.148924		3.930218	-7.891644	1.187329	-0.846149	-2.879626
30	0.871321	0.148924	1	0.826709	0.198874		2.674212	-6.529206	0.996831	-0.932767	-2.862365
31	0.826709	0.198874	2	0.835644	0.188936		2.899752	-6.792953	1.031670	-0.915014	-2.861698
32	0.835644	0.188936	3	0.833861	0.190922		2.853718	-6.739975	1.024592	-0.918540	-2.861671
33	0.833861	0.190922	4	0.834217	0.190526		2.862870	-6.750540	1.026001	-0.917835	-2.861670
34	0.834217	0.190526	5	0.834146	0.190605		2.861041	-6.748430	1.025719	-0.917976	-2.861670
35	0.834146	0.190605	6	0.834160	0.190589		2.861406	-6.748852	1.025775	-0.917947	-2.861670
36	0.834160	0.190589	7	0.834157	0.190592		2.861333	-6.748767	1.025764	-0.917953	-2.861670
37	0.834157	0.190592	8	0.834158	0.190591		2.861348	-6.748784	1.025766	-0.917952	-2.861670
38	0.834158	0.190591	9	0.834158	0.190592		2.861345	-6.748781	1.025766	-0.917952	-2.861670

Figure 5.4a   The worksheet 'Dzcalc'! in fig5-4.xls for the calculation of the various terms in the Fock matrix for the helium atom calculation using the Slater *double-zeta* basis from Table 1.3 and in the second part of the figure, the outputs from each iteration until convergence is reached after nine cycles of calculation.

3.   Then the components of the two-electron potential are constructed in cells $H$10 to $H$15 following exactly the prescription in equation 5.33, we have

$H$10 = (11|11) = **POWER**(4***PI**( ),2)***SUMPRODUCT**($B$19:$B$3019, $C$19:$C$3019,$J$19:$J$3019,$G$19:$G$3019)* ($A$20-$A$19)/3

$H$11 = (22|22) = **POWER**(4***PI**( ),2)***SUMPRODUCT**($B$19:$B$3019, $D$19:$D$3019,$K$19:$K$3019,$H$19:$H$3019)* ($A$20-$A$19)/3

$H$12 = (11|22) = **POWER**(4***PI**( ),2)***SUMPRODUCT**($B$19:$B$3019, $C$19:$C$3019,$J$19:$J$3019,$H$19:$H$3019)* ($A$20-$A$19)/3

$H$13 = (12|12) = **POWER**(4***PI**( ),2)***SUMPRODUCT**($B$19:$B$3019, $C$19:$C$3019,$K$19:$K$3019,$I$19:$I$3019)* ($A$20-$A$19)/3

$H$14 = (11|12) = **POWER**(4***PI**( ),2)***SUMPRODUCT**($B$19:$B$3019, $C$19:$C$3019,$K$19:$K$3019,$G$19:$G$3019)* ($A$20-$A$19)/3

$$\$H\$15 = (12|22) = \textbf{POWER}(4^*\textbf{PI}(\,),2)^*\textbf{SUMPRODUCT}(\$B\$19{:}\$B\$3019,$$
$$\$C\$19{:}\$C\$3019,\$K\$19{:}\$K\$3019,\$H\$19{:}\$H\$3019)^*$$
$$(\$A\$20{-}\$A\$19)/3$$

4. Next, the data are collected together to form the Fock matrix on the worksheet 'hfs'!. Most of the design details are familiar from previous applications of the procedures in Sections 3.6 and 3.7. The extra features provide for the calculation of the two-electron part of the potential over the basis set using the density matrix procedure of equations 5.46 to 5.48.

	A	B	C	D	E	F	G	H	I	J	K
1	Clementi DZ calculation on the helium atom.									K.E.	
2										2.86134	
3	S			theta =	0.78540		S-diagonal			N.E.	
4	1.00000	0.84240		omega1	1.84240		1.84240	0.00000		-6.74878	E
5	0.84240	1.00000		omega2	0.15760		0.00000	0.15760		(g|g)	-2.86167
6										1.02577	
7	U			U - adjoint			The S^1/2 matrix			ε	
8	0.70711	0.70711		0.70711	0.70711		0.73673	0.00000		-0.91795	
9	0.70711	-0.70711		0.70711	-0.70711		0.00000	2.51898			
10											
11	H			G			Fock matrix			(11|11)	0.903813
12	-1.84660	-1.88596		0.95903	0.97968		-0.88757	-0.90628		(22|22)	1.788876
13	-1.88596	-1.62831		0.97968	1.29241		-0.90628	-0.33589		(11|22)	1.174939
14										(12|12)	0.955422
15				X-canonical			X-canon transpose			(11|12)	0.904063
16				0.52095	1.78119		0.52095	0.52095		(12|22)	1.289092
17				0.52095	-1.78119		1.78119	-1.78119			
18										README	
19	theta	0.18165		Transformed Fock matrix						Density matrix (in)	
20	omega1	-0.91795		-0.93450	0.03333	==>	-0.82393	-0.51190		0.69582	0.15898
21	omega2	1.96301		-0.64710	-1.01596		-0.51190	1.86898		0.15898	0.03633
22											
23	Coefficients			Fock - diagonal			C - originals			Density matrix (out)	
24	0.98355	0.18066		-0.91795	0.00000		0.83416	-1.65777		0.69582	0.15898
25	0.18066	-0.98355		0.00000	1.96301		0.19059	1.84599		0.15898	0.03633

Figure 5.4b The output of each iteration to *self-consistency* for the Clementi *double-zeta* Slater basis calculation of the ground state energy of the helium atom and the final converged result of the worksheet 'hfs'! in fig5-4.xls. Further iterations lead to no improvement of the results and the energy of the helium atom is found to be 2.86167 a.u.

Thus, in cells \$J\$20 to \$K\$21 provision is made for the entry of the coefficients of the linear combination of equation 5.40 in the form of the density matrix elements of equation 5.42 and 5.43. Then these are applied, with

$$g_{11} = \$J\$20^*\$K\$11+2^*\$K\$20^*\$K\$15+\$K\$21^*\$K\$13$$

$$g_{12} = \$J\$20^*\$K\$15+2^*\$K\$20^*\$K\$14+\$K\$21^*\$K\$16$$

$$g_{22} = \$J\$20^*\$K\$13+2^*\$K\$20^*\$K\$16+\$K\$21^*\$K\$12$$

and the individual $\langle ij|kl\rangle$ terms linked to 'Dzcalc'! from cells \$K\$11 to \$K\$16 on 'hfs'!.

5. The output density matrix required for restarting the calculation, until self-consistency is achieved, is defined in cells \$J\$24 to \$K\$25. The matrix elements are based on the output coefficients in cells \$G\$24 and \$G\$25 for the original

basis. However, note that the factor of 2 in equation 5.44 has been omitted since the Fock operator is for one-electron only.

6. Next, we need to provide for the calculation of the totals of the energy components. These are presented in column M of 'Dzcalc'! and involve multiplying by the squares and cross terms of the input coefficients linked at 'Dzcalc'! cells $D$8 and $E$8 and allowing for the presence of two electrons except when counting the two-electron term. Thus[12],

$$T = \$M\$2 = 2^*(\$D\$8^\wedge2^*\$H\$6+2^*\$D\$8^*\$E\$8^*\$H\$7+\$E\$8^\wedge2^*\$H\$8)/\$D\$16$$

$$V_{ne} = \$M\$4 = 2^*(\$D\$8^\wedge2^*\$H\$2+2^*\$D\$8^*\$E\$8^*\$H\$3+\$E\$8^\wedge2^*\$H\$4)/\$D\$16$$

$$EE = \$M\$6 = \$D\$8^\wedge2^*K5+2^*\$D\$8^*\$E\$8^*K6+\$E\$8^\wedge2^*K7)/\$D\$16$$

and finally on adding all together, or applying equation 5.21, we find the electronic energy of the ground state of helium to be as calculated by Clementi in 1963 (63)

$$E_{He} = \$M\$11 = 2^*\$M\$8-\$M\$6 = -2.8167 \text{ a.u.}$$

7. Now, initiate the *self-consistent* field calculation by inputting a null density matrix in cells 'hfs'!j20 to $J$21. The output density matrix elements cells 'hfs'!$J$24 to $K$25 then are pasted as values only, using the EDIT/PASTE SPECIAL option to start the second iteration. The sequence to self-consistency, the condition in which the input and output density matrix elements are different, only by a small set amount, is set out in Figure 5.4b.

The converged result, returning -2.86167 Hartrees for the energy of the helium atom corresponds, to the *Hartree–Fock* limit, the 'best' result that can be obtained using a simple product of one-electron functions for the two-electron wave function. 'Complete' agreement, with the 'exact' energy of helium, -2.9037 Hartrees (70), has been reported in more extended calculations by Hylleras, Kinoshita and Pekeris (71–73) using 14, 39 and 1029 term polynomial forms for the ground state wave function.

5.5 GAUSSIAN BASIS SET CALCULATIONS FOR THE HELIUM ATOM – TWO-ELECTRON INTEGRALS OVER GAUSSIAN BASIS FUNCTIONS

A spreadsheet for the calculation of the energy of the helium atom using a Gaussian basis exhibits substantial similarities to previous ones in this chapter and in Chapter 4. Thus, we know how to calculate the potential and kinetic energy terms for the case of linear combinations of quadratic exponential functions to model the correct electronic eigenfunctions. However, we cannot follow the Hartree procedure for the calculation of the electron–electron repulsive potential since it is not possible to find a solution to a second-order differential equation involving a Gaussian function. So, it is appropriate, to begin to adopt appropriate procedures for the evaluation of integrals involving Gaussian functions.

Since the determination of these integrals dominate modern molecular orbital methodology, let us write out a general approach, which we can carry over into the next chapter on

[12]Note, the extra check to ensure normalization of the overall wavefunction in 5.40, the division by $D$16.

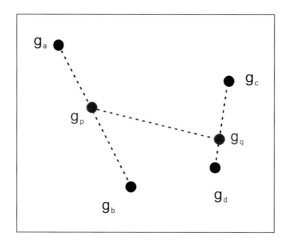

Figure 5.5 Schematic of the sequence in the reduction of the two-electron integral term in Hartree–Fock theory. There are four terms in the basic integral over the primitive Gaussian functions about the centres A, B, C and D. These are transformed into the Gaussians about the intermediate centres P and Q, using equation 1.17 to 1.19. Then, the integral is evaluated using equation 5.55.

the theory of diatoms[13]. For simplicity, only calculations involving $|\text{sto-ng:1s}\rangle$ Gaussian primitives are considered, since it is only for these cases that we can apply the simplifying transformations of equations 1.17 to 1.19 directly.

The general two-electron potential problem is set out schematically in Figure 5.5. Electron density components repel in the manner of equation 5.15[14]. The most general case is shown in the figure, with each electron density component comprising the squares of orbital components about different positions in the atomic or molecular space. For any Gaussian set based representation of the 1s-type components, the four primitive Gaussians, $|i\rangle$, $|j\rangle$, $|k\rangle$ and $|l\rangle$ about A B, C and D contribute to the two-electron integral, $\langle ij|kl\rangle$, as

$$\langle ij|kl\rangle = \iint g_i\,(\mathbf{r}_1 - \mathbf{R}_A)\,g_j\,(\mathbf{r}_1 - \mathbf{R}_B)\left(\frac{1}{r_{12}}\right)g_j\,(\mathbf{r}_2 - \mathbf{R}_C)\,g_k\,(\mathbf{r}_2 - \mathbf{R}_D)\,\mathrm{d}\mathbf{r}_1\,\mathrm{d}\mathbf{r}_2 \quad 5.50$$

Pair-wise reduction of Gaussian products following equation 1.16 to 1.19 about the intermediate positions P and Q, leads to

$$\langle ij|kl\rangle = e^{\left[\frac{\alpha\beta}{\alpha+\beta}|\mathbf{R}_A-\mathbf{R}_B|^2 - \frac{\gamma\delta}{\gamma+\delta}|\mathbf{R}_C-\mathbf{R}_D|^2\right]}\iint e^{\left(-p|r_1-\mathbf{R}_p|^2\right)}\frac{1}{r_{12}}e^{\left(-q|r_2-\mathbf{R}_Q|^2\right)}\,\mathrm{d}\mathbf{r}_1\mathrm{d}\mathbf{r}_2 \quad 5.51$$

with p the sum of α and β and q the sum of γ and δ.

---

[13]This account and the similar discussion in Chapter 6 derives from the excellent discussion on the calculation of integrals over Gaussians in *Modern Quantum Chemistry–Introduction to Advanced Electronic Structure Theory*, A. Szabo *and* N. Ostlund (47). The general procedures were set out in an early review by Shavitt (74) and reviewed, more recently, by Gill (38).

[14]Assuming that equation 5.15 has been modified to include the primitive Gaussians only, rather than the products $rR(r)$ used in Hartree's original analysis.

This integral and, as you will see in the next chapter, indeed all the integrals over Gaussian functions are evaluated by replacing each quantity in the integral by its Fourier transform.

In the end[15], after appropriate Fourier transformations on the terms in equation 5.51 and some changes of variable we arrive at the final expression for the two-electron integral over the four primitive Gaussians

$$\langle ij|kl\rangle = (ij|kl)$$

$$= \frac{2\pi^{5/2}\, e^{\left[\frac{\alpha\beta}{\alpha+\beta}|\mathbf{R}_A-\mathbf{R}_B|^2 - \frac{\gamma\delta}{\gamma+\delta}|\mathbf{R}_C-\mathbf{R}_D|^2\right]} F_0\left[(\alpha+\beta)(\gamma+\delta)(\alpha+\beta+\gamma+\delta)\,|R_P-R_Q|^2\right]}{\left[(\alpha+\beta)(\gamma+\delta)(\alpha+\beta+\gamma+\delta)^{1/2}\right]}$$

$$5.52$$

The spreadsheet, fig5-6.xls, provides for the calculation of all the integrals in the solution of the helium atom problem over a Gaussian basis set of up to six 1s primitives. The familiar integration procedures appear in Figure 5.6a, based on the worksheet oneel! in the spreadsheet, while the two-electron components occur on the worksheet, twoel!,

	A	B	C	D	E	F	G	H	I	J
1	Helium ground-state energy |sto-ng:1s⟩ calculation.							alpha(i)	alpha(j)	
2		|sto-ng⟩	Slater ζ	1.68500				1.00000	1.000	E_{He}
3		|1⟩	|2⟩	|3⟩	|4⟩	|5⟩	|6⟩	⟨i|j⟩	1.000001	-2.80780
4	α	0.1098180	0.4057710	2.2276600	0.0000000	0.0000000	0.00000	V	-3.33362	
5	d	0.4446350	0.5353280	0.1543290	0.0000000	0.0000000	0.00000	T	1.403422	ε_{1s}
6	αζ²	0.31180	1.15208	6.32483	0.00000	0.00000	0.00000	F	1.052593	-0.87760
7	d'	0.13223	0.42427	0.43868	0.00000	0.00000	0.00000			
8		[1|1]	[2|2]	[3|3]	[1|2]	[1|3]	[2|3]	[4|4]	[5|5]	[6|6]
9	d'*d'	0.01748	0.18000	0.19244	0.05610	0.05800	0.18612	0.00000	0.00000	0.00000
10	α'+α'	0.62360	2.30415	12.64966	1.46387	6.63663	7.47690	0.00000	0.00000	0.00000
11	5.5683	11.30758	1.59206	0.12377	3.14391	0.32569	0.27236	113097	113097	113097
12	sij	0.19770	0.28658	0.02382	0.17637	0.01889	0.05069	0	0	0
13		[1|1]	[2|2]	[3|3]	[1|2]	[1|3]	[2|3]	[4|4]	[5|5]	[6|6]
14		0.62360	2.30415	12.64966	1.46387	6.63663	7.47690	0	0	0
15	12.5664	20.1515	5.4538	0.9934	8.5843	1.8935	1.6807	11310	11310	11310
16	vij	-0.35233	-0.98170	-0.19117	-0.48158	-0.10983	-0.31280	0.00000	0.00000	0.00000
17		[1|1]	[2|2]	[3|3]	[1|2]	[1|3]	[2|3]	[4|4]	[5|5]	[6|6]
18		-10.57705	-5.50251	-2.34843	-4.62883	-0.58067	-1.59258	0.00000	0.00000	0.00000
19	tij	0.0924638	0.4952358	0.2259618	0.1298378	0.0168408	0.148203	0	0	0
20	0.0000	0.62360	2.30415	12.64966	1.46387	6.63663	7.47690	0.00000	0.00000	0.00000
21	0.3118	-10.577								
22	1.1521		-5.503							
23	6.3248			-2.348						

Figure 5.6a   The familiar part of the general spreadsheet, the worksheet 'oneel!', for the calculation of the energy of the helium atom using the |sto-3g⟩ basis set of Table 1.6 and the optimized choice for the Slater exponent, with the electron–electron repulsive potential as the extra term in cell $I$7 worked out on the worksheet 'twoel!', detail from which is shown in Figure 5.6b. Note, the $A$15 master formula entry allows for the He atomic number in the calculation of the nucleus–electron potential term.

[15] Reference 46 and note that F_0 is the *Error function* (75), available, fortunately, as an intrinsic function in the *Engineering function* library in EXCEL. You will need to install this extra part of the software using your EXCEL CD.

	A	B	C	D	E	F	G	H	I	J
2		|sto-ng:1s)	scaler	1.68500						
3	exponent =)	0.109818	0.405771	2.227660		0.000000	0.000000	0.000000		
4	coeff =)	0.444635	0.535328	0.154329		0.000000	0.000000	0.000000		
5	alpha =)	0.311798	1.152075	6.324828		0.000000	0.000000	0.000000		
6	d =)	0.132227	0.424268	0.438676		0.000000	0.000000	0.000000		
7	Calculation of the 2-electron integral.							#		
8	Ra	Rb	Rc	Rd		V1111 =)	1.0525928	81		
9	0	0	0	0						
10	α_i	α_j	α_k	α_l		d_i	d_j	d_k	d_l	v1111
11	1.152075	1.152075	1.152075	0.311798	4	0.424268	0.424268	0.424268	0.132227	0.215841
12	1.152075	0.311798	0.311798	1.152075	4	0.424268	0.132227	0.132227	0.424268	0.120118
13	0.311798	1.152075	0.311798	0.311798	4	0.132227	0.424268	0.132227	0.132227	0.104075
14	1.152075	1.152075	1.152075	1.152075	1	0.424268	0.424268	0.424268	0.424268	0.099466
15	1.152075	1.152075	0.311798	0.311798	2	0.424268	0.424268	0.132227	0.132227	0.089572
16	1.152075	0.311798	1.152075	6.324828	8	0.424268	0.132227	0.424268	0.438676	0.089295
17	6.324828	1.152075	1.152075	1.152075	4	0.438676	0.424268	0.424268	0.424268	0.087017
18	0.311798	0.311798	6.324828	1.152075	4	0.132227	0.132227	0.438676	0.424268	0.034317
19	6.324828	0.311798	0.311798	0.311798	8	0.438676	0.132227	0.424268	0.132227	0.032939
20	6.324828	0.311798	1.152075	1.152075	4	0.438676	0.132227	0.424268	0.424268	0.031957
21	0.311798	0.311798	0.311798	0.311798	1	0.132227	0.132227	0.132227	0.132227	0.024627

	L	M	N	O	P	Q	R	S
10	Rp	Rq	Rab2	Rcd2	Rpq2	B	fzero(B)	Integrals
11	0.000000	0.000000	0.000000	0.000000	0.000000	0.000000	1.000000	5.343596
12	0.000000	0.000000	0.000000	0.000000	0.000000	0.000000	1.000000	9.541816
13	0.000000	0.000000	0.000000	0.000000	0.000000	0.000000	1.000000	26.526982
14	0.000000	0.000000	0.000000	0.000000	0.000000	0.000000	1.000000	3.069818
15	0.000000	0.000000	0.000000	0.000000	0.000000	0.000000	1.000000	14.230619
16	0.000000	0.000000	0.000000	0.000000	0.000000	0.000000	1.000000	1.069035
17	0.000000	0.000000	0.000000	0.000000	0.000000	0.000000	1.000000	0.649351
18	0.000000	0.000000	0.000000	0.000000	0.000000	0.000000	1.000000	2.636475
19	0.000000	0.000000	0.000000	0.000000	0.000000	0.000000	1.000000	1.265313
20	0.000000	0.000000	0.000000	0.000000	0.000000	0.000000	1.000000	0.765172
21	0.000000	0.000000	0.000000	0.000000	0.000000	0.000000	1.000000	80.562204

Figure 5.6b   The worksheet for the calculation of the two-electron repulsion term, equation 5.52. The values, calculated in column R, follow from the intrinsic *Error function* defined in the *Engineering function* library in EXCEL.

represented in Figure 5.6b. This component of the potential energy appears then, as an energy term in cell $I$7 of the oneel! worksheet. The individual parts of this potential energy follow from the Gaussian primitives involvement in equation 5.52 with

$$g_a = |i\rangle = d_i \exp(-\alpha_i (\mathbf{r} - \mathbf{R}_A)^2$$

$$g_b = |j\rangle = d_j \exp(-\alpha_j (\mathbf{r} - \mathbf{R}_B)^2$$

$$g_c = |k\rangle = d_k \exp(-\alpha_k (\mathbf{r} - \mathbf{R}_C)^2$$

$$g_d = |l\rangle = d_l \exp(-\alpha_l (\mathbf{r} - \mathbf{R}_D)^2$$

assembled on 'twoel!', in the general form, displayed in Figure 5.5, and, in detail, as follows,

---

Exercise 5.3.   Calculation of the energy of the helium atom using a Gaussian basis set.

1.   Construct the worksheet 'oneel!' as displayed in Figure 5.6a in the usual manner and link the input data through to the appropriate cells of worksheet 'twoel!'.

Thus, cells $A$2 to $D$6 in 'twoel!' relate to cells in 'oneel!' as, for example, 'twoel!'$B$3 = 'oneel!'$B$4 and so on.

2. Set all the primitives about the same centre, the helium nucleus, with the entries 0.0 for R_A, R_B, R_C and R_D in cells $A$9 to $D$9 of this worksheet.

3. Then, identify the possible $ijkl$ components of the integral which are in cells $A$11 to $I$31, with the exponents of the primitives in cells $A$11 to $D$31, the coefficients in cells $F$11 to $I$31 and the 'degeneracy' of each particular choice in column E. The present example is over the $|sto\text{-}3g\rangle$ basis and so we expect 3^4 components of the total integral, which is the sum of the degeneracies in column E given in cell $H$8 [*It is probably good practice here to write these terms out in full, rather than to make use of the identities in footnote 9, page 169*].

4. The *Vijkl* entries in column J, then, are the products of these factors with the individual two-electron components calculated on the remainder of the worksheet as shown in the second-part of Figure 5.9.

   (i)   Thus, there are the trivial, in this application, entries for the various inter-site distances of Figure 5.5 in cells $L$11 to $P$31, which provide for the construction of the parts of equation 5.53 in columns Q to S.

   (ii) For example, cell $Q$11 is the argument of the error function for the Gaussian primitives identified in cells $A$11 to $I$11 with

$$\$Q\$11 = (\$A11+\$B11)^*(\$C11+\$D11)/(\$A11+\$B11 +\$C11+\$D11)^*\$P11$$

   (iii) Column R entries return the *Error function* needed to complete equation 5.53, with, for example,

$$\$R11 = \textbf{IF}(Q11<0.000001,(1\text{-}\$Q11/3),$$
$$(0.5^*\textbf{SQRT}(\textbf{PI}(\ )/\$Q11))^*\textbf{ERF}(0, \textbf{SQRT}(\$Q11)))$$

Note, the inclusion of a fixed value choice in this formula to avoid the singularity in the *Error function* for small values of its argument.

   (iv) Equation 5.53 is completed in column S, with, for example, the entry in cell $S11

$$\$S11 = 2^*\textbf{POWER}(\textbf{PI}(\ ),2.5)/((\$A11+\$B11)^*(\$C11+\$D11)^*$$
$$\textbf{POWER}((\$A11+\$B11+\$C11+\$D11),0.5))^*$$
$$\textbf{EXP}(\text{-}\$A11^*\$B11^*\$N11/(\$A11+\$B11)\text{-}\$C11^*\$D11^*\$O11/$$
$$(\$C11+\$D11))^*R11$$

and similarly in the remaining cells of the column required for the individual components of the integral over the Gaussian basis set.

   (v)   Finally, the entries in column J are the products of the coefficients of the primitive Gaussians and the equation 5.53 integrals, for example

$$\$J\$11 = \$S11^*\$F11^*\$G11^*\$H11^*\$I11^*\$E11$$

weighted, in this example, by the number of identities, column E, for each i,j,k and l.

Note, how close the Slater exponent in cell $D$2, returned by the SOLVER analysis, is to the *effective atomic number* of Hartree's calculation. Again, as chemists, we have a pleasing agreement, in the calculation, with the notion that the electrons screen each other from the full effect of the atomic number of protons in the nucleus. However, note, too, that the calculated results are not very good, especially when compared to the results of the *double-zeta* calculation in the previous section.

Table 5.1 lists Huzinaga Gaussian basis sets (45) designed specifically for helium atom energy calculations and his results are summarized in Table 5.2. We see that while Huzinaga's |sto-3g⟩ basis set is deficient as a product wave function in the form of equation 5.2 for helium.

Huzinaga's |sto-6g⟩ basis returns an energy nearer the HF limit compared to the result for the |sto-6g⟩ basis of Table 1.6. This is another example of a result that we might have expected. The data in Table 1.6 derive from the matching of the basis set to the Slater function, not the energy of the wave function. Figure 5.7 displays the Pople *et al.* result. Huzinaga's results are returned on substitution of his data into the appropriate input cells of fig5-7.xls.

Fig5-7.xls has been designed to provide for calculations with basis sets ranging up to |sto-6g⟩. This flexibility requires that, in the detail, on the worksheet 'twoel!' we avoid the #DIV/0! error message for the smaller basis sets with the entries

$$K7 = \mathbf{IF}(\mathbf{OR}(\$A7 = 0, \$B7 = 0, \$C7 = 0, \$D7 = 0), 0, (\$A7^*\$A\$5 + \$B7^*\$B\$5) / (\$A7 + \$B7))$$

$$L7 = \mathbf{IF}(\mathbf{OR}(\$A7 = 0, \$B7 = 0, \$C7 = 0, \$D7 = 0), 0, (\$C7^*\$C\$5 + \$D7^*\$D\$5) / (\$C7 + \$D7))$$

Table 5.1   The Huzinaga basis sets (45) for the calculation of the energy of the helium atom. The Slater exponent is set equal to 1.0 in any calculation.

	α	d		α	d		α	d
2	0.532149	0.82559		28.95149	0.02465		0.308364	0.31570
	4.097728	0.28317		182.4388	0.00330		0.725631	0.34783
3	0.382938	0.65722	7	0.160274	0.18067		1.802569	0.24466
	1.998942	0.40919		0.447530	0.43330		4.951881	0.11748
	13.62324	0.08026		1.297177	0.34285		15.41660	0.03844
4	0.298073	0.51380		4.038781	0.15815		55.41029	0.00939
	1.242567	0.46954		14.22123	0.04727		246.8036	0.00178
	5.782948	0.15457		62.24915	0.00971		1663.571	0.00023
	38.47497	0.02373		414.4665	0.00127	10	0.107951	0.05242
5	0.244528	0.39728	8	0.137777	0.11782		0.240920	0.24887
	0.873098	0.48700		0.347207	0.36948		0.552610	0.36001
	3.304241	0.22080		0.918171	0.36990		1.352436	0.28403
	14.60940	0.05532		2.580737	0.21021		3.522261	0.14909
	96.72976	0.00771		7.921657	0.07999		9.789053	0.05709
6	0.193849	0.26768		28.09935	0.02134		30.17990	0.01721
	0.589851	0.46844		124.5050	0.00415		108.7723	0.00412
	1.879204	0.29801		833.0522	0.00053		488.8941	0.00076
	6.633653	0.10964	9	0.129793	0.09809		3293.694	0.00010

Table 5.2   Huzinaga's results for the
helium atom for the basis sets listed in
Table 5.1, together with the result of
an |sto-6g⟩ calculation using the data
of Table 1.6.

|sto-ng⟩	E [hartrees]	ε
2	−2.7470661	−0.858911
3	−2.8356798	−0.903577
4	−2.8551603	−0.914124
5	−2.8598949	−0.916869
6	−2.8611163	−0.917688
7	−2.8614912	−0.917895
8	−2.8616094	−0.917931
9	−2.8616523	−0.917846
10	−2.8616692	−0.917952
[Table 1.6]		
6	−2.8463523	−0.8958613

$P7 = \mathbf{IF(OR}(\$A7 = 0, \$B7 = 0, \$C7 = 0, \$D7 = 0),0, 2^*\mathbf{POWER(PI}(\ ),2.5)/$
$((\$A7+\$B7)^*(\$C7+\$D7)^*\mathbf{POWER}((\$A7+\$B7+\$C7+\$D7),0.5))^*$
$\mathbf{EXP}(-\$A7^*\$B7^*\$M7/(\$A7+\$B7)-\$C7^*\$D7^*\$N7/(\$C7+\$D7))^*R7)$

$Q7 = \mathbf{IF(OR}(\$A7 = 0, \$B7 = 0, \$C7 = 0, \$D7 = 0),0,(\$A+\$B7)^*$
$(\$C7+\$D7)/(\$A7+\$B7+\$C7+\$D7)^*\$O7)$

down columns K, L, P and Q of this worksheet. Note, the remark, in the legend, to Figure 5.7, all the contributions to the two-electron integral for up to six primitive Gaussians are included.

As you see from Table 5.2, Huzinaga found that his calculated energy for helium approached the Hartree–Fock limit for his large basis sets, and you will find that this is true by making an enlarged version of fig5-7.xls and repeating the Huzinaga calculations.

5.6   A HFS-SCF CALCULATION WITH *SPLIT-BASIS* |4-31⟩ FOR HELIUM

However, it is more instructive to examine the effect of using a *split-basis* approach, for example, let us develop the equivalent of the Dunning |4−31⟩ contraction of the Huzinaga |sto-4g:1s⟩ basis of Table 1.7 and equations 1.24 and 1.25 for the helium basis set of four primitives in Table 5.1. The following analysis presents another example of the *HFS-SCF* procedure and emphasizes the utility of approximate functions in the form of equation 5.40.

The results for Huzinaga's |sto-4g:1s⟩ basis set for helium are displayed in Figure 5.8 which is based on the spreadsheet fig5-8.xls developed straightforwardly from fig4-18.xls taking into account the atomic number of helium and the two-electron potential field for the ground state electronic configuration. As you see, there is agreement to the fourth decimal figure with the results in Table 5.2.

	A	B	C	D	E	F	G	H	I	J	K
1		|sto-ng⟩	Slater ζ	1.6875	Z =⟩	2.00000		alpha(i)	alpha(j)		E_{He}
2		|1⟩	|2⟩	|3⟩	|4⟩	|5⟩	|6⟩	1.0000	1.000		-2.84635
3	α	0.06511	0.15809	0.40710	1.18506	4.23592	23.31030	⟨i|j⟩	0.999983		
4	d	0.13033	0.41649	0.37056	0.16854	0.04936	0.00916	V	-3.373776		$ε_{1s}$
5	αζ²	0.18541	0.45018	1.15928	3.37464	12.06244	66.37972	T	1.423285		-0.89586
6	d'	0.02625	0.16314	0.29506	0.29907	0.22771	0.15188	ee	1.054630		
7		[1|1]	[2|2]	[3|3]	[1|2]	[1|3]	[2|3]	[4|4]	[5|5]	[6|6]	[4|5]
8	d'*d'	0.00069	0.02661	0.08706	0.00428	0.00774	0.04814	0.08945	0.05185	0.02307	0.06810
9	α'+α'	0.37082	0.90036	2.31856	0.63559	1.34469	1.60946	6.74929	24.12489	132.759	15.43709
10	5.568	24.65937	6.51779	1.57724	10.98903	3.57102	2.72713	0	0	0	0
11	sij	0.01699	0.17347	0.13732	0.04705	0.02765	0.13127	0.02841	0.00244	0.00008	0.00625
12		[1|1]	[2|2]	[3|3]	[1|2]	[1|3]	[2|3]	[4|4]	[5|5]	[6|6]	[4|5]
13		0.37082	0.90036	2.31856	0.63559	1.34469	1.60946	6.74929	24.12489	132.759	15.43709
14	12.566	33.8882	13.9570	5.4199	19.7712	9.3452	7.8078	1.9	0.5	0.1	0.8
15	vij	-0.02334	-0.37146	-0.47187	-0.08466	-0.07237	-0.37584	-0.1665	-0.02701	-0.0022	-0.05544
16		[1|1]	[2|2]	[3|3]	[1|2]	[1|3]	[2|3]	[4|4]	[5|5]	[6|6]	[4|5]
17		-13.71624	-8.80254	-5.48539	-8.65869	-3.42485	-5.30580	-3.2150	-1.70053	-0.7249	-1.45253
18	tij	0.0047243	0.117136	0.23878	0.018537	0.013261	0.1277	0.14379	0.044086261	0.00836	0.0494595
19	0.000	0.37082	0.90036	2.31856	0.63559	1.34469	1.60946	6.74929	24.12489	132.759	15.43709
20	0.1854	-13.716									
21	0.4502		-8.803								

	A	B	C	D	E	F	G	H	I	J
1	alpha =⟩	0.18541	0.45018	1.15928		3.37464	12.0624	66.3797		
2	d =⟩	0.02625	0.16314	0.29506		0.29907	0.22771	0.15188		
3										
4	Ra	Rb	Rc	Rd		Vijkl	1.05463			
5	0	0	0	0						
6	$α_i$	$α_j$	$α_k$	$α_l$		d_i	d_j	d_k	d_l	vijkl
7	0.18541	0.18541	0.18541	0.18541		0.02625	0.02625	0.02625	0.02625	0.000140201260245573
8	0.18541	0.18541	0.45018	0.45018		0.02625	0.02625	0.16314	0.16314	0.001703999976475230
9	0.45018	0.45018	0.18541	0.18541		0.16314	0.16314	0.02625	0.02625	0.001703999976475230
10	0.45018	0.18541	0.18541	0.18541		0.16314	0.02625	0.02625	0.02625	0.000436451068393849
11	0.18541	0.18541	0.18541	0.45018		0.02625	0.02625	0.02625	0.16314	0.000436451068393849
12	0.18541	0.18541	0.45018	0.18541		0.02625	0.02625	0.16314	0.02625	0.000436451068393849
13	0.18541	0.45018	0.18541	0.18541		0.02625	0.16314	0.02625	0.02625	0.000436451068393849
14	0.45018	0.45018	0.45018	0.45018		0.16314	0.16314	0.16314	0.16314	0.022781113556988000
15	0.45018	0.18541	0.45018	0.45018		0.16314	0.02625	0.16314	0.16314	0.005621591981302140
16	0.45018	0.45018	0.18541	0.45018		0.16314	0.16314	0.02625	0.16314	0.005621591981302140
17	0.18541	0.45018	0.45018	0.45018		0.02625	0.16314	0.16314	0.16314	0.005621591981302140
18	0.45018	0.45018	0.45018	0.18541		0.16314	0.16314	0.16314	0.02625	0.005621591981302140
19	0.18541	0.45018	0.45018	0.18541		0.02625	0.16314	0.16314	0.02625	0.001408296790429120
20	0.45018	0.18541	0.45018	0.18541		0.16314	0.02625	0.16314	0.02625	0.001408296790429120

Figure 5.7   Application of the general spreadsheet to the calculation of the energy of the helium atom using the Pople, Hehre and Stewart \|sto-6g⟩ basis set of Table 1.6 and the 'best' Slater exponent, also reported elsewhere (8,9) from variation of the entry in cell 'oneel!'$D$1. For the Slater-rules exponent, 1.7, the helium energy is found to be −2.8461945 hartree with the 1s orbital energy equal to −0.8918763 H. Note, the detail shown for the $Vijkl$ term. On this spreadsheet all 1296 [6^4 integrals] are calculated, with the degeneracies over the primitives, colour-coded in the second diagram in the figure.

The next step is to renormalize the individual primitive functions of the basis set so that the most diffuse primitive Gaussian is split from the others to form the new linear combination. There are three stages, therefore, in this analysis, the renormalization of the \|sto-4g⟩ data of Table 5.1, the calculation of the two-electron term using the *split-basis* set and the straightforward calculation of the one-electron kinetic and potential energy terms, as in previous spreadsheets.

	A	B	C	D	E	F	G	H	I	J	K
1		|sto-ng⟩	Slater ζ	1.0000	Z =⟩	2.00000		alpha(i)	alpha(j)		E_He
2		|1⟩	|2⟩	|3⟩	|4⟩	|5⟩	|6⟩	1.0000	1.00000		-2.85519
3	α	1.242567	5.782948	38.47497	0.298073	0.000000	0.000000	⟨i|j⟩	0.99999		
4	d	0.469540	0.154570	0.023730	0.513800	0.000000	0.000000	V	-3.36869		ε_1s
5	αζ²	1.24257	5.78295	38.47497	0.29807	0.00000	0.00000	T	1.42766		-0.91415
6	d'	0.39384	0.41082	0.26127	0.14772	0.00000	0.00000	ee	1.02688		
7		[1|1]	[2|2]	[3|3]	[1|2]	[1|3]	[2|3]	[4|4]	[5|5]	[6|6]	[4|5]
8	d'*d'	0.15511	0.16877	0.06826	0.16180	0.10290	0.10733	0.02182	0.00000	0.00000	0.00000
9	α'+α'	2.48513	11.56590	76.94994	7.02552	39.71754	44.25792	0.59615	0.00000	0.000	0.29807
10	5.568	1.42135	0.14156	0.00825	0.29903	0.02225	0.01891	12	4189	4189	34
11	sij	0.22047	0.02389	0.00056	0.04838	0.00229	0.00203	0.26399	0.00000	0.00000	0.00000
12		[1|1]	[2|2]	[3|3]	[1|2]	[1|3]	[2|3]	[4|4]	[5|5]	[6|6]	[4|5]
13		2.48513	11.56590	76.94994	7.02552	39.71754	44.25792	0.59615	0.00000	0.000	0.29807
14	12.566	5.0566	1.0865	0.1633	1.7887	0.3164	0.2839	21.1	1256.6	1256.6	42.2
15	vij	-0.78434	-0.18337	-0.01115	-0.28940	-0.03256	-0.03048	-0.4600	0.00000	0.0000	0.00000
16		[1|1]	[2|2]	[3|3]	[1|2]	[1|3]	[2|3]	[4|4]	[5|5]	[6|6]	[4|5]
17		-5.29836	-2.45599	-0.95216	-1.83506	-0.16066	-0.57046	-10.818	0.00000	0.0000	0.00000
18	tij	0.410919	0.207248	0.0325	0.148453	0.008266	0.030615	0.11803	0	0	0
19	0.000	2.48513	11.56590	76.94994	7.02552	39.71754	44.25792	0.59615	0.00000	0.000	0.29807
20	1.2426	-5.298									
21	5.7829		-2.456								

Figure 5.8   Application of the general spreadsheet, fig5-7.xls, to reproduce the Huzinaga results for the \|sto-4g⟩ representation of the helium 1s orbital and equation 5.2. Note, the setting of the Slater exponent to 1.0.

Exercise 5.4.   Formation of a \|4–31⟩ split-basis set for helium using the data of Table 5.1 and the calculation of the energy of the helium ground state.

1. To construct the renormalized *split-basis* linear combination in the manner of equations 1.24 and 1.25, we need to calculate normalization integrals for the three primitives to be taken as one group in the calculation. So, on a new spreadsheet provide for this change as shown in the first diagram of Figure 5.9.

    (i)   Name the worksheet 'renorm'! and lay out the Simpson's rule integration procedure in columns A, B and C.
    (ii)   Enter the Huzinaga data from Table 5.1 in cells $D$2 to $G$3 and the Slater exponent equal 1.0, in cell $H$2, with the difference that the coefficient in cell $D$3 is set equal to 1.0.
    (iii)   Calculate the scaled Gaussian exponents and normalization constants in cells $D$4 to $G$5.
    (iv)   Project the individual Gaussians over the radial array in columns D to G and check the normalizations of the *split-basis*, using the **SUMPRODUCT** function in cells $D$7 and $H$7.
    (v)   Renormalize the three-component term of the *split-basis* set as in columns I to L by redefining the pre-exponential factors of these primitives, to include division by the square root of the normalization integral in cell $H$7.

2. Open the spreadsheet 5-6.xls. Use the EDIT/MOVE OR COPY SHEET sequence to copy the 'twoel!' worksheet into the new file. This is the simplest way to ensure that the sequence of spreadsheet operations required for

Renormalization of Huzinaga |sto-4g⟩ to|4-31g⟩

	A	B	C	D	E	F	G	H	I	J	K
2			α	0.298073	1.242567	5.782948	38.47497	Zeta			
3			d	1.00000	0.46954	0.15457	0.02373	1.0000			
4			α'	0.298073	1.242567	5.782948	38.47497				
5			d'	0.287509	0.393842	0.410816	0.261271		Renormalized data		
6				⟨1|1⟩					⟨234|234⟩	0.793301	0.261150	0.040093
7				1.000000					0.350323	0.665408	0.694085	0.441425
8	r	0	#	|1⟩	|2⟩	|3⟩	|4⟩	|sto-3g⟩	|2n⟩	|3n⟩	|4n⟩
9	0.00	1	1	0.28751	0.39384	0.41082	0.26127	1.06593	0.665408	0.694085	0.441425
10	0.01	2	4	0.28750	0.39379	0.41058	0.26027	1.06464	0.665325	0.693684	0.439730
11	0.02	3	2	0.28748	0.39365	0.40987	0.25728	1.06079	0.665077	0.692481	0.434684
12	0.03	4	4	0.28743	0.39340	0.40868	0.25238	1.05446	0.664664	0.690482	0.426401
13	0.04	5	2	0.28737	0.39306	0.40703	0.24567	1.04576	0.664086	0.687692	0.415070
14	0.05	6	4	0.28730	0.39262	0.40492	0.23731	1.03485	0.663344	0.684123	0.400944
15	0.06	7	2	0.28720	0.39208	0.40235	0.22748	1.02191	0.662438	0.679785	0.384329

	A	B	C	D	E	F	G	H	I	J	K
1											
2			|1⟩			|21⟩	|22⟩	|23⟩		⟨2222⟩	1.54427
3			1.00000			1.00000				⟨1112⟩	0.47625
4			0.29807			1.24257	5.78295	38.47497		⟨1212⟩	0.42707
5			1.00000			0.79330	0.26115	0.04009		⟨1122⟩	0.80410
6			0.29807			1.24257	5.78295	38.47497		⟨1222⟩	0.78558
7	README		0.28751			0.66541	0.69409	0.44143		⟨1111⟩	0.61605
8	0.00000	0.00000	0.00000	0.00000							
9	i	j	k	l		d_i	d_j	d_k	d_l	⟨ijkl⟩	Rp
10	V1111										
11	1.24257	1.24257	1.24257	1.24257		0.66541	0.66541	0.66541	0.66541	0.498158	0.000000
12	1.24257	1.24257	5.78295	1.24257		0.66541	0.66541	0.69409	0.66541	0.132877	0.000000
13	1.24257	5.78295	1.24257	1.24257		0.66541	0.69409	0.66541	0.66541	0.132877	0.000000
14	5.78295	1.24257	1.24257	1.24257		0.69409	0.66541	0.66541	0.66541	0.132877	0.000000
15	1.24257	1.24257	1.24257	5.78295		0.66541	0.66541	0.66541	0.69409	0.132877	0.000000

	A	B	C	D	E	F	G	H	I	J	K	L
1	SCF											
2	|sto_4-31g:1s⟩calculation for helium					alpha(i)	alpha(j)		⟨1|1⟩		v_{11}	-1.74245
3	|4-31g⟩	Slater ζ	1.00000			Slater	1.00000		1.00000		v_{12}	-2.01396
4	|1⟩				|21⟩	|22⟩	|23⟩		⟨2|2⟩		v_{22}	-4.80621
5	α	0.298073			1.242567	5.782948	38.474970		1.00000		t	0.44711
6	d	1.000000			0.793301	0.261150	0.040093		⟨1|2⟩		t_{12}	0.46738
7	$\alpha\zeta^2$	0.29807			1.24257	5.78295	38.47497		0.63410		t_{22}	2.92682
8	d'	0.28751			0.66541	0.69409	0.44143					
9		[a1|a1]	[a2|a2]	[a3|a3]	[a1|a2]	[a1|a3]	[a2|a3]	[b1|b1]	[b2|b2]	[b3|b3]	[b1|b2]	[b1|b3]
10	d'*d'	0.08266	0.00000	0.00000	0.00000	0.00000	0.00000	0.44277	0.48175	0.19486	0.46185	0.29373
11	α'+α'	0.59615	2.00000	2.00000	1.29807	1.29807	2.00000	2.48513	11.56590	76.94994	7.02552	39.71754
12	5.57	12.0975	1.97	1.97	3.77	3.77	1.97	1.42135	0.14	0.01	0.29903	0.02225
13	sij	1.00000	0.00000	0.00000	0.00000	0.00000	0.00000	0.62933	0.06820	0.00161	0.13810	0.00653

Figure 5.9 Detail from the spreadsheet, fig5-9.xls, showing the renormalization worksheet, individual components of the two-electron potential for the basis set written in the form of equation 1.26 and, in the third diagram, the remodelling of the worksheet 'oneel'! from fig4-18.xls.

equation 5.52 are transferred for use in the present calculation. You need, now, to remember that for the Clementi *double-zeta* calculation of fig5-4.xls, there are six two-electron terms corresponding to the different possibilities for ⟨ij|kl⟩ and that these occur, also, in the present case, but with the extra complication that one term in the Gaussian *double-zeta* basis is comprised of three Gaussian primitives. Thus, for every two-electron component involving both parts of the *split-basis* all the permutations over the three primitives have to be included as you can see on the worksheet.

3. Open fig4-18 and copy the worksheet for the calculation of the one-electron terms over four Gaussian primitives into the new spreadsheet.

4. Open the spreadsheet fig5.4.xls and copy the 'hfs!' worksheet into fig5-8.xls.
5. Make appropriate links to the 'renorm'! worksheet cells containing the *split-basis* data.
6. Modify the master formula for the nuclear-electron potential for the case of the helium atom on the one-electron worksheet and collect the various terms in the integrals to form the one-electron Hamiltonian matrix on 'hfs'!
7. Finally link all the integrals through to the 'hfs'! worksheet for the diagonalization of the Fock matrix and iterate to self-consistency.

The result of the calculations using fig5-9.xls are shown in Figure 5.10, in the same manner as in Figure 5.4b, with the iteration details to *self-consistency* and then in the second part of the diagram the worksheet 'hfs'! at *self-consistency* is displayed. The

	A	B	C	D	E	F	G	H	I	J	K
1	4-31g) split basis calculation of the energy of the helium atom.									T	
2										2.85534	
3	S			theta =	0.78540		S-diagonal			N.E.	
4	1.00000	0.63410		omega1	1.63410		1.63410	0.00000		-6.59656	E-He
5	0.63410	1.00000		omega2	0.36590		0.00000	0.36590		ee	-2.85516
6										1.02692	
7	U			U - adjoint			The S^1/2 matrix			ε	
8	0.70711	0.70711		0.70711	0.70711		0.78228	0.00000		-0.91412	
9	0.70711	-0.70711		0.70711	-0.70711		0.00000	1.65318			
10											
11	H			G			Fock matrix		g11	(11|11)	0.616051
12	-1.29534	-1.54658		0.73401	0.66070		-0.56134	-0.88588	0.73401	(22|22)	1.544268
13	-1.54658	-1.87939		0.66070	1.23110		-0.88588	-0.64829	g12	(11|22)	0.804105
14									0.66070	(12|12)	0.427072
15				X-canonical			X-canon transpose		g22	(11|12)	0.476252
16				0.55315	1.16898		0.55315	0.55315	1.23110	(12|22)	0.785578
17				0.55315	-1.16898		1.16898	-1.16898			
18										README	
19	theta	-0.03341		Transformed Fock matrix						Density matrix (in)	
20	omega1	-0.91412		-0.80053	0.37938	==>	-0.91224	0.05622		0.26399	0.30411
21	omega2	0.77004		-0.84863	-0.27774		0.05622	0.76816		0.30411	0.35034
22											
23	Coefficients			Fock - diagonal			C - originals			Density matrix (out)	
24	0.99944	-0.03340		-0.91412	0.00000		0.51380	-1.18680		0.26399	0.30411
25	-0.03340	-0.99944		0.00000	0.77004		0.59189	1.14985		0.30411	0.35033
26											
27	c₁	c₂		c₁	c₂		T	N.E.	ee	e	E - He
28	0.00000	0.00000		0.31237	0.77232					-1.99361	
29	0.31237	0.77232		0.56444	0.54181		2.57502	-6.90015	1.09093	-0.81943	-2.72978
30	0.56444	0.54181		0.49992	0.60527		2.93363	-6.49615	1.00626	-0.94191	-2.89008
31	0.49992	0.60527		0.51754	0.58826		2.83436	-6.62236	1.03225	-0.90678	-2.84580
32	0.51754	0.58826		0.51278	0.59287		2.86104	-6.58947	1.02546	-0.91612	-2.85770
33	0.51278	0.59287		0.51407	0.59163		2.85381	-6.59846	1.02731	-0.91358	-2.85448
34	0.51407	0.59163		0.51372	0.59196		2.85577	-6.59603	1.02681	-0.91427	-2.85535
35	0.51372	0.59196		0.51382	0.59187		2.85524	-6.59669	1.02694	-0.91408	-2.85511
36	0.51382	0.59187		0.51379	0.59190		2.85538	-6.59651	1.02691	-0.91413	-2.85517
37	0.51379	0.59190		0.51380	0.59189		2.85534	-6.59656	1.02692	-0.91412	-2.85516
38	0.51380	0.59189		0.51380	0.59189		2.85535	-6.59655	1.02692	-0.91412	-2.85516
39											
40	README										
41	0.513800	0.591895		0.513798	0.591891		2.855344	-6.596558	1.026919	-0.914119	-2.855156

Figure 5.10 Determination of the 'best' ground state energy for helium by varying the independent coefficient of the *split-basis* linear combination from the $|sto-4g\rangle$ Gaussian set of Table 5.1. The energy of helium is found to be -2.85516 Hartree for the coefficients of equation 5.40 equal 0.51380 and 0.59189. The orbital energy, ε_{1s}, is calculated to be -0.91412 Hartree and compares well with Huzinaga's result, Table 5.2.

minimum ground-state energy of helium -2.855161 H for c_1 equal to 0.591892 and c_2 equal to 0.513797.

As you might expect there is no difference in the result of this calculation and that presented in Figure 5.8. After all, the second calculation results simply from renormalization of the original basis set. However, it is important to remember that the *split-basis* procedure has increased the flexibility of the calculation. It allows us to vary the two components independently by varying the Slater exponent.

Exercise 5.5. **Modify fig5-9.xls to provide for distinct Slater exponents for each component of the *split-basis*. Repeat the calculation of the energy of the helium atom to investigate whether a better description of the ground state can be obtained using the electronic energy as the test parameter.**

5.7 HELIUM, SINGLET AND TRIPLET EXCITED STATES, ELECTRON SPIN AND THE ROLE OF THE EXCHANGE INTEGRAL

The Hall–Roothaan equations were developed in Section 5.3 for the case of the ground state of the helium atom, with both electrons *spin-paired* in the 1s spatial orbital. For this case of a single closed electron shell, the Fock operator takes the simple form set out in equation 5.33, because of the spin orthogonality, which renders the exchange integral, equation 5.37, equal to zero. While the Hartree–Fock–Slater theory of open-shell systems is complicated (76,46) and not amenable to simple illustration on a spreadsheet, the following demonstration of the role of the Exchange integral is possible, for the special case of two electrons when the Pople–Nesbet equations reduce to the Hall–Roothaan ones.

Consider the two possible spin configurations of the excited state of helium with one-electron in the 1s orbital and the other in the 2s orbital, the electronic configuration $1s^12s^1$. With reference to equations 5.28 and 5.29, the spin orbitals, again, are the independent products of distinct space and spin functions, with

$$\phi_{1s}{}^+ (\mathbf{r}, \sigma) = \phi_{1s} (\mathbf{r}) \, \alpha \, (\sigma) \qquad\qquad 5.53$$

and

$$\phi_{2s}^+ (\mathbf{r}, \sigma) = \phi_{2s} (\mathbf{r}) \, \alpha \, (\sigma) \ \text{ or } \ \phi_{2s}^- (\mathbf{r}, \sigma) = \phi_{2s} (\mathbf{r}) \, \beta \, (\sigma) \qquad\qquad 5.54$$

with $\alpha \, (\sigma)$ and $\beta \, (\sigma)$ the possible spin wave functions. The Pauli Principle requirement that the electronic wave function change sign on the exchange of electron coordinates can be fulfilled in the anti-symmetry of either the spatial or spin components of the wave functions.

For the two-helium excited state electron configurations, we can write three product two-electron symmetric spin functions,

$$\alpha \, (1) \, \alpha \, (2) = +\alpha \, (1) \, \alpha \, (2) \qquad\qquad 5.55$$

$$\beta \, (1) \, \beta \, (2) = +\beta \, (1) \, \beta \, (2) \qquad\qquad 5.56$$

$$\alpha \, (1) \, \beta \, (2) + \alpha \, (2) \, \beta \, (1) = +[\alpha \, (2) \, \beta \, (1) + \alpha \, (1) \, \beta \, (2)] \qquad\qquad 5.57$$

and a single anti-symmetric spin function

$$\alpha\,(1)\,\beta\,(2) - \alpha\,(2)\,\beta\,(1) = -\,[\alpha\,(2)\,\beta\,(1) - \alpha\,(1)\,\beta\,(2)] \qquad 5.58$$

The symmetric and anti-symmetric linear combinations over the space coordinates are

$$\phi\,(1)_{1s}\,\phi_{2s}\,(2) + \phi\,(2)_{1s}\,\phi_{2s}\,(1) = +\,[\phi_{1s}\,(2)\,\phi_{2s}\,(1) + \phi_{1s}\,(1)\,\phi_{2s}\,(2)] \qquad 5.59$$

$$\phi\,(1)_{1s}\,\phi_{2s}\,(2) - \phi_{1s}\,(2)\,\phi_{2s}\,(1) = -\,[\phi_{1s}\,(2)\,\phi_{2s}\,(1) - \phi_{1s}\,(1)\,\phi_{2s}\,(2)] \qquad 5.60$$

As you see in these equations, interchanging the electron coordinates is equivalent to multiplying by $+/-1$, demonstrating the exchange properties of the wave functions.

To comply with the anti-symmetry requirement of the Pauli Principle, we can form the total two-electron wave function in the spatial and spin coordinates only by taking products which involve an anti-symmetric linear combination of the one with an anti-symmetric linear combination of the other. In the standard notation this leads to four possible two-electron wave functions, three describing the triplet excited state, ϕ_t, and one the singlet, ϕ_s, with normalized Slater determinants

$$\phi_{t(1)}\,(1,2) = \sqrt{\frac{1}{2}} \begin{vmatrix} \phi_{1s}^{+}\,(1) & \phi_{2s}^{+}\,(1) \\ \phi_{1s}^{+}\,(2) & \phi_{2s}^{+}\,(2) \end{vmatrix} \qquad 5.61$$

$$\phi_{t(2)}\,(1,2) = \sqrt{\frac{1}{2}} \begin{vmatrix} \phi_{1s}^{-}\,(1) & \phi_{2s}^{-}\,(1) \\ \phi_{1s}^{-}\,(2) & \phi_{2s}^{-}\,(2) \end{vmatrix} \qquad 5.62$$

$$\phi_{t(3)}\,(1,2) = \sqrt{\frac{1}{2} \left[\sqrt{\frac{1}{2}} \begin{vmatrix} \phi_{1s}^{+}\,(1) & \phi_{2s}^{-}\,(1) \\ \phi_{2s}^{+}\,(2) & \phi_{2s}^{-}\,(2) \end{vmatrix} + \sqrt{\frac{1}{2}} \begin{vmatrix} \phi_{1s}^{-}1s\,(1) & \phi_{2s}^{+}\,(1) \\ \phi_{1s}^{-}\,(2) & \phi_{2s}^{+}\,(2) \end{vmatrix} \right]} \qquad 5.63$$

$$\phi_{s(1)}\,(1,2) = \sqrt{\frac{1}{2} \left[\sqrt{\frac{1}{2}} \begin{vmatrix} \phi_{1s}^{+}\,(1) & \phi_{2s}^{-}\,(1) \\ \phi_{2s}^{+}\,(2) & \phi_{2s}^{-}\,(2) \end{vmatrix} - \sqrt{\frac{1}{2}} \begin{vmatrix} \phi_{1s}^{-}1s\,(1) & \phi_{2s}^{+}\,(1) \\ \phi_{1s}^{-}\,(2)\phi_{2s}^{+}\,(2) \end{vmatrix} \right]} \qquad 5.64$$

Since these functions are symmetry-adapted approximations to the eigenfunctions, we need only operate on any one with the two-electron Hamiltonian and integrate to return the energy of the corresponding triplet or singlet state (47, 77, 78), which result then can be optimized in terms of the parameters in the linear combinations.

For example, applying the Hamiltonian of equation 5.8 to equation 5.61 and integrating to calculate the expectation value, leads to a one-electron contribution to the energy and then the single two-electron potential term, equation 5.36. We find two kinetic and nuclear-electron energy terms deriving from equation 5.4, in this case, different because of the different space components of the two-electron wave functions. But equation 5.36 does not reduce to a single Coulomb integral term, since the Exchange term is non-zero in the case of the parallel spin wave function, equation 5.61. Then, for example, using equation 5.68, the two-electron potential term, equation 5.36, on multiplying through of the Slater determinant, becomes

$$\left\langle \phi_{t(1)}(1,2) \left| \frac{1}{r_{12}} \right| \phi_{t(1)}(1,2) \right\rangle = \frac{1}{2} \left\langle \phi_{1s}^+(1)\, \phi_{2s}^+(1) \left| \frac{1}{r_{12}} \right| \phi_{1s}^+(2)\, \phi_{2s}^+(2) \right\rangle$$

$$+ \frac{1}{2} \left\langle \phi_{1s}^+(1)\, \phi_{2s}^+(1) \left| \frac{1}{r_{12}} \right| \phi_{1s}^+(2)\, \phi_{2s}^+(2) \right\rangle - \frac{1}{2} \left\langle \phi_{1s}^+(1)\, \phi_{2s}^+(1) \right| \qquad \text{5.65}$$

$$\times \frac{1}{r_{12}} \left| \phi_{1s}^+(2)\, \phi_{2s}^+(2) \right\rangle - \frac{1}{2} \left\langle \phi_{1s}^+(1)\, \phi_{2s}^+(1) \left| \frac{1}{r_{12}} \right| \phi_{1s}^+(2)\, \phi_{+2s}^+(2) \right\rangle$$

which simplifies to the difference of the Coulomb, J, and the Exchange, K, integrals neither of which is zero

$$J - K$$

$$= \frac{1}{2} \left[\begin{array}{l} \left\langle \phi_{1s}^+(1)\, \phi_{1s}^+(1) \left| \frac{1}{r_{12}} \right| \phi_{2s}^+(2)\, \phi_{2s}^+(2) \right\rangle - \left\langle \phi_{1s}^+(1)\, \phi_{2s}^+(1) \left| \frac{1}{r_{12}} \right| \phi_{1s}^+(2)\, \phi_{2s}^+(2) \right\rangle \\[2mm] - \left[\left\langle \phi_{1s}^+(1)\, \phi_{2s}^+(1) \left| \frac{1}{r_{12}} \right| \phi_{1s}^+(2)\, \phi_{2s}^+(2) \right\rangle + \left\langle \phi_{2s}^+(1)\, \phi_{2s}^+(1) \left| \frac{1}{r_{12}} \right| \phi_{1s}^+(2)\, \phi_{1s}^+(2) \right\rangle \right] \end{array} \right]$$

$$\text{5.66}$$

For the singlet state, equation 5.64, we find the one-electron terms, as we expect, to be the same, but the two-electron term ranges over the sixteen product terms arising from the substitution of equation 5.64 into equation 5.36 and reduces to the sum of the Coulomb and Exchange integrals

$$J + K =$$

$$\frac{1}{4} \left[\begin{array}{l} \left\langle \phi_{1s}^+(1)\, \phi_{1s}^+(1) \left| \frac{1}{r_{12}} \right| \phi_{2s}^-(2)\, \phi_{2s}^-(2) \right\rangle - \left\langle \phi_{1s}^+(1)\, \phi_{1s}^+(1) \left| \frac{1}{r_{12}} \right| \phi_{2s}^+(2)\, \phi_{2s}^-(2) \right\rangle \\[2mm] + \left\langle \phi_{1s}^+(1)\, \phi_{2s}^+(1) \left| \frac{1}{r_{12}} \right| \phi_{1s}^+(2)\, \phi_{2s}^-(2) \right\rangle - \left\langle \phi_{1s}^+(1)\, \phi_{2s}^-(1) \left| \frac{1}{r_{12}} \right| \phi_{2s}^+(2)\, \phi_{2s}^-(2) \right\rangle \\[2mm] - \left\langle \phi_{1s}^-(1)\, \phi_{1s}^+(1) \left| \frac{1}{r_{12}} \right| \phi_{2s}^-(2)\, \phi_{2s}^+(2) \right\rangle + \left\langle \phi_{1s}^-(1)\, \phi_{1s}^-(1) \left| \frac{1}{r_{12}} \right| \phi_{2s}^+(2)\, \phi_{2s}^+(2) \right\rangle \\[2mm] - \left\langle \phi_{1s}^-(1)\, \phi_{2s}^+(1) \left| \frac{1}{r_{12}} \right| \phi_{1s}^-(2)\, \phi_{2s}^+(2) \right\rangle + \left\langle \phi_{1s}^-(1)\, \phi_{2s}^-(1) \left| \frac{1}{r_{12}} \right| \phi_{1s}^+(2)\, \phi_{2s}^+(2) \right\rangle \\[2mm] + \left\langle \phi_{1s}^+(1)\, \phi_{2s}^+(1) \left| \frac{1}{r_{12}} \right| \phi_{1s}^-(2)\, \phi_{2s}^-(2) \right\rangle - \left\langle \phi_{1s}^-(1)\, \phi_{2s}^+(1) \left| \frac{1}{r_{12}} \right| \phi_{1s}^-(2)\, \phi_{2s}^+(2) \right\rangle \\[2mm] + \left\langle \phi_{2s}^+(1)\, \phi_{2s}^+(1) \left| \frac{1}{r_{12}} \right| \phi_{1s}^-(2)\, \phi_{1s}^-(2) \right\rangle - \left\langle \phi_{2s}^+(1)\, \phi_{2s}^-(1) \left| \frac{1}{r_{12}} \right| \phi_{1s}^+(2)\, \phi_{1s}^-(2) \right\rangle \\[2mm] - \left\langle \phi_{1s}^+(1)\, \phi_{2s}^-(1) \left| \frac{1}{r_{12}} \right| \phi_{2s}^-(2)\, \phi_{1s}^+(2) \right\rangle + \left\langle \phi_{1s}^-(1)\, \phi_{2s}^-(1) \left| \frac{1}{r_{12}} \right| \phi_{1s}^+(2)\, \phi_{2s}^+(2) \right\rangle \\[2mm] - \left\langle \phi_{2s}^+(1)\, \phi_{2s}^-(1) \left| \frac{1}{r_{12}} \right| \phi_{1s}^-(2)\, \phi_{2s}^+(2) \right\rangle + \left\langle \phi_{2s}^-(1)\, \phi_{2s}^-(1) \left| \frac{1}{r_{12}} \right| \phi_{1s}^+(2)\, \phi_{1s}^+(2) \right\rangle \end{array} \right]$$

$$\text{5.67}$$

For the approximate representation of the spin orbitals in the manner of equation 5.36 over two spin orbitals, which involve different spatial orbitals, the expectation value of the Fock operator of equation 5.39 takes the form

$$\iint \phi_1 (\mathbf{r})\, F (\mathbf{r}, \mathbf{r}')\, \phi_1 (\mathbf{r})\, \mathbf{dr}\, \mathbf{dr}' = \int \phi_1 (\mathbf{r}) \left[-\frac{1}{2} \nabla^2 (\mathbf{r}) - \frac{2}{r} + \int |\phi_2 (\mathbf{r}')|^2 \frac{1}{r_{12}} \mathbf{dr}' \right.$$

$$\left. + \int \phi_1 (\mathbf{r}')\, \phi_2 (\mathbf{r}') \frac{1}{r_{12}} \mathbf{dr}' \right] \phi_{1s} (\mathbf{r})\, \mathbf{dr}$$

$$= \int \phi_1 (\mathbf{r}) \left[-\frac{1}{2} \nabla^2 (\mathbf{r}) - \frac{2}{r} + 2J (\mathbf{r}, \mathbf{r}') - K (r, \mathbf{r}') \right]$$

$$\phi_{1s} (\mathbf{r})\, \mathbf{dr} \qquad\qquad 5.68$$

since the matrix elements of the J and K integrals reduce to

$$J_{11} = \tfrac{1}{2} [P_{11} (11|11) + P_{12} (11|12) + P_{21} (11|21) + P_{22} (11|22)] \qquad 5.46$$

$$J_{12} = \tfrac{1}{2} [P_{11} (11|11) + P_{12} (11|12) + P_{21} (11|21) + P_{22} (11|22)] \qquad 5.47$$

$$J_{22} = \tfrac{1}{2} [P_{11} (22|11) + P_{12} (22|12) + P_{21} (22|21) + P_{22} (22|22)] \qquad 5.48$$

$$K_{11} = \tfrac{1}{2} [P_{11} (11|11) + P_{12} (11|21) + P_{21} (12|11) + P_{22} (12|21)] \qquad 5.69$$

$$K_{12} = \tfrac{1}{2} [P_{11} (11|12) + P_{12} (11|22) + P_{21} (12|12) + P_{22} (12|22)] \qquad 5.70$$

$$K_{22} = \tfrac{1}{2} [P_{11} (21|12) + P_{12} (22|12) + P_{21} (21|22) + P_{22} (22|22)] \qquad 5.71$$

with K equal J when the two-electron state is a 'closed-shell' singlet and then the Fock operator is equation 5.39.

Exercise 5.6.   Calculation of the helium singlet and triplet state energies using an STO-nG 1s basis set.

1. Make a copy of the spreadsheet, fig5-8.xls; rename it fig5-11.xls.
2. Delete the renormalization worksheet and make sure that the input data cells in 'oneel'! contain the data for the Pople $|sto\text{-}3g\rangle$ 1s and 2s approximations, Table 1.6.
3. Check the links in 'hfs'! for the elements of the overlap and one-electron matrices.
4. Use the '$\langle ij|kl\rangle$'! as a template to generate the 'two-electron potential' worksheets over the Gaussian primitives for the 1s and 2s linear combinations used in the calculation.
5. Link the individual contributions to the two-electron potential to the appropriate cells in 'hfs'!
6. Collect the total J and K terms of the two-electron potential in cells 'hfs'!$J$2 and $J$4.
7. The density matrix elements follow from reference 47 and are as in cells 'hfs'!$J$24 to $K$25 in fig5-11.xls.

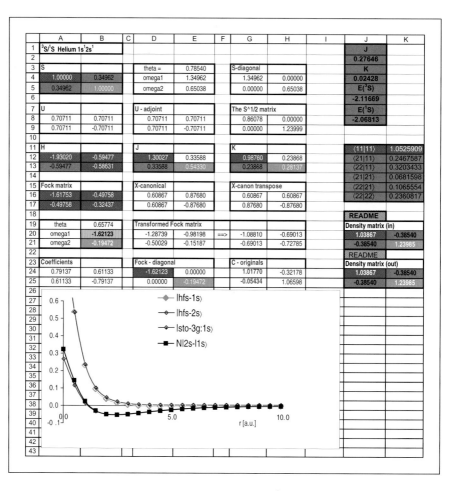

Figure 5.11 Calculation of the energies of the $^3S/^1S$ states for the possible electronic configurations of helium $1s^12s^1$. The two-electron terms are as given in equation 5.67, with J and K components as given in cells $J$2 and $J$4.

8. As usual, start the calculation with the input density matrix set to zero and then iterate to self-consistency, the condition of the spreadsheet displayed in Figure 5.11.

9. Complete Figure 5.11 with the chart of the HFS radial functions compared to the starting Schmidt-orthogonalized pair. As is to be expected the 1s function is little altered, the change to self-consistency being reflected in the differences in the 2s radial functions projections on the radial array.

The 'best' energy for the ground state in the $|$sto-3g$\rangle$ representation of Table 1.6, Figure 5.7 and fig5-7.xls, is -2.8464 Hartree, compared with the HFS limiting value -2.9037 Hartree. From the data in Figure 5.11, have found that the excited state singlet lays some 0.78 Hartree above this ground state in the same basis, while the triplet state is found at slightly lower energy at some 0.73 Hartree above the ground state result.

This is as we would expect; the exchange energy in the high spin excited state lowers the energy[16]. Moreover, we can compare the calculated energy differences with available spectra[17], where the first excited state is found 0.732 Hartree above the ground state, with the lower-spin singlet lying at 0.748 above the ground state.

While our calculations have been fairly primitive, in that the basis set is limited in its replication of the actual 1s and 2s orbitals, hopefully, there is the beginning of a theoretical explanation of the spectral measurements.

Exercise 5.7.   Calculation of the *doubly-excited* state energy for the configuration $1s^0 2s^2$.

The doubly-excited state of helium is some 160 000 cm^{-1} above the ground state (78). Apply the spreadsheet approach to obtain a 'best' estimate of this energy difference.

---

[16] I am indebted to Professor Peter Gill of the University of Nottingham, UK and to Dr Mike Schmidt of the GAMESS laboratory at Iowa State University, USA, for correspondence and help with these calculations. The calculations were checked using the GAMESS programme (79).

[17] See, for example, reference 75, Figure 17.3, page 631 and *http://spider.upac.caltech.edu/staff/rosalie/grotrian.html*.

6

One- and two-electron diatoms

The hydrogen molecule ion, H_2^+, is to molecular electronic structure theory what the hydrogen atom is to atomic theory. However, Schrödinger's equation for even this simplest molecular species is much more complicated than for any atom and it cannot be solved without simplification. The universal simplifying approximation in molecular theory is to apply the Born-Oppenheimer approximation (52,78,80) and assume that the electronic interactions can be considered against the constant potential field of a fixed nuclear framework defining the molecular geometry. For the hydrogen molecule ion this is a sufficient approximation for exact solution of its Schrödinger equation but for no other molecular species and so the details of the solution are not helpful in the construction of a general theory of the electronic structure of molecules. Therefore this final chapter is concerned only with the application of basis set theory to the molecular electronic structure problem within the *linear combination of atomic orbitals — molecular orbital [LCAO-MO]* theory. Dihydrogen, H_2, is the simplest molecule. The helium hydride ion, HeH^+, is the simplest heterogeneous diatom. H_2^+, H_2 and HeH^+ examples used in this chapter to exemplify the application of basis set theory to the *LCAO-MO* method using spreadsheet technology.

In this chapter, you will learn:

1. How to solve Schrödinger's equation for the hydrogen molecule ion using the spreadsheet for numerical integration over a cylindrical volume and the exact hydrogen 1s atomic orbitals.
2. how to solve Schrödinger's equation for the hydrogen molecule ion using |sto-ng⟩ basis sets and the simplifications, which follow for the the direct evaluation of integrals over products of 1s Gaussians.
3. How to carry out a full SCF calculation, on a spreadsheet, for the dihydrogen molecule.
4. How to carry out a full SCF calculation, on a spreadsheet, for the HeH^+ ion, the simplest heterogeneous diatomic species.
5. How to demonstrate the effect of a polarized basis set using a Gaussian-lobe approach.
6. How to extend the model calculations to larger molecules, when appropriate integration formulae have been devised for higher-order Gaussian basis sets.

For the one-electron diatomic ion, H_2^+, then the Schrödinger equation, subject to the Born-Oppenheimer approximation, takes the form

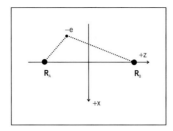

Figure 6.1 The geometry of the hydrogen molecule ion, within the Born-Oppenheimer approximation with the bond length R set at $|\mathbf{R}_A - \mathbf{R}_B|$.

$$\left[-\frac{1}{2}\nabla^2(\mathbf{r}) - \frac{1}{(\mathbf{r} - \mathbf{R}_a)} - \frac{1}{(\mathbf{r} - \mathbf{R}_b)} + \frac{1}{(\mathbf{R}_a - \mathbf{R}_b)} \right] \varphi(\mathbf{r}) = \varepsilon\varphi(\mathbf{r}) \qquad 6.1$$

for the molecular geometry drawn in Figure 6.1, with $\mathbf{R}_a$ and $\mathbf{R}_b$ constant.

In the *LCAO-MO* approximation, we try to find the eigenfunctions and eigenvalues of the stationary states from linear combinations of atomic orbital, in reality, atomic orbital-like functions. Since we know from the analysis of atomic theory that linear combinations of Slater-type functions and Gaussians functions can be used within the spirit of the variation principle to obtain good results for atoms, much of the background for the molecular orbital theory herein, has been covered. At the simplest level, the standard procedure to solve the Schrödinger equation, for the hydrogen molecule ion is to take a linear combination of two hydrogenic 1s orbital-like functions sited about each atomic position, *viz*

$$\varphi(\mathbf{r}) = c_a\phi_a(\mathbf{r} - \mathbf{R}_a) + c_b\phi_b(\mathbf{r} - \mathbf{R}_b) \qquad 6.2$$

which equation takes us back to equation 1.26 and the calculations with *double-zeta* and *split basis* sets on spreadsheets in Chapter 4.

In textbooks, justifications for this choice of *LCAO* function, often, involve considerations of the electron probability densities for different separations of the nuclei and positions of the electron. In the region of one nucleus, especially for large separations of the nuclei, the probability distribution should correspond to that for an occupied 1s atomic orbital about the nuclear position. At very large separations, we do not know whether to describe the electron in terms of one atomic orbital or the other. Thus, the most appropriate function is the form given in equation 6.2.

The reality, of course, is that any complete set of square integrable functions, such as all the atomic orbitals of any atom, provide for an exact description of the eigenfunction solutions to the Schrödinger equation for a molecule. However, since such an approach is not practicable even on the largest modern computers, we settle for an approximation, which leads to good results when comparisons with experimental data are made. This is the essence of the *LCAO-MO* approximation and it would be normal to extend equation 6.2 and take linear combinations of valence atomic orbitals chosen from all the atoms in a molecule. For example, for simple heteronuclear diatomics we might choose

$$\varphi(\mathbf{r}) = \sum_i c_{iA}\phi_{iA}(\mathbf{r} - \mathbf{R}_A) + \sum_j c_{jB}\phi_{jB}(\mathbf{r} - \mathbf{R}_B) \qquad 6.3$$

with the linear combinations ranging over the valence atomic orbitals of the atoms.[1]

However, to return to the case of the hydrogen molecule ion and equation 6.2, since there cannot be any difference in the electron probability about either nucleus because of the symmetry of the molecular framework, we know that the coefficients follow as $[2 +/- 2S]^{-0.5}$ with Δs_{ab} the overlap integral between the atomic functions.

With this definition of a suitable trial function, we can proceed directly to the determination of the two expectation values since there are two linear combinations, just as in Chapter 4, but now with the components of the molecular Hamiltonian in the numerator, i.e.

$$\varepsilon = \frac{\left\langle \varphi(\mathbf{r}) \left| -\frac{1}{2}\nabla^2(\mathbf{r}) - \frac{1}{(\mathbf{r} - \mathbf{R}_a)} - \frac{1}{(\mathbf{r} - \mathbf{R}_b)} + \frac{1}{(\mathbf{R}_a - \mathbf{R}_b)} \right| \varphi(\mathbf{r}) \right\rangle}{\langle \varphi(\mathbf{r}) | \varphi(\mathbf{r}) \rangle} \qquad 6.4$$

As you see, the main difference is the need to calculate an extra potential term to represent the attraction between the electron and a second nucleus. We can calculate the expectation values for any approximate trial function exactly as in the case of the hydrogen atom, then optimize the result using the variation theorem using any available disposable parameter available.

6.1 CALCULATIONS USING HYDROGEN 1S ORBITALS

The simplest *LCAO-MO* function, in the form of equation 6.3, using the atomic orbitals of the hydrogen atom, is the linear combination, compare equation 6.1, using the 1s function of Table 1.1

$$\varphi(\mathbf{r}) = e^{-\zeta(\mathbf{r} - \mathbf{R}_a)} +/- e^{-\zeta(\mathbf{r} - \mathbf{R}_b)} \qquad 6.5$$

and the matrix elements of the Hamiltonian in equation 6.4 follow straightforwardly from the discussion in Chapter 4 on the hydrogen atom calculations.

---

[1] It is appropriate, perhaps, to mention here that imprudent application of linear combinations like equation 6.3 in basis set theory can lead to error. The interaction energy, E_{int}, between two atoms A and B, separated by a distance R, is defined by

$$E_{int}(R) = E_{AB}(R) - (E_A + E_B)$$

when E_A and E_B are the total energies of the isolated atoms A and B and E_{AB} is the extra energy due to the atoms being within R of each other. It turns out that the application of equation 6.3 to the calculation of $E_{int}(R)$ using two small basis sets leads to the error known as the *Basis Set Superimposition Error* [BSSE] (41). This error arises because the incomplete basis set on the first atom effectively increases the size of the basis set on the other and *vice versa*. Thus the energies of the isolated atoms are calculated to be lower than one would expect, which, in turn, leads to an incorrect interaction energy.

A method to overcome this error is to use both basis sets in the calculation of the energy of each atom on its own so that the phantom component of the atomic energies can be determined and subtracted in what is known as the *counterpoise* method. BSSE can lead to over-estimations of molecular binding energies and so to the prediction of incorrect molecular geometries and charge density distributions.

Exercise 6.1a.   Calculation of the $1\sigma_g^+$ LCAO-MO in the hydrogen molecule ion, using hydrogen 1s atomic orbitals.

The detail of this calculation is set out in the spreadsheet fig6-2.xls. As you can see, in Figure 6.2, the design involves the cylindrical integration procedure described in Section 2.4.

1. Open fig2-7.xls and use the EDIT/MOVE OR COPY SHEET sequence to make three copies of the integration procedure.
2. Name the worksheets 'SN!', 'V!' and 'T!' to identify the calculations of orthonormality, potential energy and kinetic energy.
3. Name the default blank worksheet of the new spreadsheet RESULTS.
4. Modify the coordinate meshes in V and T to link to those in SN, with, for example, the cell entries

    'V!'$A$13 = 'T!'$A$13 = 'SN!'$A13
    'V!'$B$12 = 'T!'$B$12 = 'SN!'B12

5. Construct one- and two-variable tables on the worksheets to facilitate calculation of the different integrals for normalization of the atomic functions, overlap of the atomic functions, the nucleus-electron potential terms for the linear combination of equation 6.3 and the kinetic energy terms. Thus,

    (i) On worksheet 'SN!', redefine the projections of the normalization and overlap integrals over the two-dimensional array using the switches of cells $B$6 and the linked bond length in $B$3. Note, too, the linked effective atomic number in $B$9 to provide for variation of the calculation and the identification of the 'best' molecular orbital energies.

    'SN!'$B$13 = B$11*(($B$9^3/PI())^0.5*EXP(-($B$9*SQRT (SUMSQ($A13, (B$12-$B$3/2)))))*($B$9^3/PI()) ^0.5*EXP(-($B$9*SQRT(SUMSQ ($A13,(B$12+$B$6*$B$3/2))))))

    This means that the full normalization expression in Table 1.1 is required, since Z need not be of unit value.

    (ii) The master formula, $D$2, is, as expected, for the copy of fig2-7.xls, with

    'SN!'$D$2 = SUM($GB13:$GB113)*($A$13-$A$14)/3

    and note, too, that the atomic functions are normalized over all coordinates in this sequence.

    (iii) On worksheet 'V!', provide a similar integration scheme, but this time, involving a two-variable table using the switches $B$5 and $B$7, integration elements, for example,

    'V!'$B$13 = ($B$9^3/PI())^0.5*EXP(-$B$9*SQRT(SUMSQ ($A13,(B$13-B$3/2))))/SQRT(SUMSQ($A13, (B$13-$B$5*$B$3/2)))*B$12*($B$9^3/PI())^0.5*EXP (-$B$9*SQRT(SUMSQ($A13,(B$13-$B$7*$B$3/2))))

'V!'\$C\$3 = -**SUM**(\$GB13:\$GB113)*(\$A\$14-\$A\$15)/
(3*'SN!'\$D\$4)

(iv) On worksheet T!, the kinetic energy components of the Hamiltonian are worked out and this requires the factor, equations 4.12 and 4.13, reflecting the effect of the Laplacian operator on the 1s function, with, for example,

'T!'\$B\$13 = (\$B\$9^3/**PI**( ))^0.5***EXP**(-\$B\$9***SQRT**(**SUMSQ**
(\$A13,(B\$13-\$B\$3/2))))*(\$B\$9^2-2*\$B\$9/**SQRT**
(**SUMSQ**(\$A13,(B\$13-\$B\$3/2))))*B\$12*(\$B\$9^3/**PI**( ))^
0.5***EXP**(-\$B\$9***SQRT**(**SUMSQ**
(\$A13,(B\$13-\$B\$6*\$B\$3/2))))

switching on \$B\$6 to return $T_{11} = T_{22}$ and T_{12} in

'T!'\$E\$3 = -**SUM**(\$GB13:\$GB113)*(\$A\$14-\$A\$15)/3/(2*'SN!'\$D\$4)

Since very good calculations for the hydrogen molecule ion and molecule problems are available[2], it is worthwhile to write out some of the details of the calculations for comparison with the present analysis. Substitution of equation 6.5 into equation 6.4 leads to the identification of the following relations

$$\left(-\frac{1}{2}\nabla^2 - \frac{1}{R_a}\right)|1s_a\rangle = \varepsilon_a|1s_a\rangle \qquad 6.6$$

and similarly for the second 'atomic' Hamiltonian

$$\left(-\frac{1}{2}\nabla^2 - \frac{1}{R_b}\right)|1s_b\rangle = \varepsilon_b|1s_b\rangle \qquad 6.7$$

so that

$$H_{aa} = H_{bb} + \varepsilon_{H_a} + J' \qquad 6.8$$

which defines the equivalent in molecular orbital theory of the 'Coulomb' integral of the Heitler–London Valence Bond theory, with

$$J' = \left\langle 1s_a \left| -\frac{1}{R_b} + \frac{1}{R} \right| 1s_a \right\rangle \qquad 6.9$$

The 'Exchange' integral equivalent follows from the definition

$$H_{ab} = \varepsilon_H S + K' \qquad 6.10$$

so that

$$K' = \left\langle 1s_a \left| -\frac{1}{R_a} + \frac{1}{R} \right| 1s_b \right\rangle \qquad 6.11$$

[Compare the definitions in equation 5.38.] Thus, equation 6.4 can be rewritten in two forms,

$$\varepsilon_{+/-} = \frac{H_{aa} +/- H_{ab}}{1 +/- S} \qquad 6.12$$

[2]Reference 82 contains a detailed survey of these calculations.

	A	B	C	D	E
1	Normalization/overlap of atomic functions.				
2		R	S/N	0.99999	
3		2.49	1	0.460694	
4			-1	0.999988	
5		+/-			
6		-1			
7					
8		Z*			
9		1			
10		sto-3g Integral components.			
11		1	4	2	4
12	y/z	-7.500	-7.417	-7.333	-7.250
13	7.500	3.1344E-11	1.4225E-10	8.0653E-11	1.8282E-10

	A	B	C	D	E
1	Calculation of the potential energy.				
2		R			
3		2.49	-0.999238	1	-1
4		+/-	1	-0.999238	-0.289294
5		1	-1	-0.391974	-0.289294
6		+/-			
7		1			
8		Z*			
9		1			
10		sto-3g Integral components.			
11		1	4	2	4
12	y/z	-7.500	-7.417	-7.333	-7.250
13	7.500	2.7207E-12	1.2415E-11	7.0782E-12	1.6133E-11

	A	B	C	D	E
1	Calculation of the kinetic energy.				
2		R			
3		2.49		T	0.499238
4				1	0.499238
5		+/-		-1	0.058944
6		1			
7					
8		Z*			
9		1			
10		sto-3g Integral components.			
11		1	4	2	4
12	y/z	-7.500	-7.417	-7.333	-7.250
13	7.500	2.5903E-11	1.1742E-10	6.6497E-11	1.5056E-10

Figure 6.2   The three worksheets required for the calculation of the orbital energies using the simple *LCAO-MO* approximation of equation 6.3. The integration array is defined on the first worksheet and linked to the others as described in the exercise.

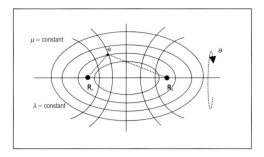

Figure 6.3   The confocal elliptical coordinate system used in early calculations to solve Schrödinger's equation for the dihydrogen molecule and its ion.

with which expression the spreadsheet results have been calculated so that

$$\varepsilon_{+/-} = \varepsilon_H + \frac{J' +/- K'}{1 +/- S} \qquad 6.13$$

Exact forms (83,81,82) for the integrals, J, K and S for the two-centre problems of dihydrogen and other homogeneous diatomic molecular species were achieved in early calculations by transforming to the elliptical coordinates system defined in Figure 6.3. In this coordinate system the nuclear positions are the two fixed points of the coordinate system the foci of the ellipses separated by the 'bond length' R with a point P defined by the three coordinates

$$\lambda = \frac{\mathbf{r_a} + \mathbf{r_b}}{R} \qquad 6.14$$

$$\mu = \frac{\mathbf{r_a} - \mathbf{r_b}}{R} \qquad 6.15$$

and θ the apex angle of the triangle defining the electron position in Figure 6.1. In equations 6.14 and 6.15 $\mathbf{r_a}$ and $\mathbf{r_b}$ are vectors from each nucleus to the electron in Figure 6.3. In this way, we arrive at the results

$$S = e^{-R} \left(1 + R + \frac{R^2}{3} \right) \qquad 6.16$$

$$J' = e^{-2R} \left(1 + \frac{1}{R} \right) \qquad 6.17$$

and

$$K' = \frac{S}{R} - e^{-R} (1 + R) \qquad 6.18$$

Exercise 6.1b.   Completion of the design of fig6-2.xls.

6.   The TABLE macros on the various worksheets contain all the matrix elements of the Hamiltonian needed to calculate the orbital energies and the approximate

molecular orbitals. So collect these together on the worksheet 'RESULTS!' with the links,

S_{11} = 'RESULTS!'\$B\$7 = 'SN!'\$D\$4 = 'RESULTS!'\$C\$8 = S_{22}

S_{12} = 'RESULTS!'\$C\$7 = 'SN!'\$D\$3 = 'RESULTS!'\$B\$8 = S_{21}

T_{11} = 'RESULTS!'\$B\$11 = 'T!'\$E\$4 = 'RESULTS!'\$C\$12 = T_{22}

T_{12} = 'RESULTS!'\$B\$12 = 'T!'\$D\$5 = 'RESULTS!'\$C\$11 = S_{21}

$V_{11}(a)$ = 'RESULTS!'\$B\$15 = 'T!'\$E\$4 = 'RESULTS!'\$C\$16 = $V_{22}(a)$

$V_{12}(a)$ = 'RESULTS!'\$B\$16 = 'T!'\$D\$5 = 'RESULTS!'\$C\$15 = $V_{21}(a)$

$V_{11}(b)$ = 'RESULTS!'\$B\$19 = 'T!'\$E\$4 = 'RESULTS!'\$C\$20 = $V_{22}(b)$

$V_{12}(b)$ = 'RESULTS!'\$B\$20 = 'T!'\$D\$5 = 'RESULTS!'\$C\$19 = $V_{21}(b)$

7. Then form the Hamiltonian matrix as the sums over the potential and kinetic energy terms, with

$$H_{11} = \$E\$7 = \$B\$11+\$B\$15+\$B\$19$$
$$H_{12} = \$F\$7 = \$C\$11+\$C\$15+\$C\$19$$
$$H_{21} = \$E\$8 = \$B\$12+\$B\$16+\$B\$20$$
$$H_{22} = \$F\$8 = \$C\$12+\$C\$16+\$C\$20$$

8. The final steps in the calculation are the determination of the orbital energies and the coefficients for the two possible linear combinations in the form of equation 6.5. The H_{ij}, above, are components in the secular equation

$$\begin{vmatrix} H_{11} - \varepsilon & H_{12} - \varepsilon S \\ H_{21} - \varepsilon S & H_{22} - \varepsilon \end{vmatrix} = 0$$

compare, equation 5.41. The usual textbook procedure is to multiply out to the quadratic equation to obtain the roots

ε_1 = \$F\$11 = (2*(\$E\$7-\$E\$8*\$C\$7)-**SQRT**((2*(\$E\$7-\$E\$8*\$C\$7))^2-4*(1-\$C\$7^2)*(\$E\$7^2-\$E\$8^2)))/(2*(1-\$C\$7^2))

ε_2 = \$F\$12 = (2*(\$E\$7-\$E\$8*\$C\$7)-**SQRT**((2*(\$E\$7-\$E\$8*\$C\$7))^2+4*(1-\$C\$7^2)*(\$E\$7^2-\$E\$8^2)))/(2*(1-\$C\$7^2))

Then, substitute back into the simultaneous equations to obtain the coefficients for each of the two possible molecular orbitals, for ε_1

c_1 = \$F\$15 = **SQRT**(1/(**POWER**((\$F\$7-\$F\$11*\$C\$7)/(\$E\$7-\$F\$11),2)-2*((\$F\$7-\$F\$11*\$C\$7)/(\$E\$7-\$F\$11))*\$C\$7+1))

c_2 = \$F\$16 = -F15*(F7-F11*C7)/(E7-F11)

and, similarly, for the c_1 and c_2 values for the ε_2 in cells \$G\$15 and \$G\$16.

Figure 6.4 shows the comparison between total energies based on calculated values using equations 6.16 to 6.18 and the spreadsheet results for different bond lengths, R.

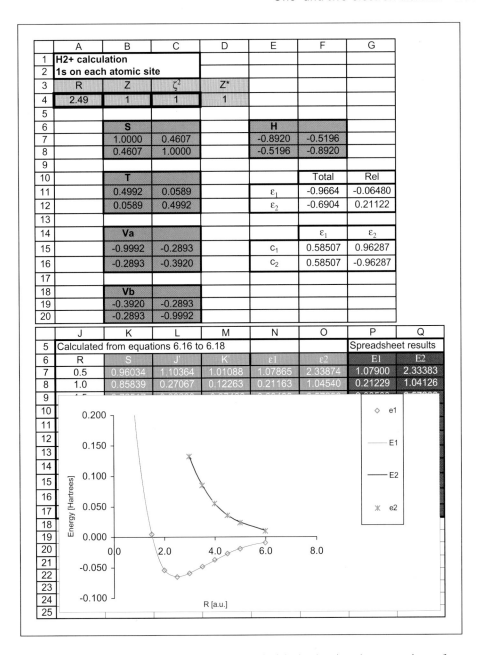

Figure 6.4 The 'RESULTS!' worksheet in fig6-2.xls showing the comparison of the spreadsheet based calculation of the $1\sigma_g$ and $1\sigma_u$ and the results of accurate calculations in elliptical coordinates.

As you can see there is good agreement between the numerically calculated results and those obtained using equations 6.16 to 6.18. Pleasingly the coefficients of the linear combinations for the two molecular orbitals are returned as $[2 + / - S)]^{-0.5}$.

Close inspection of Figure 6.4 and, indeed, your own energy search reveals that the basic calculation has returned a stable molecular species exhibiting an electronic energy 0.0665 H lower than that of a 1s electron in hydrogen for an equilibrium bond length of 2.4 a.u. The experimental values for these parameters are 0.102 a.u. and 2.00 a.u.

Exercise 6.2.   **Use the scaler in cell 'RESULTS!' $C$4 to optimize the results for the hydrogen molecule ion.**

It is proper to present this calculation using the canonical orthonormalization procedure described in Section 3.6, since this would be the standard approach in a molecular orbital calculation. The only modifications of fig6-2 and 6-4.xls required are the additions of a canonical worksheet, taken from any of the previous spreadsheets, using the EDIT/MOVE -COPY option in the EDIT menu and the creation of suitable links to the RESULTS! worksheet cells.

This change is shown in Figure 6.5, which displays the extra 'canonical'! worksheet in fig6-5.xls. Note that small changes in the results are obtained because of the different approach to the determination of the eigenfunctions of the Fock matrix. These differences lead to closer agreement with the analytical results.

	A	B	C	D	E	F	G	H
1	Canonical orthonormalization for the H_2^+ problem based on H-1s orbitals.							
2								
3	S			theta =	0.785398			
4	0.999988	0.460694		omega1	1.460682			
5	0.460694	0.999988		omega2	0.539293			
6								
7	U			U - adjoint			S-diagonal	
8	0.70711	0.70711		0.70711	0.70711		1.46068	0.00000
9	0.70711	-0.70711		0.70711	-0.70711		0.00000	0.53929
10							The S^1/2 matrix	
11							0.82741	0.00000
12							0.00000	1.36172
13	Fock matrix			X-canonical			X-canon transpose	
14	-0.891974	-0.519643		0.5850691	0.9628806		0.58507	0.58507
15	-0.519643	-0.891974		0.5850691	-0.9628806		0.96288	-0.96288
16								
17	theta	0.00000		Transformed Fock matrix				
18	omega1	-0.966410		-0.82589	-0.35851	==>	-0.96641	0.00000
19	omega2	-0.690404		-0.82589	0.35851		0.00000	-0.69040
20								
21	Coefficients			Fock - diagonal			C - originals	
22	1.0000000	0.0000000		-0.966410	0.000000		0.58507	-0.96288
23	0.0000000	-1.0000000		0.000000	-0.690404		0.58507	0.96288

Figure 6.5   The extra worksheet 'canonical'! added to fig6-2 and 6-4.xls providing for application of the canonical orthonormalization procedure. Note the slight differences in the results compared with the earlier ones calculated by solving the simultaneous equation problem directly.

> **Exercise 6.3.** Improve the 'user' friendliness of fig6-5.xls by assembling all the primary data of interest on the CANONICAL! worksheet.

6.2 |sto-3g) CALCULATIONS FOR H$_2$$^+$

It is clear that the use of a Gaussian basis set to solve Schrödinger's equation for the hydrogen molecule ion will involve the calculation of many components of the various integrals required. The calculation is set out in fig6-6.xls and details from the various worksheets are presented in Figure 6.6a and 6.6b. In all the worksheets, links to master entries for the input data and the 'x' and 'y' meshes ensure that interactive calculations can be performed for different bond lengths, basis sets and scaling parameter for the Slater exponents. A further detail is the provision for the checking of the convergence of the calculations in the panel $H$26 to $I$33 on the 'RESULTS!' worksheet. Apart from this feature, all the design details in fig6-6.xls have been explained for previous spreadsheets and so the instructions, for this exercise, are confined to the writing out, here, of only the major details.

> **Exercise 6.4.** Solution of Schrödinger's equation for H$_2$$^+$ using an |sto-ng) basis.
>
> 1. Make a copy of fig2-7.xls and use the EDIT/MOVE OR COPY SHEET facility to create six copies of the worksheet for integration over a cylindrical array. Create a primary worksheet and label all the worksheets as in Figures 6.6a and 6.6b.
> 2. Construct links on the worksheets to the primary input data on 'RESULTS!' and the meshes for 'x' and 'y' axes on 'S!'
> 3. Then, to provide for the analysis of one-electron species AB as well as H$_2$$^+$ create one and two-variable tables on the various worksheets, with,
>
> (i)   on sheet 'S!'
>
> $$`S!'\$B\$16 = \mathbf{EXP}(-\$B\$8^*\mathbf{SUMSQ}(\$A16,(B\$15-\$D\$1)))^*$$
> $$\mathbf{EXP}(-\$B\$10^*\mathbf{SUMSQ}(\$A16,(B\$15-\$G\$1)))^*B\$14$$
>
> etc., the master inputs for the two-variable table S in cells $B$8 to $B$11 and the master formula for this overlap integral table in cell $D$8, with,
>
> $$S!\$D\$8 = B\$9\$^*\$B\$11^*\mathbf{SUM}(\$GB\$16:\$GB\$116)^*(\$A\$16-\$A\$17)/3$$
>
> (ii) on sheet 'Viia!'
>
> $$Viia!\$B\$20 = \mathbf{EXP}(-\$B\$8^*\mathbf{SUMSQ}(\$A20,(B\$19-\$D\$1)))^*$$
> $$\mathbf{EXP}(-\$B\$10^*\mathbf{SUMSQ}(\$A20,(B\$19-\$G\$1)))^*$$
> $$\mathbf{SQRT}(1/\mathbf{SUMSQ}(\$A20,(B\$19-\$D\$1)))^*B\$18$$

etc., the master inputs for the two one-variable tables in cells $B$8 to $B$11 and the master formulae in cell $D$8 and $D$12, with,

$$V_{iia} = \text{'Viia!'}\$D\$8 = -\$B\$9^*\$B\$11^*\$B\$2^* \\ \textbf{SUM}(\$GB\$20:\$GB\$120)^*(\$A\$20-\$A\$21)/3$$

$$V_{jjb} = \text{'Viia!'}\$D\$12 = -\$B\$9^*\$B\$11^*\$E\$2^* \\ \textbf{SUM}(\$GB\$20:\$GB\$120)^*(\$A\$20-\$A\$21)/3$$

(iii)   on sheet 'Viiab!'

$$\text{'Viiab!'}\$B\$20 = \textbf{EXP}(-\$B\$8^*\textbf{SUMSQ}(\$A20,(B\$19-\$G\$1)))^* \\ \textbf{EXP}(-\$B\$10^*\textbf{SUMSQ}(\$A20,(B\$19-\$G\$1)))^* \\ \textbf{SQRT}(1/\textbf{SUMSQ}(\$A20,(B\$19-\$D\$1)))^*B\$18$$

etc., the master inputs for the two one-variable tables in cells $B$8 to $B$11 and the master formulae in cell $D$8 and $D$12, with,

$$V_{iib} = \text{'Viiab!'}\$D\$8 = -\$B\$9^*\$B\$11^*\$E\$2^*\textbf{SUM}(\$GB\$20:\$GB\$120)^* \\ (\$A\$20-\$A\$21)/3$$

$$V_{jja} = \text{'Viiab!'}\$D\$12 = -\$B\$9^*\$B\$11^*\$B\$2^*\textbf{SUM}(\$GB\$20:\$GB\$120)^* \\ (\$A\$20-\$A\$21)/3$$

(iv)   on sheet 'Vija!'

$$\text{'Vija!'}\$B\$16 = \textbf{EXP}(-\$B\$8^*\textbf{SUMSQ}(\$A16,(B\$15-\$D\$1)))^* \\ \textbf{EXP}(-\$B\$10^*\textbf{SUMSQ}(\$A16,(B\$15-\$G\$1)))^* \\ \textbf{SQRT}(1/\textbf{SUMSQ}(\$A16,(B\$15-\$D\$1)))^*B\$14$$

etc., the master inputs for the one-variable table in cells $B$8 to $B$11 and the master formulae in cell $D$8, with,

$$V_{ija} = \text{'Vija!'}\$D\$8 = -\$B\$9^*\$B\$11^*\$B\$2^*\textbf{SUM}(\$GB\$20:\$GB\$120)^* \\ (\$A\$20-\$A\$21)/3$$

(v)   on sheet 'Vijb!'

$$\text{'Vijb!'}\$B\$16 = \textbf{EXP}(-\$B\$8^*\textbf{SUMSQ}(\$A16,(B\$15-\$D\$1)))^* \\ \textbf{EXP}(-\$B\$10^*\textbf{SUMSQ}(\$A16,(B\$15-\$G\$1)))^* \\ \textbf{SQRT}(1/\textbf{SUMSQ}(\$A16,(B\$15-\$D\$1)))^*B\$14$$

etc., the master inputs for the one-variable table in cells $B$8 to $B$11 and the master formulae in cell $D$8, with,

$$V_{ijb} = \text{Vijb!}\$D\$8 = -\$B\$9^*\$B\$11^*\$B\$2^*\textbf{SUM}(\$GB\$20:\$GB\$120)^* \\ (\$A\$20-\$A\$21)/3$$

(vi)   on sheet 'Tii!'

$$\text{'Tii!'}\$B\$20 = \textbf{EXP}(-\$B\$8^*\textbf{SUMSQ}(\$A20,(B\$19-\$D\$1)))^* \\ \textbf{EXP}(-\$B\$10^*\textbf{SUMSQ}(\$A20,(B\$19-\$G\$1)))^* \\ (4^*\$B\$10^\wedge2^*\textbf{SUMSQ}(\$A20,(B\$19-\$G\$1))) \\ -6^*\$B\$10)^*B\$18$$

etc., the master inputs for the two one-variable tables in cells $B$8 to $B$11 and the master formulae in cell $D$8 and $D$12, with,

$$T_{ii} = \text{`Tii!'} \$D\$8 = -0.5*\$B\$9*\$B\$11*\textbf{SUM}(\$GB\$20:\$GB\$120)*$$
$$(\$A\$20-\$A\$21)/3$$

$$T_{jj} = \text{`Tjj!'} \$D\$12 = -0.5*\$B\$9*\$B\$11*\textbf{SUM}(\$GB\$20:\$GB\$120)*$$
$$(\$A\$20-\$A\$21)/3$$

(vii)   on sheet 'Tij!'

$$\text{Tij!}\$B\$15 = \textbf{EXP}(-\$B\$8*\textbf{SUMSQ}(\$A15,(B\$14-\$D\$1)))*$$
$$\textbf{EXP}(-\$B\$10*\textbf{SUMSQ}(\$A15,(B\$14-\$G\$1)))*$$
$$(4*\$B\$10\char94 2*\textbf{SUMSQ}(\$A15,(B\$14-\$G\$1))$$
$$-6*\$B\$10)*B\$13$$

etc., the master inputs for the one-variable table in cells $B$8 to $B$11 and the master formulae in cell $D$8 and $D$12, with,

$$T_{ij} = \text{`Tij!'} \$D\$8 = -0.5*\$B\$9*\$B\$11*\textbf{SUM}(\$GB\$15:\$GB\$115)*$$
$$(\$A\$15-\$A\$16)/3$$

4. Gather together the various integral components over the basis sets as links onto the 'RESULTS!' worksheet between cells $A$11 to $J$22.
5. Make the appropriate sums of the overlap, potential and kinetic energy integral components to form the overlap, potential, kinetic energy matrices and then the Hamiltonian matrix, as before, in cells $H$27 to $I$28.
6. Solve the secular determinant and generate the eigenvalues and coefficients of the *LCAO*-MOS over sto-3g representations of equation 6.3.
7. Note, the detail of the convergence test panel, which generate the mesh intervals on the 'x' and 'y' axes of the integration array on the worksheets.

Note the inputs for the bond length, cell $B$24, and Slater exponents, cells $F$24 and $H$24, in the calculation. The first is the bond length found in the neutral diatomic molecule. The choice of 1.24 as the scaler of the Slater exponent for the Gaussian primitives, Table 1.10, is to model better the molecular environment of the electron in the presence of the second nucleus. Remember that the scaler is squared, as in equations 1.21 and 1.22, because the Gaussian primitive is a quadratic exponential.

The experimental data, for the hydrogen molecule ion, is that the bond length is 2.00 a.u. and the dissociation energy [the negative of the 1σ orbital energy] is 0.106 a.u.

Exercise 6.5.   Optimize the calculation of the orbital energy in the hydrogen molecule ion and determine, too, the equilibrium bond length for this level of approximation.

Finally Figure 6.6c presents the standard format for the molecular calculation based on the canonical orthonormalization procedure standard in modern programs.

Sheet S!

	A	B	C	D	E	F	G
1		Z	Ra	-0.7	Z	Rb	0.7
2		1	scaler	1.24	1	scaler	1.24
3	α	0.109818	0.405771	2.22766	0.109818	0.405771	2.22766
4	d	0.444635	0.535328	0.154329	0.444635	0.535328	0.154329
5	α'	0.168856157	0.62391349	3.42525002	0.16885616	0.62391349	3.42525002
6	d'	0.083474036	0.2678387	0.27693436	0.08347404	0.2678387	0.27693436
7							
8	α	0.168856	S	0.1675	0.16886	0.62391	3.42525
9	d	0.083474		0.16885616	0.1675	0.1359	0.0138
10	α	0.168856		0.62391349	0.1359	0.1555	0.0180
11	d	0.083474		3.42525002	0.0138	0.0180	0.0008
12							
14		1	4	2	4	2	4
15	y/z	-10.00	-9.92	-9.83	-9.75	-9.67	-9.58
16	7.500	1.02678E-23	7.19389E-23	6.27079E-23	2.17623E-22	1.87927E-22	6.46095E-22

Sheet Viia!

	A	B	C	D	E	F	G
8	α	0.168856	Viia	-0.129584	0.168856	0.623913	3.425250
9	d	0.083474		0.168856	-0.129584	-0.177021	-0.040231
10	α	0.168856		0.623913	-0.177021	-0.360654	-0.114513
11	d	0.083474		3.425250	-0.040231	-0.114513	-0.069738
12			Vjjb	-0.129584	0.168856	0.623913	3.425250
13				0.168856	-0.129584	-0.177021	-0.040231
14				0.623913	-0.177021	-0.360654	-0.114513
15				3.425250	-0.040231	-0.114513	-0.069738

Sheet Viiab!

	A	B	C	D	E	F	G
8	α	0.168856	Viib	-0.105893	0.168856	0.623913	3.425250
9	d	0.083474		0.168856	-0.105893	-0.116130	-0.013492
10	α	0.168856		0.623913	-0.116130	-0.199129	-0.036206
11	d	0.083474		3.425250	-0.013492	-0.036206	-0.017014
12			Vjja	-0.105893	0.168856	0.623913	3.425250
13				0.168856	-0.105893	-0.116130	-0.013492
14				0.623913	-0.116130	-0.199129	-0.036206
15				3.425250	-0.013492	-0.036206	-0.017014

Sheet Vija!

	A	B	C	D	E	F	G
8	α	0.168856	Viia	-0.1041	0.16886	0.62391	3.42525
9	d	0.083474		0.16885616	-0.1041	-0.1333	-0.0292
10	α	0.168856		0.62391349	-0.1029	-0.1623	-0.0383
11	d	0.083474		3.42525002	-0.0103	-0.0152	-0.0012

Sheet Vijb

	A	B	C	D	E	F	G
8	α	0.168856	Vijb	-0.1040625	0.1688562	0.6239135	3.4252500
9	d	0.083474		0.1688562	-0.1041	-0.1029	-0.0103
10	α	0.168856		0.6239135	-0.1333	-0.1623	-0.0152
11	d	0.083474		3.4252500	-0.0292	-0.0383	-0.0012

Figure 6.6a   Details from the worksheets for the calculation of the overlap and potential energy integrals for H_2^+ using an $|$sto-3g$\rangle$ basis. These are in fig6-6abc.xls.

Sheet Tii!

	A	B	C	D	E	F	G
8	α	0.168856	Tii	0.0501	0.16885616	0.62391349	3.42525002
9	d	0.083474		0.16885616	0.0501	0.0703	0.0091
10	α	0.168856		0.62391349	0.0703	0.2682	0.0803
11	d	0.083474		3.42525002	0.0091	0.0803	0.1224
12			Tjj	0.05008	0.16886	0.62391	3.42525
13				0.16886	0.0501	0.0703	0.0091
14				0.62391	0.0703	0.2682	0.0803
15				3.42525	0.0091	0.0803	0.1224

Sheet Tij!

	A	B	C	D	E	F	G
8	α	0.168856	Tij	0.03776	0.16886	0.62391	3.42525
9	d	0.083474		0.16886	0.0378	0.0448	0.0053
10	α	0.168856		0.62391	0.0448	0.0862	0.0089
11	d	0.083474		3.42525	0.0053	0.0089	-0.0053

Sheet 'results'!

	A	B	C	D	E	F	G	H	I	J
1	sto-3g calculations for 1e-diatomics.									
2										
3		Z	Ra	-1.0000000	Z	Rb	1.0000000			
4		1	scaler	1.2400000	1	scaler	1.2400000			
5	α	0.1098180	0.405771	2.2276600	0.1098180	0.4057710	2.2276600			
6	d	0.4446350	0.535328	0.1543290	0.4446350	0.5353280	0.1543290			
7	α'	0.1688562	0.623913	3.4252500	0.1688562	0.6239135	3.4252500			
8	d'	0.0834740	0.267839	0.2769344	0.0834740	0.2678387	0.2769344			
9										
10	Integral components over the basis sets.									
11	t11	0.0501	0.0703	0.0091	0.0703	0.2682	0.0803	0.0091	0.0803	0.1225
12	t12	0.0277	0.0267	0.0027	0.0267	0.0129	-0.0040	0.0027	-0.0040	-0.0005
13	t22	0.0501	0.0703	0.0091	0.0703	0.2682	0.0803	0.0091	0.0803	0.1225
14										
15	v11a	-0.1296	-0.1769	-0.0401	-0.1769	-0.3603	-0.1142	-0.0401	-0.1142	-0.0694
16	v22a	-0.0889	-0.0871	-0.0094	-0.0871	-0.1431	-0.0253	-0.0094	-0.0253	-0.0119
17	v12a	-0.0830	-0.0627	-0.0052	-0.0992	-0.0728	-0.0036	-0.0209	-0.0123	0.0000
18	v12b	-0.0830	-0.0992	-0.0209	-0.0627	-0.0728	-0.0123	-0.0052	-0.0036	0.0000
19	v11b	-0.0889	-0.0871	-0.0094	-0.0871	-0.1431	-0.0253	-0.0094	-0.0253	-0.0119
20	v22b	-0.1296	-0.1769	-0.0401	-0.1769	-0.3603	-0.1142	-0.0401	-0.1142	-0.0694
21										
22	S	0.1410	0.1037	0.0099	0.1037	0.0823	0.0061	0.0099	0.0061	0.0000
23										
24	R	2	ε_H	-0.4613	scaler A	1.24	scaler B	1.24		
25	S									
26	1.0000	0.4628		H			Convergence check			
27	0.4628	1.0000		-0.94910	-0.62833		README			
28	T			-0.62833	-0.94910		#/z	100		
29	0.7603	0.0911					#/y	100		
30	0.0911	0.7603		Eigenvalues	Total	Relative	y[max]	10		
31	V[A]			$|1\rangle$	-1.0784	-0.1170	z[min]	10		
32	-1.2216	-0.3597		$|2\rangle$	-0.5971	0.3643	Δy	-0.1		
33	-0.3597	-0.4877					Δz	0.1		
34	V[B]			LCAO	$|1\rangle$	$|2\rangle$				
35	-0.4877	-0.3597		c_1	0.5846	-0.9647				
36	-0.3597	-1.2216		c_2	0.5846	0.9647				

Figure 6.6b   Details from the remaining worksheets for the calculation of the kinetic energy integral for H_2^+ and the 'RESULTS!' worksheet returning the eigenvalues and the $LCAO$ coefficients. Note the 'convergence check' panel on the worksheet.

	A	B	C	D	E	F	G	H
1	Canonical orthonormalization for the H_2^+ problem							
2	in an $|$sto-3g$\rangle$ basis.							
3	S			theta =	0.785398			
4	1.000000	0.462783		omega1	1.462783			
5	0.462783	1.000000		omega2	0.537217			
6								
7	U			U - adjoint			S-diagonal	
8	0.70711	0.70711		0.70711	0.70711		1.46278	0.00000
9	0.70711	-0.70711		0.70711	-0.70711		0.00000	0.53722
10							The S^1/2 matrix	
11							0.82682	0.00000
12							0.00000	1.36435
13	Fock matrix			X-canonical			X-canon transpose	
14	-0.949098	-0.628335		0.5846488	0.9647397		0.58465	0.58465
15	-0.628335	-0.949098		0.5846488	-0.9647397		0.96474	-0.96474
16								
17	theta	0.00000		Transformed Fock matrix				
18	omega1	-1.078378		-0.92224	-0.30945	==>	-1.07838	0.00000
19	omega2	-0.597082		-0.92224	0.30945		0.00000	-0.59708
20								
21	Coefficients			Fock - diagonal			C - originals	
22	1.0000000	0.0000000		-1.078378	0.000000		0.584649	-0.964740
23	0.0000000	-1.0000000		0.000000	-0.597082		0.584649	0.964740

Figure 6.6c.   Application of the $S^{-1/2}$ orthonormalization routine to the *LCAO*-MO theory calculation of the bonding and antibonding orbitals of H_2^+.

6.3   CALCULATIONS USING GAUSSIAN BASIS SETS WITH THE EXACT EVALUATION OF INTEGRALS USING FOURIER TRANSFORMS

The calculation for the hydrogen molecule ion makes plain the basic problem with attempting to extend the numerical approach to molecular calculations. Even, for the simplest molecular species, we have to build a spreadsheet with multiple worksheets devoted to the calculation of the components of the potential and kinetic energies and, this, without any consideration of the extra need in neutral species to evaluate the two-electron potential term. It is appropriate, therefore, to extend the direct approach of Chapter 5, for the calculation of the two-electron integral components for helium, to the calculation of all the integrals in any calculation.

There are four types of integral to be calculated (38,46,73). These are the overlap integral, S, and the kinetic energy integral, T, both of which involve application of equations 1.16 to 1.19 and a change of integration variable for the integral over the product Gaussian form. The other two integrals are the nuclear attraction integral, V_n and, for the cases other than one-electron systems, the electron–electron repulsion integral, V_{ee}.

For the diatom problem, the two-centre overlap integral, in the $|$sto-3g$\rangle$ Gaussian basis of Table 1.6, contains components over the primitive functions, $g_{1s}(\mathbf{r}_1 - \mathbf{R})$, about the nuclear positions defined by $\mathbf{R}$;

$$\int = \int g_{1s}(\mathbf{r}_1 - \mathbf{R_a})\, g_{1s}(\mathbf{r} - \mathbf{R_b}) \qquad \qquad 6.19$$

which, using equations 1.16 to 1.19, transforms to

$$\int = \tilde{K}\int g_{1s}(\mathbf{r}_1 - \mathbf{R_p})\, d\mathbf{r}_1 = \tilde{K}\int e^{-\rho|\mathbf{r} - \mathbf{R_p}|^2} \qquad \qquad 6.20$$

This integral is known when we make the change of variable $\mathbf{r} = \mathbf{r}_1 - \mathbf{R_p}$ and $d\mathbf{r} = d\mathbf{r}_1$ and carry out the integrations over the angular coordinates so that

$$S = 4\pi\tilde{K}\int_0^\infty r^2 e^{-\rho r^2} = \left[\frac{\pi}{(\alpha_a + \alpha_b)}\right]^{3/2} e^{\left(-\frac{\alpha_a\alpha_b}{\alpha_a+\alpha_b}|\mathbf{R_a}-\mathbf{R_b}|^2\right)} \qquad 6.21$$

The components, t, of the kinetic energy integral are only slightly more complicated. The action of the Laplacian follows from equations 4.21 to 4.23 and so the components

$$t = \int g_{1s}(\mathbf{r} - \mathbf{R_a})\left(-\frac{1}{2}\nabla_1^2\right) g_{1s}(\mathbf{r}_1 - \mathbf{R_b})\, d\mathbf{r}_1 \qquad 6.22$$

take the form

$$t = \left[\left(\frac{\alpha_a\alpha_b}{\alpha_a + \alpha_b}\right)\left(3 - \frac{2\alpha_a\alpha_b}{\alpha_a + \alpha_b}|\mathbf{R_a} - \mathbf{R_b}|^2\right)\right]\left[\frac{\pi}{\alpha_a + \alpha_b}\right]^{3/2} e^{\left(-\frac{\alpha_a\alpha_b}{\alpha_a+\alpha_b}|\mathbf{R_a}-\mathbf{R_b}|^2\right)} \qquad 6.23$$

The components of the nuclear attraction integrals, v_n, take the form

$$v_n = \int g_{1s}(\mathbf{r} - \mathbf{R_a})\left(-\frac{Z}{|\mathbf{r}_1 - \mathbf{R_c}|}\right) g_{1s}(\mathbf{r}_1 - \mathbf{R_b})\, d\mathbf{r}_1 \qquad 6.24$$

which, after considerable effort (46), involving the Fourier transformation procedure, reduce to the result

$$v_n = \left[-\frac{2\pi}{\alpha_a + \alpha_b}Z_c\right]F_0\left[(\alpha_a + \alpha_b)|\mathbf{R_p} - \mathbf{R_c}|^2\right]e^{\left(-\frac{\alpha_a\alpha_b}{\alpha_a+\alpha}|\mathbf{R_a}-\mathbf{R_b}|^2\right)} \qquad 6.25$$

In equations 6.24 and 6.25 the positions $\mathbf{R_p}$ and $\mathbf{R_c}$ locate the product Gaussian of the primitives in equation 6.24, while $F_0[\ ]$ is related, as before, to the *Error function*, $erf[\ ]$ of pure mathematics (75), through the identity

$$F_0[t] = \left(\frac{\pi}{4t}\right)^{1/2} erf[t^{1/2}] \qquad 6.26$$

As you might expect, the use of exact formulae for the different integrations leads to a dramatic reduction in size of the spreadsheet file.

Exercise 6.6.   Solution of the Schrödinger equation for the hydrogen molecule ion problem using Gaussian basis sets and the direct evaluation of integrals.

1. Make a copy of fig6-5.xls and rename it fig6-7.xls. The changes required in this exercise are simply to provide for the direct determination of all the integrals, as indicated in Figure 6.7a. Using the notation $|*a\rangle$ and $|*b\rangle$ to identify Gaussian primitives from the atomic sites, the important formulae entries are as follows

2. Provide for the Boys transformations of equations 1.16 to 1.19 in cells $V$10 to $V$20.

3. The various transforms required for the integrations are the formulae in cells $B$23 to $V$31, with, for example,

$$t[A] = \$B\$23 = B\$19^*(B\$21\text{-}B\$18)\text{^}2$$

$$F_0(t)[A] = \$B\$24 = \textbf{IF}(B\$23{<}0.000001,(1\text{-}B\$23/3),$$
$$(0.5^*\textbf{SQRT}(\textbf{PI}(\,)/B\$23))^*\textbf{ERF}(0,\text{SQRT}(B\$23)))$$

$$t[B] = \$B\$25 = B\$19^*(B\$22\text{-}B\$18)\text{^}2$$

$$F_0(t)[B] = \$B\$26 = \textbf{IF}(B\$25{<}0.000001,(1\text{-}B\$25/3),$$
$$(0.5^*\textbf{SQRT}(\textbf{PI}(\,)/B\$25))^*\textbf{ERF}(0,\text{SQRT}(B\$25)))$$

4. Then, calculate the integrals in cells $B$28 to $V$31, with, for example,

$$s = \$B\$28 = \textbf{POWER}(\textbf{PI}(\,)/B\$19,1.5)^*B\$17^*B\$20$$

$$t = \$B\$29 = B\$12^*B\$14/B\$19^*(3\text{-}2^*B\$12^*B\$14/B\$19^*$$
$$(B\$10\text{-}B\$11)\text{^}2)^*\textbf{POWER}(\textbf{PI}(\,)/B\$19,1.5)^*B\$17^*B\$20$$

$$v[A] = \$B\$30 = \text{-}2^*\textbf{PI}(\,)/B\$19^*\$D\$1^*B\$17^*B\$24^*B\$20$$

$$v[B] = \$B\$31 = \text{-}2^*\textbf{PI}(\,)/B\$19^*\$G\$1^*B\$17^*B\$26^*B\$20$$

5. Finally, collect the various components of the integrals on the 'RESULTS!' worksheet and determine the total energy, including the nuclear repulsion term in cells $G$12 and $G$13, with

$$E+ = \text{'RESULTS!'}\$G\$12 = \$F\$12+1/\$A\$2\text{-}(\$B\$12+\$B\$15)$$

$$E- = \text{'RESULTS!'}\$G\$13 = \$F\$13+1/\$A\$2\text{-}(\$B\$12+\$B\$15)$$

and construct the included chart of the variation in total energy with bond length as in Figure 6.7b.

6. As usual now, add the canonical worksheet and make this the primary location of the input and control data in the calculation as displayed in Figure 6.7c.

6.4   CALCULATIONS INVOLVING THE TWO-ELECTRON TERMS; THE |sto-3g⟩ HF-SCF RESULTS FOR DIHYDROGEN

The extension of the one-electron calculations to the case of the simplest molecule dihydrogen with two electrons is beyond our numerical methodology on a spreadsheet

	A	B	C	D	E	F	G
1	R		Z = ⟩	1		Z = ⟩	1
2	2.00	|sto-3g:1s⟩	scaler	1.24	|sto-3g:1s⟩	scaler	1.24
3		|1a⟩	|2a⟩	|3a⟩	|1b⟩	|2b⟩	|3b⟩
4	α	0.109818	0.405771	2.227660	0.109818	0.405771	2.227660
5	d	0.444635	0.535328	0.154329	0.444635	0.535328	0.154329
6	α'	0.168856	0.623913	3.425250	0.168856	0.623913	3.425250
7	d'	0.083474	0.267839	0.276934	0.083474	0.267839	0.276934
8							
9		[1a|1a]	[1a|2a]	[1a|3a]	[1a|1b]	[1a|2b]	[1a|3b]
10	Ra	0.00	0.00	0.00	0.00	0.00	0.00
11	Rb	0.00	0.00	0.00	2.00	2.00	2.00
12	alpha[a]	0.168856	0.168856	0.168856	0.168856	0.168856	0.168856
13	d[a]	0.083474	0.083474	0.083474	0.083474	0.083474	0.083474
14	alpha[b]	0.168856	0.623913	3.425250	0.168856	0.623913	3.425250
15	d[b]	0.083474	0.267839	0.276934	0.083474	0.267839	0.276934
16		[1a|1a]	[1a|2a]	[1a|3a]	[1a|1b]	[1a|2b]	[1a|3b]
17	kappa	1.000000	1.000000	1.000000	0.713400	0.587686	0.525349
18	Rp	0.000000	0.000000	0.000000	1.000000	1.574010	1.906037
19	alpha[p]	0.337712	0.792770	3.594106	0.337712	0.792770	3.594106
20	d[p]	0.006968	0.022358	0.023117	0.006968	0.022358	0.023117
21	R_A	0.000000	0.000000	0.000000	0.000000	0.000000	0.000000
22	R_B	2.000000	2.000000	2.000000	2.000000	2.000000	2.000000
23	t[A]	0.000000	0.000000	0.000000	0.337712	1.964092	13.057308
24	F_o(t)[A]	1.000000	1.000000	1.000000	0.897974	0.602333	0.245255
25	t[B]	1.350849	3.171079	14.376425	0.337712	0.143862	0.031732
26	F_o(t)[B]	0.686069	0.491802	0.233733	0.897974	0.954047	0.989522
27		[1a|1a]	[1a|2a]	[1a|3a]	[1a|1b]	[1a|2b]	[1a|3b]
28	s	0.197700	0.176372	0.018892	0.141039	0.103651	0.009925
29	t	0.050074	0.070314	0.009120	0.027680	0.026679	0.002735
30	v[a]	-0.129639	-0.177198	-0.040413	-0.083049	-0.062725	-0.005207
31	v[b]	-0.088941	-0.087146	-0.009446	-0.083049	-0.099351	-0.021008

Figure 6.7a   Detail from the worksheet, 1e-integrals! in fig6-7abc.xls for the direct calculation of the integrals required to solve Schrödinger's equation for the hydrogen molecule ion using Gaussian basis sets.

	A	B	C	D	E	F	G	H	I	J	K	L
1	R		Z = ⟩	1		Z = ⟩	1		R	E+	E-	
2	2	|sto-3g:1s⟩	scaler	1.24	|sto-3g:1s⟩	scaler	1.24		0.5	0.8866	2.5746	
3		|1a⟩	|2a⟩	|3a⟩	|1b⟩	|2b⟩	|3b⟩		1.0	0.0764	1.1749	
4	α	0.109818	0.405771	2.227660	0.109818	0.405771	2.227660		1.5	-0.0882	0.6269	
5	d	0.444635	0.535328	0.154329	0.444635	0.535328	0.154329		2.0	-0.1161	0.3616	
6	α'	0.168856	0.623913	3.425250	0.168856	0.623913	3.425250		2.5	-0.1038	0.2157	
7	d'	0.083474	0.267839	0.276934	0.083474	0.267839	0.276934		3.0	-0.0810	0.1307	
8		S			H				4.0	0.04027	0.04914	
9		1.00000	0.46278		-0.95437	-0.62938						
10		0.46278	1.00000		-0.62938	-0.95437						
11		T			Eigenvalues	$ε_i$						
12		0.76003	0.09107		|1⟩	-1.0827	-0.11611					
13		0.09107	0.76003		|2⟩	-0.6049	0.36163					
14		V[A]			Coefficients	|1⟩	|2⟩					
15		-1.22662	-0.36022		c_1	0.5846	0.96474					
16		-0.36022	-0.48779		c_2	0.5846	-0.96474					
17		V[B]										
18		-0.48779	-0.36022									
19		-0.36022	-1.22662									

Figure 6.7b   The 'RESULTS!' worksheet in fig6-7abc.xls linking to the integral components in 1e-integrals! calculated using equations 6.21, 6.23 and 6.25.

	A	B	C	D	E	F	G	H
1	Canonical orthonormalization with exact calculation of integral over Gaussian primitives.							
2	|sto-3g⟩ basis for the hydrogen molecule ion.							
3								
4	R		Z = ⟩	1.00		Z = ⟩	1.00	
5	2	|sto-3q:1s⟩	scaler	1.24	|sto-3q:1s⟩	scaler	1.24	
6		|1a⟩	|2a⟩	|3a⟩	|1b⟩	|2b⟩	|3b⟩	
7	α	0.109818	0.405771	2.22766	0.109818	0.405771	2.22766	
8	d	0.444635	0.535328	0.154329	0.444635	0.535328	0.154329	
9	α'	0.168856	0.623913	3.425250	0.168856	0.623913	3.425250	
10	d'	0.083474	0.267839	0.276934	0.083474	0.267839	0.276934	
11								
12	S			theta =	0.785398			
13	1.000001	0.462778		omega1	1.462779			
14	0.462778	1.000001		omega2	0.537223			
15								
16	U			U - adjoint			S-diagonal	
17	0.70711	0.70711		0.70711	0.70711		1.46278	0.00000
18	0.70711	-0.70711		0.70711	-0.70711		0.00000	0.53722
19							The S^1/2 matrix	
20							0.82682	0.00000
21							0.00000	1.36434
22	Fock matrix			X-canonical			X-canon transpose	
23	-0.954368	-0.629376		0.5846495	0.9647339		0.58465	0.58465
24	-0.629376	-0.954368		0.5846495	-0.9647339		0.96473	-0.96473
25								
26	theta	0.00000		Transformed Fock matrix				
27	omega1	-1.082695		-0.92594	-0.31353	==>	-1.08270	0.00000
28	omega2	-0.604948		-0.92594	0.31353		0.00000	-0.60495
29								
30	Coefficients			Fock - diagonal			C - originals	
31	1.0000000	0.0000000		-1.082695	0.000000		0.584649	-0.964734
32	0.0000000	-1.0000000		0.000000	-0.604948		0.584649	0.964734

Figure 6.7c   Calculation using exact integral results over the Gaussian primitives and canonical orthonormalization with $S^{-1/2}$ diagonalization for $H_2{}^+$.

not only because of size considerations for even the minimum basis set representation but principally, because of the complexity of even the simplest two-electron integrals for molecules.

The general problem is set out schematically in Figure 5.5 and in equations 5.51 and 5.53. We know that the two-electron terms over the Gaussian primitives require the evaluation of equation 5.55 over the Gaussian basis set.

$$\langle ij|kl\rangle = \frac{2\pi^{5/2}}{[(\alpha + \beta)(\gamma + \delta)(\alpha + \beta + \gamma + \delta)^{1/2}]} e^{\left[-\frac{\alpha\beta}{\alpha+\beta}|R_A - R_B|^2 - \frac{\gamma\delta}{\gamma+\delta}|R_C - R_D|^2\right]} .$$

$$F_0[(\alpha + \beta)(\gamma + \delta)(\alpha + \beta + \gamma + \delta)|R_P - R_Q|^2]$$

5.53

However, the simplifications, in the case of the helium atom, with each Gaussian primitive about the atomic nucleus cannot be invoked for the hydrogen molecule, because one basis set is centred about one atomic centre, while the other is centred about the second atomic position.

To carry out the full Hartree–Fock calculation for dihydrogen, we have to assemble the two-electron term in the molecular Hamiltonian over the Gaussian sets. For the |sto-3g⟩ representation of the hydrogenic 1s orbitals in the LCAO-MO function, this means the determination of the potential terms V_{aaaa}, V_{aaab}, V_{abab}, V_{aabb}, V_{abbb} and V_{bbbb}, with the subscripts identifying all the primitives in each |sto-3g⟩ group on the atomic sites A and B.

Thus, we have to extend the design of the two-electron worksheet of fig5-6.xls to the present case involving Gaussian primitives about different sites, the general case represented in Figure 5.5. Details from these worksheets in fig6-8.xls are shown in Figure 6.8 and you can see that in each copy of the basic worksheet design the significant new features are the different values of the $\mathbf{R}_a$, $\mathbf{R}_b$, $\mathbf{R}_c$, and $\mathbf{R}_d$ parameters, reflecting the origins of the components in the $\langle ij|kl\rangle$ integrals.

For the one-electron problem, we would proceed, at this point, to transform the core Hamiltonian matrix to obtain the approximate eigenfunctions and eigenvalues for the basis set chosen as, for example, in Figure 6.6c and fig6-7abc.xls, since step 2 above would not involve the calculation of any two-electron integrals. However, for dihydrogen and the myriad of other many-electron molecules we need, too, to form the Fock matrix by adding to the core Hamiltonian matrix the electron–electron repulsion potential term due to the two-electron integrals over the Gaussian primitives, in the following manner.

Dihydrogen is a 'closed-shell' molecule and so it is adequate to describe it by a single Slater determinant to describe the double occupation of the $1\sigma_g^+$ space orbital. Thus, as in Section 5.2, the total charge density, $\rho(\mathbf{r})$, for all closed shell electron configurations is just

$$\rho(\mathbf{r}) = 2\sum_i^{n/2} |\varphi(\mathbf{r})|^2 \mathbf{dr} \qquad 5.42$$

which we interpret as the probability of finding an electron in the volume element $\mathbf{dr}$ about the position $\mathbf{r}$. For the LCAO-MO of equation 6.2, this can be rewritten to include the density matrix, $P_{\mu\nu}$, from equation 5.43, as

$$\rho(\mathbf{r}) = 2\sum_{i=1}^{N/2} |\phi_i(\mathbf{r})|^2 = 2\sum_{i=1}^{N/2}\sum_j c_{ji}\zeta_j(\mathbf{r})\sum_k c_{ki}\zeta_k(\mathbf{r})$$

$$= \sum_j\sum_k \left[2\sum_{i=1}^{N/2} c_{ji}c_{ki}\right]\phi_j(\mathbf{r})\phi_k(\mathbf{r}) \qquad 5.43$$

$$= \sum_j\sum_k P_{jk}\phi_j(\mathbf{r})\phi_k(\mathbf{r})$$

with the individual terms ranging over the two Gaussian sto-3g approximations to the hydrogen 1s atomic orbitals about $\mathbf{R}_a$ and $\mathbf{R}_b$.

Thus, as indicated in the README box on the spreadsheet, the *HFS-SCF* calculation is started with the null-density matrix and *self-consistency* is achieved when the output density matrix substituted into the input cells returns the same values for the parameters calculated or leads to agreement with some other appropriate criterion. This requires the restructuring of the Hamiltonian matrix into the Fock form developed in equation 5.45 and the determination of the roots of the secular equation, equation 5.49, as set out for the helium $1s^2$ state in equations 5.51 and 5.52.

Exercise 6.7.   The SCF calculation of the energy of dihydrogen with bond length, orbital energies and LCAO-MO functions.

1. Make a copy of fig6-7.xls and rename it fig6-8.xls.
2. Insert copies of the two-electron worksheets from fig5-14.xls. Since $|sto\text{-}3g\rangle$ basis sets are located at each hydrogen site, the only difference in the permutations leading to the $\langle ij|kl\rangle$ terms rests with the values of $\mathbf{R}_a$, $\mathbf{R}_b$, $\mathbf{R}_c$ and $\mathbf{R}_d$. As with the helium calculation, there are 3^4 individual terms worked out on each worksheet. Details from the individual two-electron worksheets are shown in Figure 6.8a.
3. As previously, link all the data through to the 'Fock!' worksheet and construct the one-electron Hamiltonian matrix in cells \$E\$16 to \$F\$17, with

$$h_{11} = \$E\$16 = \$B\$15 + \$B\$19 + \$B\$23$$

$$h_{12} = \$F\$16 = \$C\$15 + \$C\$19 + \$C\$23$$

$$h_{21} = \$E\$17 = \$B\$16 + \$B\$20 + \$B\$24$$

$$h_{22} = \$F\$18 = \$C\$16 + \$C\$20 + \$C\$24$$

4. Transform this into the Fock matrix form of equation 5.46 with, in cells \$E\$20 to \$F\$21

$$F_{11} = \$E\$16 + \$E\$12^*(\$C\$26 - 0.5^*\$C\$26) + \$F\$12^*(\$C\$27 - 0.5^*\$C\$27) + \\ \$E\$13^*(\$C\$27 - 0.5^*\$C27) + \$F\$13^*(\$C\$29 - 0.5^*\$C\$28)$$

$$F_{12} = \$F\$16 + \$E\$12^*(\$C\$27 - 0.5^*\$C\$27) + F12^*(\$C\$28 - 0.5^*\$C\$28) + \\ \$E\$13^*(\$C\$28 - 0.5^*\$C\$29) + \$F\$13^*(\$C\$30 - 0.5^*\$C\$30)$$

$$F_{21} = \$E\$17 + \$E\$12^*(\$C\$27 - 0.5^*\$C\$27) + \$F\$12^*(\$C\$28 - 0.5^*\$C\$29) + \\ \$E\$13^*(\$C\$28 - 0.5^*\$C\$28) + \$F\$13^*(\$C\$30 - 0.5^*\$C\$30)$$

$$F_{22} = \$F\$17 + \$E\$12^*(\$C\$29 - 0.5^*\$C\$28) + \$F412^*(\$C\$30 - 0.5^*\$C\$30) + \\ \$E\$13^*(\$C\$30 - 0.5^*\$C\$30) + \$F\$13^*(\$C\$31 - 0.5^*\$C\$31)$$

and link these formulae to cells \$A\$24 to \$B\$25 on the 'scf'! worksheet.
5. Note that the starting choice for the coefficients of equation 6.2 are based on the assumption there are no two-electron terms, i.e. the input density matrix, cells \$E\$11 to \$F\$12 are set to zero from the link to 'canonical' !\$G\$2 to 'scf'!\$G\$3. Remember, too, that as in the case of the ground state of helium there is only one doubly-occupied orbital, so the output density matrix is comprised of the coefficients of the $1\sigma_g{}^+$ orbital only.

Figure 6.8c displays the calculations in the form resulting from the application of the canonical orthonormalization procedure of Section 3.6 and Table 6.1 shows the full iterative process for the dihydrogen calculation within the *HFS-SCF* approximation based on the spreadsheet design in fig6-8.xls.

V_{aaaa}

	A	B	C	D	E	F	G	H	I	J	K
3						sto-3g	scaler	1.24000E+00	sto-3g	scaler	1.24000E+00
4						(g11)	(g12)	(g13)	(g21)	(g22)	(g23)
5				exponent		1.09818E-01	4.05771E-01	2.22766E+00	1.09818E-01	4.05771E-01	2.22766E+00
6				coefficient		4.44635E-01	5.35328E-01	1.54329E-01	4.44635E-01	5.35328E-01	1.54329E-01
7				alpha		1.68856E-01	6.23913E-01	3.42525E+00	1.68855E-01	6.23911E-01	3.42524E+00
8				d		8.34740E-02	2.67839E-01	2.76934E-01	8.34738E-02	2.67838E-01	2.76934E-01
9	R = ⟩	1.4000									
10	Ra	Rb	Rc	Rd		V_{aaaa}	0.77461				
11	0	0	0	0							
12	a	b	c	d		d_a	d_b	d_c	d_d	V_{aaaa}	Rp
13	0.16886	0.16886	0.16886	0.16886		0.08347	0.08347	0.08347	0.08347	0.01812	0.0000
14	0.16886	0.16886	0.16886	0.62391		0.08347	0.08347	0.08347	0.26784	0.01915	0.0000
15	0.16886	0.16886	0.62391	0.16886		0.08347	0.08347	0.26784	0.08347	0.01915	0.0000

V_{baaa}

Ra	Rb	Rc	Rd	V_{baaa}	0.44411			
1.40000	0.00000	0.00000	0.00000					
a	b	c	d	d_a	d_b	d_c	d_d	V_{baaa}	Rp
0.16886	0.16886	0.16886	0.16886	0.08347	0.08347	0.08347	0.08347	0.01495	0.7000
0.16886	0.62391	0.16886	0.16886	0.08347	0.26784	0.08347	0.08347	0.01465	0.2982
0.16886	0.16886	0.16886	0.62391	0.08347	0.08347	0.08347	0.26784	0.01562	0.7000

V_{baba}

	A	B	C	D	E	F	G	H	I	J	K
10	Ra	Rb	Rc	Rd		V_{baba}	0.29703				
11	1.4	0	1.4	0							
12	a	b	c	d		d_a	d_b	d_c	d_d	V_{baba}	Rp
13	0.16886	0.16886	0.16886	0.16886		0.08347	0.08347	0.08347	0.08347	0.01302	0.7000
14	0.16886	0.16886	0.16886	0.62391		0.08347	0.08347	0.08347	0.26784	0.01235	0.7000
15	0.16886	0.62391	0.16886	0.16886		0.08347	0.26784	0.08347	0.08347	0.01235	0.2982

V_{bbaa}

	A	B	C	D	E	F	G	H	I	J	K
10	Ra	Rb	Rc	Rd		V_{bbaa}	0.56968				
11	1.40000	1.40000	0.00000	0.00000							
12	a	b	c	d		d_a	d_b	d_c	d_d	V_{bbaa}	Rp
13	0.16886	0.16886	0.16886	0.16886		0.08347	0.08347	0.08347	0.08347	0.01631	1.4000
14	0.16886	0.16886	0.16886	0.62391		0.08347	0.08347	0.08347	0.26784	0.01656	1.4000
15	0.16886	0.16886	0.16886	3.42525		0.08347	0.08347	0.08347	0.27693	0.00194	1.4000

V_{bbba}

Ra	Rb	Rc	Rd	V_{bbba}	0.44411			
1.40000	1.40000	1.40000	0.00000					
a	b	c	d	d_a	d_b	d_c	d_d	Vbbba	Rp
0.16886	0.16886	0.16886	0.16886	0.08347	0.08347	0.08347	0.08347	0.01495	1.4000
0.16886	0.16886	0.16886	0.62391	0.08347	0.08347	0.08347	0.26784	0.01346	1.4000
0.16886	0.62391	0.16886	0.16886	0.08347	0.26784	0.08347	0.08347	0.01562	1.4000

V_{bbbb}

	B	C	D	E	F	G	H	I	J	K
10	Rb	Rc	Rd		V_{bbbb}	0.77461				
11	0.00000	0.00000	0.00000							
12	c	d		d_a	d_b	d_c	d_d	V_{bbbb}	Rp
13	0.16886	0.16886	0.16886		0.08347	0.08347	0.08347	0.08347	0.01812	0.0000
14	0.16886	0.16886	0.62391		0.08347	0.08347	0.08347	0.26784	0.01915	0.0000
15	0.16886	0.62391	0.16886		0.08347	0.08347	0.26784	0.08347	0.01915	0.0000

Figure 6.8a   The two-electron potential term worksheets for the dihydrogen calculation; note how the colour coding of the entries for the exponents and coefficients of the Gaussian primitives identify individual $\langle ij|kl \rangle$ integrals.

Computational Quantum Chemistry

	A	B	C	D	E	F	G	H	I
1		R=)	1.40000						
2			Z=)	1.00000		Z=)	1.00000		
3		sto-3g	scaler	1.24000		scaler	1.24000		
4		|g11)	|g12)	|g13)	|g21)	|g22)	|g23)		
5	α	0.10982	0.40577	2.22766	0.10982	0.40577	2.22766		
6	d	0.44464	0.53533	0.15433	0.44464	0.53533	0.15433	G array	
7	α'	0.16886	0.62391	3.42525	0.16886	0.62391	3.42524	0.00000	0.00000
8	d'	0.08347	0.26784	0.27693	0.08347	0.26784	0.27693	0.00000	0.00000
9									
10		S			README				
11		1.00000	0.65932		Density matrix [in]			Density matrix [out]	
12		0.65932	1.00000		0.00000	0.00000		0.60266	0.60266
13					0.00000	0.00000		0.60266	0.60266
14		T							
15		0.76003	0.23646		H core			Eigenvalues	
16		0.23646	0.76003		-1.12041	-0.95838		ε_1	-1.25280
17					-0.95838	-1.12041		ε_2	-0.47561
18		V[A]							
19		-1.22662	-0.59742		F			c_1[out]	0.54893
20		-0.59742	-0.65383		-1.12041	-0.95838		c_2[out]	0.54893
21					-0.95838	-1.12041			
22		V[B]							
23		-0.65382818971	-0.59741805677						
24		-0.59741805677	-1.22661299504						
25									
26		(1111)	0.77460836003						
27		(2111)	0.44410926761						
28		(2121)	0.29702966414						
29		(2211)	0.56967737648						
30		(2221)	0.44410876229						
31		(2222)	0.77460679832						

Figure 6.8b   The 'Fock!' worksheet for the *HFS-SCF* calculations on dihydrogen. These results are for the equilibrium experimental bond length of 1.4 a.u. found for H_2, but show the situation before iteration, with the density matrix set to zero. Again, note, the calculation follows Exercise 6.1.

Table 6.1.   Iteration of the spreadsheet calculation, fig6-8.xls, to self-consistency based on the criterion that the input coefficients of the linear combination of equation 6.2 and those returned by the calculation should be the same. The calculation follows the procedure set out in Exercise 6.1.

	B	C	D	E	F	G	H
37	Iteration	c_1[in]	c_2[in]	c_1[out]	c_2[out]	ε_1	ε_2
38	1	1.000000	0.000000	0.519611	0.577939	-0.875903	0.059688
39	2	0.519600	0.577900	0.546134	0.551731	-0.577936	0.670016
40	3	0.546100	0.551700	0.548665	0.549203	-0.578200	0.670267
41	4	0.548700	0.549200	0.548908	0.548960	-0.578200	0.670300
42	5	0.548908	0.548960	0.548932	0.548936	-0.578203	0.670270
43	6	0.548932	0.548936	0.548934	0.548934	-0.578203	0.670270
44	7	0.548934	0.548934	0.548934	0.548934	-0.578203	0.670270

As you see, self-consistency is achieved after about six iterations based on the criterion that the coefficients in the linear combination of equation 6.2, the starting *LCAO*-MO and the output coefficients, the *LCAO*-MO after the calculation cycle should be the same. You should devise a criterion based on the convergence of the density matrix elements and apply this to the calculation using the 'scf! worksheet.

	A	B	C	D	E	F	G	H
1	Dihydrogen calculation in an |sto-3g⟩ basis					Density matrix (in)		
2	Direct integrals and S$^{-1/2}$ orthogonalization.						0.60266	0.60266
3							0.60266	0.60266
4		R = ⟩		1.4				
5			Z = ⟩	1.00000			Z = ⟩	1.00000
6	README	sto-3g	scaler	1.24000	sto-3g	scaler	1.24000	
7		|g11⟩	|g12⟩	|g13⟩	|g21⟩	|g22⟩	|g23⟩	
8	α	0.10982	0.40577	2.22766	0.10982	0.40577	2.22766	
9	d	0.44464	0.53533	0.15433	0.44464	0.53533	0.15433	
10	α'	0.16886	0.62391	3.42525	0.16886	0.62391	3.42524	
11	d'	0.08347	0.26784	0.27693	0.08347	0.26784	0.27693	
12								
13	S			theta =	0.78540	Density matrix (out)		
14	1.00000	0.65932		omega1	1.65932	0.60266	0.60266	
15	0.65932	1.00000		omega2	0.34068	0.60266	0.60266	
16								
17	U			U - adjoint			S-diagonal	
18	0.70711	0.70711		0.70711	0.70711		1.65932	0.00000
19	0.70711	-0.70711		0.70711	-0.70711		0.00000	0.34068
20							The S^1/2 matrix	
21							0.77631	0.00000
22							0.00000	1.71327
23	Fock matrix			X-canonical			X-canon transpose	
24	-0.36554	-0.59389		0.54893	1.21146		0.54893	0.54893
25	-0.59389	-0.36554		0.54893	-1.21146		1.21146	-1.21146
26								
27	theta	0.00000		Transformed Fock matrix				
28	omega1	-0.57820		-0.52666	0.27664	==>	-0.57820	0.00000
29	omega2	0.67027		-0.52666	-0.27663		0.00000	0.67027
30								
31	Coefficients			Fock - diagonal			C - originals	
32	1.00000	0.00000		-0.57820	0.00000		0.54893	-1.21146
33	0.00000	-1.00000		0.00000	0.67027		0.54893	1.21146

Figure 6.8c 'scf'! worksheet involving the input of density matrix data in the \|sto-3g⟩ calculations on dihydrogen.

6.5   THE STANDARD FORM FOR THE RESULTS OF HFS-SCF CALCULATIONS

Our calculations have been based on the determination of integrals over the primitive Gaussian components of the *LCAO*-MOs defined in equation 6.2 and this is the normal manner in which modern molecular orbital theory programs are written. For interpretation and applications of molecular orbital theory, however, it is customary to present results in terms of the actual molecular orbitals determined after the SCF calculation has been completed. This is not an easy step in the molecular orbital theory of more complicated molecular structures, since it requires the collecting together of all the individual one-electron Hamiltonian, h, Coulomb, J, and exchange, K, integrals for each molecular orbital. In particular, the transformations over the two-electron integral components can involve a considerable number of steps for large molecules, since for a generalized equation 6.3,

$$\varphi_i(\mathbf{r}) = \sum_{k=1}^{K} c_{ki} \phi_k(\mathbf{r} - \mathbf{R}_k) \qquad 6.27$$

we require the integrals

$$h_{ij} = \langle \varphi_i | h | \varphi_j \rangle = \sum_{k(i)} \sum_{k(j)} c_{k(i)} c_{k(j)} [t + v_{n/e}]_{k(i)k(j)} \qquad 6.28$$

and

$$\varphi_i \varphi_j \varphi_k \varphi_l = \sum_{k(i)} \sum_{k(j)} \sum_{k(k)} \sum_{k(l)} c_{k(i)} c_{k(j)} c_{k(k)} c_{k(l)} (ij|kl) \qquad 6.29$$

For the case of dihydrogen and the single occupied $1\sigma_g^+$ molecular orbital. We need

$$h_{11} = \langle 1\sigma_g^+ | h | 1\sigma_g^+ \rangle = -1.2528 \text{ a.u.} \qquad 6.30$$

$$h_{22} = \langle 1\sigma_g^- | h | 1\sigma_g^- \rangle = -0.4756 \text{ a.u.} \qquad 6.31$$

$$J_{11} = \langle 1\sigma_g^+ 1\sigma_g^+ | 1\sigma_g^+ 1\sigma_g^+ \rangle = 0.6746 \text{ a.u.} \qquad 6.32$$

$$J_{22} = [1\sigma_g^- 1\sigma_g^- | 1\sigma_g^- 1\sigma_g^-] = 0.6975 \text{ a.u.} \qquad 6.33$$

$$J_{12} = [1\sigma_g^+ 1\sigma_g^+ | 1\sigma_g^- 1\sigma_g^-] = 0.6636 \text{ a.u.} \qquad 6.34$$

$$K_{12} = \langle 1\sigma_g^+ 1\sigma_g^- | 1\sigma_g^- 1\sigma_g^+ \rangle = 0.1813 \text{ a.u.} \qquad 6.35$$

Equations 6.30 and 6.31 are the sums of the kinetic and nuclear attraction energies for an electron in the two possible *LCAO* functions. Similarly, the '*J*' integrals are the Coulomb terms for the two electron in the same molecular orbital and in the different molecular orbitals, while K_{12} is the exchange interaction between an electron in first linear combination and the other electron, with the same spin, in the second possible linear combination of the $|$sto-3g$\rangle$ representations of the 1s orbitals in hydrogen.

Finally, for the closed-shell electron configuration, the ground state of H_2, we can use these parameters to write the orbital energies in the form

$$\varepsilon_i = h_{ii} + \sum_j 2J_{ij} - K_{ij} \qquad 6.36$$

remembering that for the closed-shell configuration both two-electron terms are equal.

Exercise 6.8.   Modification of fig6-8abc.xls for the presentation of the SCF calculation in standard form.

1. Make a copy of fig6-8abc.xls; rename it fig6-9.xls.
2. Provide for the compilation of the integrals defined in equations 6.32 to 6.38 with the formulae entries in cells \$E\$23 to \$K\$28,

   a. h_{11} = \$F\$28 = 2*(\$E\$16 + \$F\$16)*\$I\$19^2
   b. h_{22} = \$G\$28 = 2*(\$E\$16*\$H\$24^2 + \$F\$16*\$H\$24*\$H\$25)
   c. J_{11} = \$H\$28 = SUM(\$H\$29:\$H\$34) i.e. the sum of the contributing $\langle ij|kl \rangle$ listed immediately below this entry and, similarly,
   d. J_{22} = \$I\$28 = SUM(\$I\$29:\$I\$34)

   e. $J_{12} = \$J\$28 = \textbf{SUM}(\$J\$29{:}\$J\$34)$

   f. $K_{12} = \$K\$28 = \textbf{SUM}(\$K\$29{:}\$K\$34)$

3.   Iterate to self-consistency, to reproduce the results displayed in Figure 6.9 and fig6-9.xls.

6.6   THE |sto-3g⟩ HFS-SCF CALCULATION FOR HEH⁺

The helium hydride two-electron ion has been detected in mass spectra and Szabo and Ostlund (47) report it to be of interest in astrophysics, for example, as the decay product in the β emission of the heavy dihydrogen species HT and in proton scattering experiments involving helium atoms. There are, too, accurate calculations by Wolniewicz (84), which suggest an equilibrium bond length of 1.4632 a.u. and an electronic binding energy of 0.0749 a.u.

To repeat the Wolniewicz calculation, as a spreadsheet exercise, it is necessary only to enter the appropriate parameters into the two-electron dihydrogen program fig6-8abc.xls and then iterate to self-consistency. The end result is set out in Figure 6.10, which reproduces the results page of the diatom spreadsheet when self-consistency has been established. Then in Table 6.2 the output of each iteration in the calculation is given from the initial guess at the density matrix that there be no mixing of the |sto-3g⟩ functions centred about each atomic position, for the bond length set to the Wolniewicz value 1.4632 a.u.

	A	B	C	D	E	F	G	H	I	J	K	L
1		R =)	1.40000									
2			Z =)	1.00000		Z =)	1.00000					
3		sto-3g	scaler	1.24000	sto-3g	scaler	1.24000					
4		|g11)	|g12)	|g13)	|g21)	|g22)	|g23)					
5	α	0.10982	0.40577	2.22766	0.10982	0.40577	2.22766					
6	d	0.44464	0.53533	0.15433	0.44464	0.53533	0.15433	G array				
7	α'	0.16886	0.62391	3.42525	0.16886	0.62391	3.42524	0.75487	0.36450			
8	d'	0.08347	0.26784	0.27693	0.08347	0.26784	0.27693	0.36450	0.75487			
9												
10		S			README							
11		1.00000	0.65932		Density matrix [in]			Density matrix [out]				
12		0.65932	1.00000		0.60266	0.60266		0.60266	0.60266			
13					0.60266	0.60266		0.60266	0.60266			
14												
15		T			H core			Eigenvalues				
16		0.76003	0.23646		-1.12041	-0.95838		ϵ_i	-0.57820			
17		0.23646	0.76003		-0.95838	-1.12041		ϵ_j	0.67027			
18		V[A]										
19		-1.22662	-0.59742		F			c_i[out]	0.54893			
20		-0.59742	-0.65383		-0.36554	-0.59389		c_j[out]	0.54893			
21					-0.59389	-0.36554						
22		V[B]										
23		-0.65382818971	-0.59741805677		E-elect	E-total	|ψ_1⟩	|ψ_2⟩	N1	N2	S	
24		-0.59741805677	-1.22661299504		-1.83100	-1.11672	0.54893	-1.21147	1.00000	1.00001	0.00000	
25							0.54893	1.21147				
26		(1111)	0.77460836003									
27		(2111)	0.44410926761		E-nuc	⟨ψ_1|h|ψ_1⟩	⟨ψ_2|h|ψ_2⟩	J11	J22	J12	K12	
28		(2121)	0.29702966414		0.71429	-1.25280	-0.47561	0.67459	0.69750	0.66357	0.18126	
29		(2211)	0.56967737648					0.07033	1.66851	0.34257	0.34257	(aaaa)
30		(2221)	0.44410876229					0.16130	-3.82645	0.00000	0.00000	(baaa)
31		(2222)	0.77460679832					0.10788	2.55921	-0.52544	0.00000	(baba)
32								0.10345	2.45417	0.50387	-0.50387	(bbaa)
33								0.16130	-3.82644	0.00000	0.00000	(bbba)
34								0.07033	1.66850	0.34257	0.34257	(bbbb)

Figure 6.9   The 'Fock!' worksheet of Figure 6.8 modified to display the SCF results in terms of the molecular orbitals of the calculation. These are the converged results for the equilibrium bond length 1.4 a.u. in dihydrogen.

	B	C	D	E	F	G	H	I	J	K	L
1	R =)	1.463200									
2		Z =)	2.000000		Z =)	1.000000					
3	sto-3g	scaler	2.092500	sto-3g	scaler	1.240000					
4	(g11)	(g12)	(g13)	(g21)	(g22)	(g23)					
5	0.109818	0.405771	2.227660	0.109818	0.405771	2.227660					
6	0.444635	0.535328	0.154329	0.444635	0.535328	0.154329	G array				
7	0.480844	1.776691	9.753935	0.168856	0.623913	3.425250	1.194109	0.296613			
8	0.182986	0.587136	0.607075	0.083474	0.267839	0.276934	0.296613	0.921319			
9											
10	S			README							
11	1.000001	0.450770		Density matrix [in]			Density matrix [out]		Best		
12	0.450770	1.000001		1.286163	0.540163		1.286143	0.540172	1.286163	0.540163	
13				0.540163	0.226858		0.540172	0.226869	0.540163	0.226858	
14	T										
15	2.164313	0.167013		H core			Eigenvalues				
16	0.167013	0.760033		-2.652745	-1.347205		ε_1	-1.597448			
17				-1.347205	-1.731828		ε_2	-0.061668			
18	V[A]										
19	-4.139827	-1.102913		F			c_1[out]	0.336801			
20	-1.102913	-1.265246		-1.458636	-1.050592		c_2[out]	0.801917			
21				-1.050592	-0.810509						
22	V[B]										
23	-0.677230	-0.411305		E-elect	E-total	$\|\psi_1\rangle$	$\|\psi_2\rangle$		N1	N2	S
24	-0.411305	-1.226615		-4.227525	-2.860658	0.801917	-0.782265		1.000002	1.000005	-0.000001
25						0.336801	1.068446				
26	(1111)	1.307152									
27	(2111)	0.437279		E-nuc	$\langle\psi_1\|h\|\psi_1\rangle$	$\langle\psi_2\|h\|\psi_2\rangle$	J11	J22	J12	K12	
28	(2121)	0.177267		1.366867	-0.907465	-0.994626	1.032625	0.755960	0.748962	0.073246	
29	(2211)	0.605703					0.540561	0.489488	0.514391	0.514391	(1111)
30	(2221)	0.311795					0.303795	-0.894609	-0.325518	-0.325518	(2111)
31	(2222)	0.774808					0.051724	0.495337	-0.160065	0.062407	(2121)
32							0.088368	0.846258	0.486701	-0.273463	(2211)
33							0.038210	-1.189980	0.133145	-0.004878	(2221)
34							0.009967	1.009466	0.100307	0.100307	(2222)

Figure 6.10 The spreadsheet-based repeat of Wolniewicz's calculation for HeH$^+$ for the bond length 1.4632 a.u. The iteration sequence to convergence is given in Table 6.2.

Table 6.2 The HFS-SCF calculation from the initial guess of the density matrix elements to self-consistency in 10 iterations. Note, again, these data relate to the calculation carried out following Exercise 6.1.

	D	E	F	G	H	I
	Iteration	P_{11}	P_{12}	P_{22}	ε_1	ε_2
40						
41	0	1.00000000	0.00000000	0.00000000	-2.04940000	-0.70720000
42	1	1.56085092	0.38291312	0.09393752	-1.54147740	-0.06385048
43	2	1.31221677	0.52759983	0.21213079	-1.59240411	-0.06133382
44	3	1.28814257	0.53922573	0.22572376	-1.59706663	-0.06163913
45	4	1.28629185	0.54010239	0.22678414	-1.59742310	-0.06166657
46	5	1.28615282	0.54016815	0.22686389	-1.59744987	-0.06166865
47	6	1.28614239	0.54017308	0.22686987	-1.59745188	-0.06166881
48	7	1.28614161	0.54017345	0.22687032	-1.59745188	-0.06166881
49	8	1.28614155	0.54017348	0.22687035	-1.59745203	-0.06166882
50	9	1.28614155	0.54017348	0.22687035	-1.597452	-0.0616688
51	10	1.28614155	0.54017348	0.22687036	-1.597452	-0.0616688

Note, the use of the value 2.0925 as the Slater exponent for helium in the calculation; this is the value suggested by Szabo and Ostlund (47), chosen there was no accepted value for helium in molecular environments but, also, chosen to obtain some agreement, in their calculation, Wolniewicz's data.

It is appropriate, too, to compare the results of the present calculations with those reported by Szabo and Ostlund (47) from their dedicated FORTRAN computer program.

	A	B	C	D	E	F	G	H
1	The HeH⁺ calculation following Szabo and Ostlund (46)							
2								
3	The S array			The G array				
4	1.0000014	0.4507704		1.1941085	0.2966132			
5	0.4507704	1.0000014		0.2966132	0.9213189		R	1.4632
6							Z_1	2.0000
7	The X array			The F array			ζ_1	2.0925
8	0.5870640	0.9541298		-1.458636	-1.050592		Z_2	1.0000
9	0.5870640	-0.9541298		-1.050592	-0.810509		ζ_2	1.2400
10								
11	The H array			The F' array			Density matrix (in)	
12	-2.6527447	-1.3472051		-1.506208	-0.363039		1.2861630	0.5401631
13	-1.3472051	-1.7318284		-0.363039	-0.152906		0.5401631	0.2268579
14								
15				The C' array				
16	⟨1111⟩	1.3071516		0.9698410	0.2437383			
17	⟨2111⟩	0.4372793		0.2437383	-0.9698410			
18	⟨2121⟩	0.1772671					E electronic	-4.22752548
19	⟨2211⟩	0.6057033		The E array				
20	⟨2221⟩	0.3117945		-1.5974482	0.0000000		E total	-2.86065834
21	⟨2222⟩	0.7746084		0.0000000	-0.0616684			
22								
23				The C array			Density matrix (out)	
24				0.80191675	-0.7822643		1.2861433	0.5401724
25				0.33680075	1.06844426		0.5401724	0.2268692

Figure 6.11 The calculation of the *self-consistent* results for the case of HeH⁺ using the |sto-3g⟩ basis of Table 1.6 to represent both the hydrogen and helium atoms when scaled. The output is to be compared directly with the results in Appendix B of reference 47.

To provide another example of the small discrepancies likely between different calculations a further worksheet has been added to fig6-9.xls so that the calculation results are presented in the same form as the output from the FORTRAN program. This worksheet, Figure 6.11, presents the spreadsheet output in the form of the sample output in reference 47 and exemplifies the information output by modern molecular orbital theory programs during the cycles to self-consistency.

The present calculation on the HeH⁺ ion, illustrates several features of the SCF methodology and facilitates the kind of interpretations that chemist's like to use to explain MO theory calculations. For example, it is useful to reconstruct Table 6.2, starting with the null matrix as the initial input to the calculation and to monitor the values of the coefficients in equation 6.2 during the cycle to self-consistency. You will find that initially, the 1σ molecular orbital is comprised largely of helium character, because there is no screening due to the two-electron term in the calculation with the null matrix. Then, as the iterative cycle proceeds and there is more mixing of the two atomic orbitals on the different atoms, this leads to a lowering of the electron−electron repulsion, and really to a very primitive example of the *nephalauxetic* effect due to the sharing of electron density over surrounding atomic orbitals in transition metal complex theory.

Then, too, in keeping with the different electronegativities of hydrogen and the helium *cation*, we expect and find that for the converged results, the bonding molecular orbital contains more helium character, while the anti-bonding molecular orbital contains more hydrogen character!

6.7   POLARIZATION FUNCTIONS, GAUSSIAN LOBES AND HIGHER-ORDER GAUSSIAN BASIS SETS

Chapter 1 ended with a short survey of modern methodology in the application of Gaussian basis set theory to molecular orbital theory calculations. In 'real' calculations it is standard to use a variety of larger basis sets tailored to particular applications (49) with the minimum basis $|sto-3g\rangle$ sets widely used for general use and then these more complicated basis sets being applied to obtain better agreement with particular experimental results.

Calculations using Gaussian lobe functions (47, 48, 85) are not used in modern program packages. However, since it is not possible to carry out a spreadsheet calculation with higher order Gaussian functions, because the function library of the EXCEL software does not include the derivatives of the Error Function, a simple Gaussian lobe calculation illustrates well the effect of polarization functions.

In the calculation associating three single Gaussians with each atomic site provides for the effect of polarization. The central s-orbital is approximated by the $|sto-1g\rangle$ function at each site and the other Gaussians, then, taken as the 'out of phase' linear combinations, sited about each atomic site, approximate to a 'p' polarizing function at each site. All cross terms in the calculations of the various integrals are included, even though this is not appropriate and I recommend that you investigate the results returned in calculations in which you exclude cross terms to take account of the nodal planes in actual p-orbital functions, if you wish to develop this calculation.

So, let us repeat the HeH$^+$ calculation using three Gaussian primitives associated with each atomic position, but with two of these at each site used to model p-type polarization. Whitten (48) proposed the Gaussian sets listed in Table 6.3 to represent the 1s and 2p of hydrogen. Hoyland (85) carried out more extensive calculations on H$_2$$^+$ and H$_2$.

In this present calculation, therefore, the two extra s-type Gaussian primitives are added to simulate the effect of polarization by constructing sp-hybrid functions $[\phi_a(r)$ and $\phi_b(r)]$ at each atomic site, with for the bond length R

$$\phi_a(\mathbf{r}) = g_{1s}\left(r - \frac{\mathbf{R}}{2}\right) + c_1\left[g_{1s}\left(\mathbf{r} - \frac{\mathbf{R}}{2} - \Delta\mathbf{R}\right) - g_{1s}\left(\mathbf{r} - \frac{\mathbf{R}}{2} + \Delta\mathbf{R}\right)\right] \quad 6.37$$

$$\phi_b(\mathbf{r}) = g_{1s}\left(r + \frac{\mathbf{R}}{2}\right) + c_2\left[g_{1s}\left(\mathbf{r} + \frac{\mathbf{R}}{2} - \Delta\mathbf{R}\right) - g_{1s}\left(\mathbf{r} + \frac{\mathbf{R}}{2} + \Delta\mathbf{R}\right)\right] \quad 6.38$$

Table 6.3   Whitten's $|sto-1g\rangle$ basis sets for the 1s and 2p orbitals in hydrogen.

$|\rangle$	α	d	N	R_i
$|sto-1g:1s\rangle$	0.28294	1.00000	1.00000	0.00000
$|sto-1g:2p\rangle$	0.045989	1.00000	0.39214	+/−0.93262

	A	B	C	D	E	F	G	H
1	|sto-1g⟩ polarized Gaussian lobe calculation for HeH+ ion							
2								
3	S			theta =	0.78540		S-diagonal	
4	1.00000	0.41232		omega1	1.41232		1.41232	0.00000
5	0.41232	1.00000		omega2	0.58768		0.00000	0.58768
6								
7	U			U - adjoint			The S^1/2 matrix	
8	0.70711	0.70711		0.70711	0.70711		0.84146	0.00000
9	0.70711	-0.70711		0.70711	-0.70711		0.00000	1.30446
10								
11	Fock matrix			X-canonical			X-canon transpose	
12	-0.79401	-0.99651		0.59500	0.92239		0.59500	0.59500
13	-0.99651	-1.20788		0.59500	-0.92239		0.92239	-0.92239
14								
15	theta	-0.15618		Transformed Fock matrix				
16	omega1	-1.45007		-1.06536	0.18678	==>	-1.41431	0.22714
17	omega2	0.02821		-1.31161	0.19497		0.22714	-0.00755
18								
19	Coefficients			Fock - diagonal			C - originals	
20	0.98783	-0.15554		-1.41431	0.00000		0.44429	-1.00371
21	-0.15554	-0.98783		0.00000	-0.00755		0.73123	0.81862
22								
23		R		ΔR			c1	c2
24		1.46320		0.50000			0.00000	0.00000
25								
26	E(electronic)			E(total)			ε_1	ε_2
27	-3.8744			-2.50755			-1.45007	0.02821
28	README				README			
29	Density matrix (in)			Density matrix (out)			⟨1111⟩	0.744258698
30	0.39478	0.64975		0.39478	0.64975		⟨2111⟩	0.286879471
31	0.64975	1.06940		0.64975	1.06940		⟨2121⟩	0.175502151
32							⟨2211⟩	0.61727257
33							⟨2221⟩	0.443569292
34							⟨2222⟩	1.255936554

Figure 6.12a   Canonical orthonormalization of the Fock matrix for the HeH$^+$ calculation using the |sto-1g:1s⟩ basis proposed by Whitten (48). Without polarization, $c_1 = c_2 = 0.0$ [cells $G$23 and $H$23] the electronic energy is found to be −3.87700473 Hartree, while the total energy is −2.51013759 Hartree. A slightly better result is returned using Stewart's |sto-1g⟩ basis (32).

Figure 6.12a, b and c illustrates the calculation for HeH$^+$ based on the functions of equations 6.37 and 6.38.

The modifications to fig6-8abc.xls required to include the polarization functions can be examined in fig6-12abcd.xls and some of the details for the calculation of the one-electron terms in the Fock matrix are displayed in Figures 6.12b. Note that all the input data are linked throughout the spreadsheet to the worksheet inputs on 'scf'! and 'Fock'!.

The two-electron terms for the |sto-1g*⟩ basis involve the taking of the possible four-term components of the multiplications of equations 6.37 and 6.38 in the calculation

	A	B	C	D	E	H	I	J
1		R	1.46320	ΔR	0.50000	c1/c2	0.00000	
2		|sto-1g:sp⟩	Z = ⟩	1.00000		|sto-1g:sp⟩	Z = ⟩	2.00000
3			scaler	1.24000			scaler	2.09250
4		|s1⟩	|p1+⟩	|p1⟩		|s2⟩	|p2+⟩	|p2-⟩
5	α	0.28294	0.04599	0.04599		0.28294	0.04599	0.04599
6	d	1.00000	0.39214	0.39214		1.00000	0.39214	0.39214
7	α'	0.43505	0.07071	0.07071		1.23887	0.20137	0.20137
8	d'	0.38178	0.00000	0.00000		0.83691	0.00000	0.00000
9	R_i	-0.73160	-1.23160	-0.23160		0.73160	0.23160	1.23160
10		[s1|s1]	[p1-|p1-]	[p1+|p1+]	[s1|p1-]	[s2|s2]	[p2+|p2+]	[p2-|p2-]
11	Ra	-0.73160	-1.23160	-0.23160	-0.73160	0.73160	0.23160	1.23160
12	Rb	-0.73160	-1.23160	-0.23160	-1.23160	0.73160	0.23160	1.23160
13	alpha[a]	0.43505	0.07071	0.07071	0.43505	1.23887	0.20137	0.20137
14	d[a]	0.38178	0.00000	0.00000	0.38178	0.83691	0.00000	0.00000
15	alpha[b]	0.43505	0.07071	0.07071	0.07071	1.23887	0.20137	0.20137
16	d[b]	0.38178	0.00000	0.00000	0.00000	0.83691	0.00000	0.00000
17								
18	kappa	1.00000	1.00000	1.00000	0.98491	1.00000	1.00000	1.00000
19	Rp	-0.73160	-1.23160	-0.23160	-0.80151	0.73160	0.23160	1.23160
20	alpha[p]	0.87010	0.14143	0.14143	0.50576	2.47774	0.40273	0.40273
21	d[p]	0.14576	0.00000	0.00000	0.00000	0.70042	0.00000	0.00000
22	R_A	-0.73160	-0.73160	-0.73160	-0.73160	-0.73160	-0.73160	-0.73160
23	R_B	0.73160	0.73160	0.73160	0.73160	0.73160	0.73160	0.73160
24	t[A]	0.00000	0.03536	0.03536	0.00247	5.30472	0.37364	1.55219
25	$F_o(t)[A]$	1.00000	0.98834	0.98834	0.99918	0.38435	0.88826	0.65579
26	t[B]	1.86284	0.54508	0.13121	1.18875	0.00000	0.10068	0.10068
27	$F_o(t)[B]$	0.61453	0.84454	0.95793	0.71278	1.00000	0.96743	0.96743
28		[s1|s1]	[p1|p1]	[p1+|p1+]	[s1|p1-]	[s2|s2]	[p2+|p2+]	[p2-|p2-]
29	S	1.00000	0.00000	0.00000	0.00000	1.00000	0.00000	0.00000
30	t	0.65257	0.00000	0.00000	0.00000	1.85830	0.00000	0.00000
31	V[a]	-1.05254	0.00000	0.00000	0.00000	-0.68266	0.00000	0.00000
32	V[b]	-1.29363	0.00000	0.00000	0.00000	-3.55233	0.00000	0.00000

Figure 6.12b   Construction of the one-electron components of the Fock matrix of Figure 6.12a. Note that some columns are hidden. The positions of the primitive Gaussians are set in the active cells of row 9, but the results displayed are for the polarization term set to zero, so that, in this case, these positions are not important.

of the ⟨ijkl⟩ components. Some detail is shown in Figure 6.12c, which is from the 'Vbaaa'! worksheet of fig6-11abcd.xls, but, in this case, with the worksheet condition shown for the choice that the polarization is turned on fully, i.e. the entries 1.00 in cells 'scf'!$G$24 and $H$24.

Note, as mentioned above that all terms are included in these two-electron calculations over the Gaussian primitives, which is in conflict with the nodal properties of actual p orbitals.

The converged condition of the worksheet 'canonical'! in fig6-11abcd.xls is shown, finally, in Figure 6.12d. As you see there are significant improvements in the estimates of the total energy and the electronic energy for the helium hydride ion, but there is also an unwelcome larger increase in the energy of the anti-bonding molecular orbital returned as the value in cell $H$27. Since the calculation is intended only to illustrate the effect of a polarization term, it is not necessary to improve the design of the calculation to mirror, in detail, the action of added 2p orbitals, but, if you are interested it is appropriate to examine the contributions from cross terms ignoring the nodal structure of actual p functions.

	A	B	C	D	E	F	G	H	I	J	K
1	V2111 Fock term		R = ⟩	1.463200		delta-R	0.500000				
2			Z	1.00000E+00			Z	2.00000E+00			
3			|sto-1g:sp⟩	scaler	1.240000	|sto-1g:sp⟩	scaler	2.092500			
4			|s1⟩	|p1+⟩	|p1-⟩		|s2⟩	|p2+⟩	|p2-⟩		
5			0.282940	0.045989	0.045989	α	0.282940	0.045989	0.045989		
6			1.000000	0.392140	0.392140	d	1.000000	0.392140	0.392140		
7			0.435049	0.070713	0.070713	α'	1.238869	0.201365	0.201365		
8			0.381780	-0.038324	0.038324	d'	0.836911	0.084012	-0.084012		
9			-0.731600	-1.231600	-0.231600	R|	0.731600	0.231600	1.231600		
10											
11				V2111		0.37476					

	a	b	c	d		da	db	dc	dd	⟨ijkl⟩	V...
12	a	b	c	d		da	db	dc	dd	⟨ijkl⟩	V...
13	1.238869	0.435049	0.435049	0.435049		0.836911	0.381780	0.381780	0.381780	0.28688	6.160
14	1.238869	0.070713	0.435049	0.435049		0.836911	-0.038324	0.381780	0.381780	-0.05664	12.116
15	1.238869	0.435049	0.435049	0.070713		0.836911	0.381780	0.381780	-0.038324	-0.05509	11.784
16	1.238869	0.435049	0.070713	0.435049		0.836911	0.381780	-0.038324	0.381780	-0.05509	11.784
17	0.201365	0.435049	0.435049	0.435049		0.084012	0.381780	0.381780	0.381780	0.20942	44.795

	L	M	N	O	P	Q	R	S	T	U	V	W	X	Y	Z
	R_a	R_b	R_c	R_d	R_p	R_q	Rab2	Rcd2	Rpq2	A	B	fzero	A	B	Err f
13	0.7316	-0.7316	-0.7316	-0.7316	0.3513	-0.7316	2.1410	0.0000	1.1727	6.1600	0.6714	0.8149	0.000058	0.466550	0.864036
14	0.7316	-1.2316	-0.7316	-0.7316	0.6256	-0.7316	3.8542	0.0000	1.8420	12.1161	0.9629	0.7539	0.000816	0.394123	0.882806
15	0.7316	-0.7316	-0.7316	-1.2316	0.3513	-0.8015	2.1410	0.2500	1.3290	11.7838	0.5162	0.8516	0.000000	0.330771	0.899875
16	0.7316	-0.7316	-1.2316	-0.7316	0.3513	-0.8015	2.1410	0.2500	1.3290	11.7838	0.5162	0.8516	0.000000	0.330771	0.899875
17	0.2316	-0.7316	-0.7316	-0.7316	-0.4268	-0.7316	0.9278	0.0000	0.0929	44.7949	0.0341	0.9887	0.000000	0.363425	0.890999

Figure 6.12c   Part of the two-electron worksheet 'V2111'! for the case that the polarization term is fully on, the condition $c_1 = c_2 = 1.00$ in cells 'canonical' !$G$24 and $H$24 with the offset at $+/-0.5$ atomic units of length, cell 'scf' !$D$24.

We have seen, Section 6.3, how the Boys' method (14) provides for the exact calculations of integrals over Gaussian products. In those calculations for diatoms, it was convenient to apply the Fourier transformation procedure only for the calculation of the nuclear attraction and two-electron integral terms over the Gaussian sets. In general, (14,73,38) it is normal to calculate all the integrals over the Gaussian primitives using appropriate Fourier transformations of the Fundamental Integral (38), I, with

$$I = \iint_{\substack{r_1 \\ r_2}} e^{-\alpha|\mathbf{r}_1-\mathbf{R}_a|^2} e^{-\beta|\mathbf{r}_1-\mathbf{R}_b|^2} f\left(|\mathbf{r}_1 - \mathbf{r}_2|\right) e^{-\gamma|\mathbf{r}_2-\mathbf{R}_c|^2} e^{-\delta|\mathbf{r}_2-\mathbf{R}_d|^2}\, d\mathbf{r}_1\, d\mathbf{r}_2 \qquad 6.39$$

which is a generalization of equation 5.51, for the Gaussian primitives defined in Figure 5.5. For the case that all the components are s-type Gaussians, the integral reduces directly to a general form of equation 5.52, viz

$$I = e^{\left[\frac{\alpha\beta}{\alpha+\beta}|\mathbf{R}_A-\mathbf{R}_B|^2 - \frac{\gamma\delta}{\gamma+\gamma}|\mathbf{R}_C-\mathbf{R}_D|^2\right]} \iint e^{\left(-p|\mathbf{r}_1-\mathbf{R}_p|^2\right)} f\,|\mathbf{r}_1 - \mathbf{r}_2|\, e^{\left(-q|\mathbf{r}_2-\mathbf{R}_Q|^2\right)} d\mathbf{r}_1 d\mathbf{r}_2 \qquad 6.40$$

The Fourier integral transform[3] is a standard device for the evaluation of integrals of the kinds in equation 6.40. Each function of the coordinates, $f(\mathbf{r})$, in the equation can be written as its Fourier transform

$$f(r) = \left(\frac{1}{2\pi}\right)^{-3} \iiint F(\mathbf{k})\, e^{-ik\cdot r}\, d\mathbf{k} \qquad 6.41$$

[3] See, for example, reference 89.

	A	B	C	D	E	F	G	H
1	⎮sto-1g⟩ polarized Gaussian lobe calculation for HeH+ ion							
2								
3	S			theta =	-0.77647		S-diagonal	
4	1.01068	0.52589		omega1	0.49409		0.49409	0.00000
5	0.52589	1.02946		omega2	1.54605		0.00000	1.54605
6								
7	U			U - adjoint			The S^1/2 matrix	
8	0.71339	-0.70077		0.71339	-0.70077		1.42264	0.00000
9	-0.70077	-0.71339		-0.70077	-0.71339		0.00000	0.80425
10								
11	Fock matrix			X-canonical			X-canon transpose	
12	-0.83694	-1.09960		1.01490	-0.56359		1.01490	-0.99694
13	-1.09960	-1.28789		-0.99694	-0.57374		-0.56359	-0.57374
14								
15	theta	-0.15366		Transformed Fock matrix				
16	omega1	0.11952		0.24683	1.10258	==>	0.08305	-0.23548
17	omega2	-1.43738		0.16797	1.35864		-0.23548	-1.40091
18								
19	Coefficients			Fock - diagonal			C - originals	
20	0.98822	-0.15305		0.08305	0.00000		1.08920	0.40162
21	-0.15305	-0.98822		0.00000	-1.40091		-0.89738	0.71957
22								
23		R		ΔR			c1	c2
24		1.46320		0.50000			1.00000	1.00000
25								
26	E(electronic)			E(total)			ε_1	ε_2
27	-3.8908			-2.52397			-1.46249	0.12404
28	README				README			
29	Density matrix (in)			Density matrix (out)			⟨1111⟩	0.750635939
30	0.32259	0.57798		0.32259	0.57798		⟨2111⟩	0.374755083
31	0.57798	1.03555		0.57798	1.03555		⟨2121⟩	0.249318092
32							⟨2211⟩	0.659197095
33							⟨2221⟩	0.528250304
34							⟨2222⟩	1.285611125

Figure 6.12d   The converged condition of the polarized calculation for the helium hydride ion, with the polarizing term turned fully on [c_1/c_2 set to 1.00 in cells $G$24 and $H$24].

which requires knowledge of $F(\mathbf{k})$ for each of the functions of position needed to construct the overlap, kinetic energy, nuclear attraction energy and the two-electron energy terms of the molecular Hamiltonian. These are the Gaussian itself, the reciprocal of distance and the three-dimensional Dirac delta function, for which, the transforms are (38,47,74)

$$e^{-\alpha r^2} = \left(\frac{1}{2\pi}\right)^3 \iiint \left[\left(\frac{\pi}{\alpha}\right)^{3/2} e^{\frac{-k^2}{4\alpha}}\right] e^{-i\mathbf{k}\cdot\mathbf{r}} \, d\mathbf{k} \qquad 6.42$$

$$\frac{1}{r} = \left(\frac{1}{2\pi}\right)^3 \iiint \left(\frac{4\pi}{k^2}\right) e^{-i\mathbf{k}\cdot\mathbf{r}} \, d\mathbf{k} \qquad 6.43$$

$$\delta(r_1 - r_2) = \left(\frac{1}{2\pi}\right)^3 \iiint (1) e^{-i\mathbf{k}\cdot\mathbf{r}} \, d\mathbf{k} \qquad 6.44$$

which last equation, of course, relates the eigenfunctions of position and momentum for the free particle.

Substitution of the appropriate transforms into the fundamental integral equation leads to integrable forms for all the integrals of interest. The setting of $f(r_1 - r_2)$ to be the Dirac delta function reduces the fundamental integral to an overlap integral. The nuclear attraction integral is recovered by setting one of the exponents in the fundamental integral equal to infinity and another equal to zero. The zero exponent returns a constant value of unity, while the infinitely large exponent fixes a point in space to coincide with the nuclear position.

The kinetic energy integral involves the Laplacian action of partial double differentiation with respect to the coordinates of the Gaussian primitives. As we know from several examples this returns the result of the differentiation times a product over Gaussian primitives for integration over the coordinates, for example, equation 4.32.

The generality of this approach, partly exploited in this chapter for the calculations on dihydrogen and the helium hydride ion is that the final formulae for integrations over 1s Gaussian primitives, can be applied to evaluate the similar integrals over higher order Gaussians, for example, which would have been required had we made a proper polarized basis in fig6-11abcd.xls. Consider such a p-type Gaussian primitive and the identity

$$x e^{(-\alpha r^2)} = \left(\frac{1}{2\alpha}\right) \frac{\partial}{\partial x} (e^{-\alpha r^2}) \qquad 6.45$$

The nuclear integral between such a 2p Gaussian primitive and a 1s one takes the form

$$\left\langle I_{2p} \left| \frac{1}{\mathbf{R}_c} \right| G_{1s} \right\rangle = \iiint \frac{1}{\mathbf{R}_c} (X - X_A) e^{(-\alpha_A |\mathbf{r} - \mathbf{R}_A|^2 - \alpha_B |\mathbf{r} - \mathbf{R}_B|^2)} \, d\mathbf{r} \qquad 6.46$$

for the 2s Gaussian at $\mathbf{R}_A$, the 1s Gaussian at $\mathbf{R}_B$ and the interaction with the nucleus at $\mathbf{R}_c$.

After suitable working the integral and all the others needed for a full calculation reduce to forms involving derivatives of the *Error function*. For example, equation 6.45 becomes

$$\int = \frac{\pi}{\alpha_A(\alpha_A + \alpha_B)} \frac{\partial}{\partial X} \left\{ F_0 \left[(\alpha_A + \alpha_B)|\mathbf{r} - \mathbf{R}_p|^2 e^{\left(\frac{-\alpha_A \alpha_B}{\alpha_A + \alpha_B}|\mathbf{R}_A - \mathbf{R}_B|^2\right)} \right] \right\}$$

$$= \frac{2\pi}{\alpha_A + \alpha_B} \left\{ (X - X_p) F_1[(\alpha_A + \alpha_B)|\mathbf{r} - \mathbf{R}_p|^2] \right. \qquad 6.47$$

$$\left. - \frac{\alpha_B}{\alpha_A + \alpha_B}(X_A - X_B) F_0[(\alpha_A + \alpha_B)|\mathbf{r} - \mathbf{R}_p|^2] \right\} e^{\left(-\frac{\alpha_A \alpha_B}{\alpha_A + \alpha_B}|\mathbf{R}_A \mathbf{R}_B|^2\right)}$$

Unfortunately, such derivatives are not available in the EXCEL function library, therefore, within the context of the objectives of this book, I hope this an acceptable end point. After all to proceed much further, anyway, we would have to start working with more complicated structures than diatoms with just two electrons. This would involve the calculation of many more integrals even without using other than 1s Gaussians. The storage and management of these rapidly become a significant problem (38); as we have seen even the simplest diatom problem in a small 1s Gaussian basis generates a large number of integrals.

So, perhaps, it is best to accept that the next step is to mount one of the electronic structure packages, now available, on your own PC, for example, as listed in footnote 18 on page 24. Then you can start your own studies of molecular electronic structure and properties, with I hope a deeper insight into the detail of calculation. Just as we have seen in the simple examples set out in this book, in the end, the problems reduce to the calculation and management of integrals, whether they be over only 1s Gaussians or their higher-order cousins!

6.8   EPILOGUE

I started this manuscript with a quotation from Professor Hartree's famous book on numerical integration. With the development of the methodology of the Hartree–Fock–Slater approximation on a spreadsheet, perhaps to finish, I can be allowed the temerity to quote a well-known comment by one of the giants of quantum mechanics, Professor P.A.M. Dirac (69),

The underlying physical laws necessary for the mathematical theory of a large part of physics and the whole of chemistry are thus completely known, and the difficulty is only that the exact application of these laws leads to equations much too complicated to be soluble.

Forty years of new technology have reduced the absoluteness of that statement to a considerable extent. As you see, even with your own PC you can begin to do meaningful calculations!

References

1. D.R. Hartree, *Proc. Cambridge Phil. Soc.*, **24** (1928) 89.
2. V. Fock, *Zeits. Phys.*, **61** (1930) 126.
3. V. Fock, *Zeits. Phys.*, **62** (1930) 795.
4. F. Herman and S. Skillman, *Atomic Structure Calculations* [Prentice Hall, Inc. New Jersey, 1963].
5. J.C. Slater, *Phys. Rev.*, **81** (1951) 385.
6. J.C. Slater, *Phys. Rev.*, **36** (1930) 57.
7. J.C. Slater, *Phys. Rev.*, **35** (1930) 210.
8. E. Clementi and D.L. Raimondi, *J. Chem. Phys.*, **38** (1963) 2686.
9. E. Clementi, *J. Chem. Phys.*, **40** (1964) 1944–1945.
10. E. Clementi and C. Roetti, *Atomic Data and Nuclear Data Tables*, **14** (1974) 177.
11. C.F. Bunge, J.A. Barrientos and A.V. Bunge, *ibid*, **53** (1993) 113.
12. B. Webster, *Chemical Bonding Theory* [Blackwell, Oxford, 1990].
13. J.L. Reed, *J. Chem. Educ.*, **76** (1999) 802.
14. S.F. Boys, *Proc. Roy. Soc. (London)*, **A200** (1950) 542.
15. R. McWeeny, *Proc. Camb. Phil. Soc.*, **45** (1949) 315.
16. R. McWeeny, *Nature*, **166** (1950) 21.
17. R. McWeeny, *Acta Crystallogr.*, **6** (1953) 631.
18. I. Shavitt, *Israel J. Chem.*, **33** (1993) 357.
19. J.V.L. Longstaff and K. Singer, *Proc. Roy. Soc.*, **A258** (1960) 421.
20. J.M. Foster and S.F. Boys, *Rev. Mod. Phys.*, **32** (1960) 303.
21. B.J. Ranzil, *Rev. Mod. Phys.*, **32** (1960) 239, 245.
22. C.M. Reeves, *J. Chem. Phys.*, **39** (1963) 1.
23. C.M. Reeves and M.C. Harrison, *J. Chem. Phys.*, **39** (1963) 11.
24. C.M. Reeves and R. Fletcher, *J. Chem. Phys.*, **42** (1965) 4073.
25. K. O-ohata, H. Taketa and S. Huzinaga, *J. Phys. Soc. (Japan)*, **21** (1966) 2306.
26. W.E. Palke and W.N. Lipscomb, *J. Am. Chem. Soc.*, **88** (1966) 2384.
27. M.D. Newton and W.E. Palke, *J. Chem. Phys.*, **45** (1966) 2329.
28. U. Kaldor and I. Shavitt, *J. Chem. Phys.*, **45** (1966) 888, **48** (1968) 191.
29. R.M. Pitzer, *J. Chem. Phys.*, **46** (1967) 4871, **47** (1967) 965.
30. W.E. Palke and R.M Pitzer, *J. Chem. Phys.*, **46** (1967) 3948.
31. S. Aung, R.M. Pitzer and S.I. Chan, *J. Chem. Phys.*, **49** (1968) 2071.
32. R.F. Stewart, *J. Chem. Phys.*, **50** (1969) 2485.
33. W.J. Hehre, R.F. Stewart and J.A. Pople, *J. Chem. Phys.*, **51** (1969) 2657.
34. I. Shavitt and M. Karplus, *J. Chem. Phys.*, **36** (1962) 550, **43** (1965) 398.
35. V.R. Saunders, in *Methods in Computational Physics* [edited by G.H.F. Diercksen, B.T. Sutcliffe and A. Veillard, Reidel, Dordrecht, 1975].
36. D. Hegarty and G. van der Velde, *Int. J. Quantum Chem.*, **23** (1983) 1135.
37. D. Hegarty, in *Advanced Theories and Computational Approaches to the Electronic Structure of Molecules* [edited by C.E. Dykstra, Reidel, Dordrecht, 1984].
38. P.W. Gill, *Adv. Quantum Chem.*, **25** (1994) 14.
39. E. Steiner, *The Determination and Interpretation of Molecular Wave Functions* [Cambridge University Press, Cambridge, 1976].

40. E.R. Davidson and D. Feller, *Chem. Rev.*, **86** (1986) 681.
41. S. Huzinaga, J. Andzelm, M. Klobukowski, E. Radzio-Andzelm, Y. Sakai and H. Tatewaki, *Gaussian Basis Sets for Molecular Calculations* [Elsevier, Amsterdam, 1984].
42. J.P. Lowe, *Quantum Chemistry* [Academic Press, New York and London, 1978].
43. K. Schwarz, *Phys. Rev. B*, **5** (1972) 2466.
44. J.C. Slater and K.H. Johnson, *Phys. Rev. B*, **5** (1972) 844.
45. S. Huzinaga, *J. Chem. Phys.*, **42** (1965) 1293–1302.
46. T.H. Dunning, *J. Chem. Phys.*, **53** (1970) 2823.
47. A. Szabo and N.S. Ostlund, *Modern Quantum Chemistry: Introduction to Advanced Electronic Structure Theory* [Macmillan, New York, 1982].
48. J.L. Whitten, *J. Chem. Phys.*, **44** (1966) 359.
49. A.A. Frost, *J. Chem. Phys.*, **47** (1967) 3707.
50. F. Jensen, *Introduction to Computational Chemistry* [John Wiley, Chichester, 1999].
51. J. Almlöf and P.R. Taylor, *Adv. Quantum Chem.*, **22** (1991) 301, *J. Chem. Phys.*, **86** (1984) 4070.
52. T.H. Dunning, *J. Chem. Phys.*, **90** (1989) 1007.
53. E. Steiner, *The Chemistry Maths Book* [Oxford Science Publications, Oxford, 1996].
54. D.R. Hartree, *Numerical Analysis* [The Clarendon Press, Oxford, 1952].
55. R.S. Mulliken, C.A. Rieke, D. Orloff and H. Orloff, *J. Chem. Phys.*, **17** (1949) 1248.
56. J.E. Huheey, *Inorganic Chemistry* [Harper & Row, New York, 1978].
57. G.G. Hall, *Proc. Roy. Soc.*, **A208** (1951) 328.
58. C.J. Roothaan, *Rev. Mod. Phys.*, **23** (1951) 69.
59. W.H. Press, B.P. Flannery, A.A. Teukolsky and W.T. Vetterling, *Numerical Recipes The Art of Scientific Computing* [Cambridge University Press, Cambridge, 1990].
60. D.R. Hartree, *The Calculation of Atomic Structures* [John Wiley and Sons, New York, 1957].
61. J.C. Slater, *The Quantum Theory of Atomic Structure* [McGraw-Hill, New York, 1960].
62. C. Froese Fischer, *The Hartree–Fock Method for Atoms* [Wiley-Interscience, 1977].
63. E. Clementi, IBM Tech. Rept., RJ-256, August 1963.
64. J. Mitroy, *Aust. J. Phys.*, **52** (1999) 973–997.
65. E.R. Cohen and B.N. Taylor, *J. Phys. Chem. Ref. Data*, **2** (1973), 663.
66. *The CRC Handbook of Chemistry and Physics* [CRC Press, Florida].
67. R.L. Snow and J.L. Bills, *J. Chem. Educ.*, **52** (1975) 506.
68. M.J. Dewar and J.J. Kelemen, *J. Chem. Educ.*, **48** (1971) 494.
69. J.M. André, D.H. Mosley, M.C. André, B. Champagne, E. Clementi, J.G. Fripiat, L. Leherte, L. Pisani, D.P. Vercuteren and M. Vracko, *Exploring Aspects of Computational Chemistry* [Presses Universitaires de Namur, Namur, Belgique, 1997].
70. S. Chandrasekhar and G. Herzberg, *Phys. Rev.*, **98** (1955) 1050.
71. E.A. Hylleras, *Z. Phyzik*, **54** (1929) 347.
72. T. Kinoshita, *Phys. Rev.*, **105** (1957) 1490.
73. C.L. Pekeris, *Phys. Rev.*, **115** (1959) 1216.
74. I. Shavitt, *Meth. Computat. Phys.*, **2** (1963) 1.
75. M. Ambramowitz and I. Segun, *Handbook of Mathematical Functions* [Dover Publications, New York, 1968].
76. J.A. Pople and R.K. Nesbet, *J. Chem. Phys.*, **22** (1954) 571.
77. C.M. Quinn, D.B. Redmond and P.W. Fowler, *J. Math. Chem.*, **23** (1998) 263.
78. R.P. Mortimer, *Physical Chemistry*, 2nd edn. [Harcourt-Academic Press, New York, 2000].
79. M.W. Schmidt, K.K. Baldridge, J.A. Boatz, S.T. Elbert, M.S. Gordon, J.H. Jenson, S. Koseki, N. Matsunaga, K.A. Nguyen, S.J. Su, T.L. Windus, M. Dupois and J.A. Montgomery, *J. Computat. Chem.*, **14** (1993) 1347.
80. M. Born and J.R. Oppenheimer, *Ann. Phys.*, **84** (1927) 457.
81. D.A. McQuarrie, *Quantum Chemistry* [Oxford University Press, Oxford, 1983].
82. I. N. Levine, *Quantum Chemistry*, 4th edn. [Prentice Hall, Englewood Cliffs, NJ, 1991].
83. R.S. Mulliken, C.A. Rieke, D. Orloff and H. Orloff, *J. Chem. Phys.*, **17** (1949) 1248.
84. L. Wolniewicz, *J. Chem. Phys.*, **43** (1965) 1807.
85. J.R. Hoyland, *J. Chem. Phys.*, **40** (1964) 3504.

Index